# SEEING AN INVISIBLE WORLD

# SEEING AN INVISIBLE WORLD

## HISTORY OF THERMAL IMAGING

HARRY SCHLEMMER

## Contents

## Foreword

The intention of this book is, to make the technically interested reader familiar with the evolutionary history of imaging devices which are solely based on the thermal radiation emitted by the observed objects themselves. Furthermore, it also tries to convey some of the fascination of "seeing an invisible world" even in the darkest night without any illumination.

We shall follow the gathering of knowledge from early experiments in ancient times on to theoretical insights and experimental results of the "Scientific Revolution" at the beginning of the 19th century, and further to the technical advances of modern thermal imaging technology during the last decades.

As the development of modern thermal imaging was pushed by military applications for the most part, the background of almost 150 years of military-optical technology also plays a role. German institutes and companies were deeply involved in thermal imaging from the very beginning. Their research and development work is presented in greater detail, the more so as plenty of information about these activities is accessible to the author. Wherever possible, however, the author has tried to value the international matrix as thoroughly as possible.

## Military optics from Zeiss to Hensoldt

As an interesting example for the development in Germany, in the following a brief search for the roots of military optics is conducted at the companies Zeiss, Goerz and Hensoldt, and the mystery is also sorted out, how essential contributions to thermal imaging technology may be provided by a company named HENSOLDT today.

**1846** Carl Zeiß founded a precision mechanics and optics workshop in Jena

**1847** Moritz Hensoldt founded an optical factory in Sonneberg, 5 years later a re-foundation was established in Wetzlar under the name Engelbert & Hensoldt

**1880** Hensoldt delivered military optics at first to the British army and afterwards also to the German army

**1888** Carl Paul Goerz founded the Optische Anstalt C. P. Goerz and began his own production of objectives

**1891** Goerz acquired orders for military optics from the Imperial Army and Naval Office

**1894** Zeiss established a department "Tele" for production of military optics

**1903** C. P. Goerz AG including a special military department was founded

**1910** Goerz absorbed the Fraunhofer'sche Glaswerk in Munich and therewith created an independent optics imperium alongside Zeiss/Schott

**1926** Zeiss absorbed the Goerz AG and incorporated it into Zeiss-IKON

**1928** Zeiss absorbed the Hensoldt AG and continued the business operations under the name M. Hensoldt & Söhne Optische Werke AG

**1945** Parts of Zeiss were resettled into the American occupation zone and started business as OPTON Optische Werke Oberkochen GmbH

**1954** Zeiss opened the department "Sonderoptik" for military optics which was renamed to "Sondertechnik" in 1991

**~1947** Eltro GmbH, Gesellschaft für Strahlungstechnik was founded in Switzerland; after being resettled to Heidelberg it was entered into possession of the companies AEG and Hughes

**1995** Zeiss-Eltro Optronik GmbH (ZEO) based in Oberkochen was founded by merging of the Zeiss-department "Sondertechnik" with the DASA-owned Eltro GmbH, the owners were Zeiss and DASA[1]

**1997** Zeiss Optronik GmbH was fully reassumed by Zeiss and renamed to Carl Zeiss Optronics GmbH later; the Hensoldt Systemtechnik GmbH (HST) was formed from the military part of the Hensoldt AG and parts of Leica

**2007** Hensoldt Systemtechnik was absorbed by Carl Zeiss Optronics GmbH

**2012** Cassidian Optronics GmbH (later: Airbus DS Optronics GmbH) became the new name after the divestment of Carl Zeiss Optronics to EADS[2]

**2017** HENSOLDT Optronics GmbH is the new company after the divestment of the Airbus DS Optronics GmbH to the American investor KKR

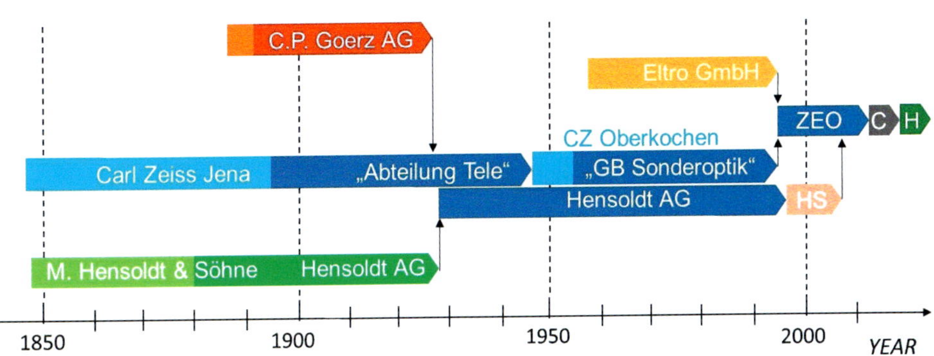


*1 – Time line of German military optics*

*Along the time axis the continued existence of companies is presented which can be viewed as the roots of today's activities in military optics. A darker part of a beam indicates that a military department had existed in that time. GB Sonderoptik was the military department of Carl Zeiss in Oberkochen which was renamed Sondertechnik in 1991. Further abbreviations are: HST = Hensoldt Systemtechnik GmbH in Wetzlar; ZEO = Zeiss-Eltro Optronik GmbH, Zeiss Optronik GmbH and Carl Zeiss Optronics GmbH; C = Cassidian Optronics GmbH and Airbus DS Optronics GmbH; H = HENSOLDT Optronics GmbH*

As the chronological table shows, the oldest lines are reaching back to the Gründerzeit in the middle of the 19th century. The production records during the time of the Great War were followed by an economical depression in the 1920s in which time the Zeiss foundation proved a noteworthy capability of surviving, and could absorb several former competitors, like Goerz and Hensoldt, and their know-how as well. After World War II, together with the partition of Germany also the partition of the Zeiss-group took place. At the western site at Carl Zeiss in Oberkochen, after a 10 years interruption, forced by Law No. 25 of the Allied Control Authority, military optics including thermal imaging could be resumed. By international cooperation in the frame of NATO and participation in the Common Module systems, thermal imaging evolved into a technological flagship in Germany again.

The idea for this book arose from a sequence of presentations in an internal technology seminar and the author wishes to thank all helpful colleagues for the provided material and - not least - for their encouraging interest. All errors and mistakes in the pages to come are my own, of course.

# Short ABCs of Thermal Imaging

## Thermal radiation

*2 – Radiation emitted by various objects*
*In the table, the wavelength of maximum emission of thermal radiation of objects in common use is given for their typical temperatures. The curves show the distribution of thermal radiation over wavelength of two examples. Note: The curve for room temperature is magnified by a factor of 1000 for the sake of an easier presentation.*

All objects whose temperature is higher than the absolute temperature zero at -273.15°C are emitting thermal radiation according to their surface temperatures. Due to the Brownian molecular motion caused by temperature, also electrically charged particles are moving and therewith generate electromagnetic fields – thermal radiation.

When the sun shines, it emits electromagnetic radiation mostly in a wavelength region of approximately 0.4 to 0.8 µm (1 µm = 1/1000 mm). We perceive this effect as light because the human eye is optimized to sunlight, and is serving us as a sensor for this wavelength band.

With decreasing temperature, the maximum of the radiation is shifted more and more to longer wavelengths: A red-hot piece of iron (approx. 1000°C) emits only just within the visible wavelength range, while the radiation emitted by an object at room temperature is by far in the region of the invisible infrared radiation at wavelength between 3 to 12 µm. For further decreasing temperatures the infrared radiation of an object is shifted to even longer wavelength, becomes weaker evermore, and vanishes completely when the temperature approaches the absolute zero point at -273.15°C.

| OBJECT | TEMPERATURE | WAVELENGTH OF MAXIMUM RADIATION |
|---|---|---|
| Surroundings | 20°C | 10 µm |
| Soldering iron | 300°C | 5 µm |
| Red-hot iron | 1000°C | 1.8 µm |
| Incandescent lamp | 3000°C | 0.9 µm |
| Sun | 6000°C | 0.56 µm |

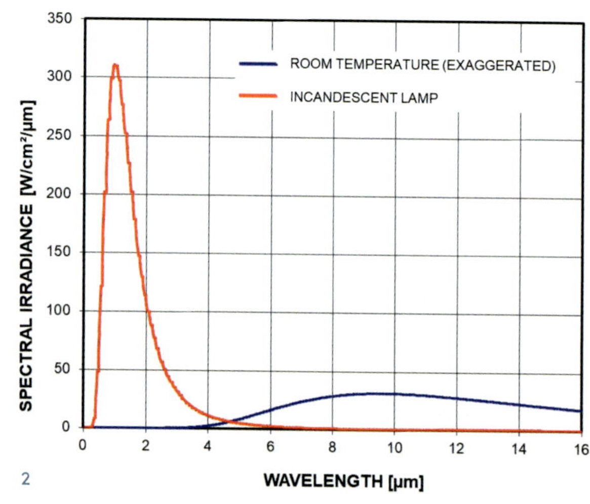

2

## Thermal imagers

A thermal imager (TI) is a special camera for infrared (IR) radiation in the region of 3 to 12 µm, as is caused by the temperature level of the surrounding. A specialty optics is imaging the infrared scene onto a matrix detector. Each detector element generates an electrical signal proportional to the power of the incident radiation. After amplification, the signal is converted to a standard video signal that can be displayed on a customary monitor. In the ideal case, each detector element creates an image point.

So far, the basic principle is not too different compared to a video camera, alas some technological and cost-pushing peculiarities come in addition.

Because customary optical glass is not transparent for infrared radiation, the lenses of the infrared optical system must be made from exotic materials like monocrystalline germanium or silicon. Zinc sulfide, zinc selenide and a few chalcogenide glasses, which are a melt of the said materials, also belong to the sparse choice of materials. Thus, the number of available materials can be considered very limited, especially when compared to the thick glass catalogues of the big international suppliers like SCHOTT, OHARA or CORNING, each of whom having more than 100 special glasses in his portfolio. Therefore, an intensive search was begun very early to find additional means to improve the image quality by a special shaping of the lens surfaces. During the 1990s, when for visual optics only spherical surfaces were used, infrared lenses with so-called aspheric surfaces were already in use in the infrared. By means of appropriate parabolic or hyperbolic surfaces, spherical and coma errors can be corrected. Since mid of the 1990s, one was capable to form additionally a circular structure on these surfaces which has the effect of a diffractive element. Therewith, color errors and thermally induced deviations can be compensated in addition. The aspheric contour as well as the diffractive structure can be manufactured using the method of diamond-turning. In this manufacturing process, a fine diamond tip is guided over the surface while the lens is rotating with high, but carefully controlled speed. In principle, the procedure is the same as that at a conventional turning lathe, only that it is more smooth and exact by an order of magnitude.

All mentioned materials have in common, that they are quite uncooperative in use. Some have a high thermal expansion, the optical properties of others are depending strongly on temperature, or the machining is difficult because the material is too hard or too soft. Moreover, a few materials are even hazardous to health. Therefore, besides the high material costs, machining and mounting of the infrared lenses in a mechanical structure additionally requires considerable expenditure.

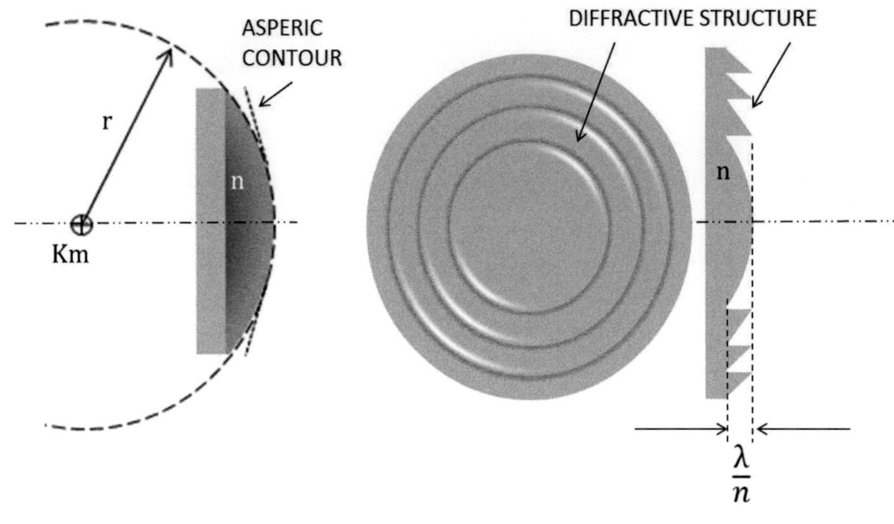


3 – *Design principles for aspheric (left) and diffractive surfaces*

*Aspheric surfaces: Compared to the commonly used spherical surfaces (shown as a broken circular arc) shaped like a spherical shell with radius r and centre of curvature (Km) the slope of the aspheric surface levels out towards the rim of the lens. Spherical surfaces typically have a refractive power that is too high in the marginal zone. This can be corrected by an aspheric contour that is depicted in the sketch in an exaggerated way.*

*Diffractive surfaces: Circular fringes are worked into the surface with a depth of λ/n (n = refractive index) corresponding to an optical path difference of one wavelength. Due to interference of the light bundles from the different zones an additional refractive power is generated which can be scaled by changing the number and the width of the fringes in the way that the colour error and the thermal influence can be corrected.*

Also for the detectors in the commonly used IR spectral regions exotic materials are required like, for example, cadmium mercury telluride for the so-called long-wavelength IR range (LWIR) from 8 to 12 µm or indium antimonide for the also common mid IR range (MWIR) from 3 to 5 µm. These exotic materials make IR detectors elaborate and costly, and have entailed that only detectors with relatively small numbers of elements have been technically feasible over a long time of history. In the early days, only one or a few detector elements per sensor were available to realize thermal pointers that could detect major heat sources. The creation of images was thus nearly impossible or only with major effort - the infrared scene could only be recorded by stepwise scanning in horizontal and vertical direction.

In the thermal imagers of the 1st and 2nd generation, addressed in the main chapters of this book, so-called line detectors were installed already, comprising a reasonable number of detector elements arranged in a column. As these detectors could cover just a strip of the image, the non-covered parts of the image had to be shifted to the detector by a scanner, e.g. by a rotating mirror. When this scanning process was repeated approximately fifty times per second, one could get a flicker-free image. The number of detector elements in the image strip determined the number of lines in the image; the resolution of the image along the lines was determined by the size of the elements. Altogether, this resulted in a typical optical-mechanical construction in which the optical system was partitioned into a telescope in front of the scanner and an imager between scanner and detector.

Only since the turn of the millennium, practical two-dimensional matrix detectors are available which can cover the whole image field. While the first of these

1

# EARLY HISTORY: LEARNING FROM NATURE

"LUX E DEO ET UMBRA"

("Light comes from God as well as shadow")

*Latin epigram*[3]

*8 – Principle design of the human eye*

*The incoming light is imaged by the cornea (C) and the eye lens (L) through the vitreous body (G) onto the retina (R). Approx. 2/3 of the optical power is generated by the cornea. Due to deformation of the eye lens, the eye can be focussed to varying distances (accommodation). The amount of light getting into the eye is regulated by opening and closing of the iris (I). The optical image of the scenery on the retina is upside down and laterally inverted. The fovea (F), an area where the resolution is highest, is located at the center of the field of view.*

Vision is by far the most effective sensory perception of the classical five senses (vision, hearing, smelling, tasting and feeling) that we humans have got. Vision contributes decisively to our ability to get a "picture" of the surrounding world. Consequently, our eyes are noteworthy optical instruments which are wonderfully designed for their purpose. They are ideally suited to effectively obtain visual sensory impressions under the typical living and environmental conditions, without being too complex[4].

## 1.1 Seen in the Light of Day: The Human Eye

The human eye comprises two sensor systems in the light sensitive retina, one for vision at daytime and a second for vision under twilight conditions. For daytime vision, chiefly the so-called cones are employed. The eye can adapt quite fast to moderate changes of brightness by changing the diameter of the pupil. When the brightness reduces more and more, the vision system is switched to night vision, for which the so-called rods are used as receptors. Altogether, that makes possible that we can view on a bright summer day without being dazzled as well as under the low-light level conditions of a half-moon night.

By means of the so-called accommodation, during which the eye lens is formed thicker controlled by a circular muscle, the eye can be focused quickly to every point in front of the nose of the observer that is farther away than approximately 20 cm. Therefore, a car driver, for example, can do a fast glance at the instruments on his dashboard and can observe the distant traffic at the next moment.

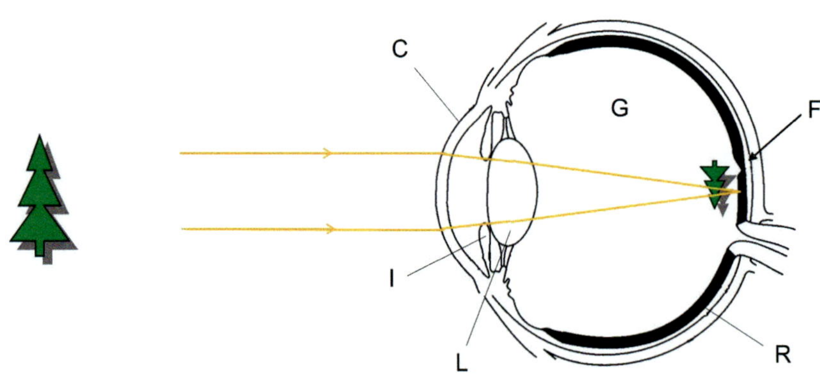


This capability is optimized in the way that in the center of the field of view (called the fovea) the achievable resolution has its maximum; the cones are packed exceedingly tight in this area. To improve the ability of discrimination, the human eye is designed to detect variant colors. For this purpose, the eye comprises not one but three types of cones for daytime vision. Each type is sensitive to another color (S-type: blue, M-type: green and L-type: red).

The high amount of information in the center of the field of view is sensibly supplemented by a less sharp but faster perception of dynamic processes at the peripheral field of view.

**Optical data of the human eye**

| | |
|---|---|
| Focal length | |
| at far point | 16.7 mm |
| at near point | 14.8 mm |
| Field of view, one eyed | 170° x 140° (h x v) |
| Field of macula lutea | Ø 0.5° |
| Field of fovea | Ø 5° |
| Distance for easy reading | 250 mm |
| Pupil diameter | |
| in bright light | 2 mm |
| in darkness | 8 mm |

**Properties of the human eye**

| | |
|---|---|
| **Daytime vision** | |
| Receptors | 5 million cones (S, M & L type) |
| Distribution | S (12%), M (54%), L (33%) |
| Spectral sensitivity | 380 nm – 780 nm |
| S-cones | blue sensitive around 440 nm |
| M-cones | green sensitive around 535 nm |
| L-cones | red sensitive around 565 nm |
| Visual acuity | 1 arc minute |
| Spatial resolution | 0.1 mm (at distance 250mm) |
| **Night vision** | |
| Receptors | 120 million rods |
| Spectral sensitivity | 380 nm – 650 nm<br>blue-green sensitive,<br>max. at 510 nm |
| Visual acuity | 10 arc minutes |
| Spatial resolution | 1 mm (at distance 250mm) |
| **Functional ranges** | |
| Day sight | with cones<br>(from dawn to dusk) |
| Sight at half-light | cones & rods<br>(day light to full moon brightness) |
| Night sight | with rods<br>(darker than full moon brightness) |
| Dynamic range | 17 orders of magnitude<br>(stimulus threshold<br>to absolute dazzling) |

Because the sun is the natural light source for all vision processes, evolution had adapted the eyes of the diurnal humans to the spectral distribution of solar radiation. The human eye has the highest sensitivity exactly in the wavelength region where the radiation power of the sun is maximum.[5]

However, the visual spectral range between 380 nm (blue) and 780 nm (red) used by men is only a fraction of the whole solar spectrum reaching roughly from 300 nm to 5000 nm. To obtain pictorial information from the other wavelength regions, humans must rely on technical auxiliaries. Typically, optoelectronic ("optronic") cameras are used today, which, by means of special detectors and optical systems, can obtain images in spectral bands that would be inaccessible otherwise. Already in classic 'infrared photography', visible pictures were recorded on specially prepared films in the near infrared (NIR) range at about 800 - 1100 nm that could be viewed by the eye. Today modern optronic cameras are available for the ultraviolet spectral range at wavelength beneath 400 nm as well as for the NIR and the shortwave infrared range SWIR[6] (at about 1100 - 3000 nm). The obtained image information is displayed in appropriate grey levels on a suitable monitor, and by this is transformed into the visible image.

The imaging processes discussed so far are fully depending on the radiation of the sun or of other, artificial light sources. In addition, there is the possibility to obtain images by means of the thermal radiation that is emitted by an object itself due to its surface temperature. The surrounding bodies approximately have room temperature and emit most of the thermal radiation in the longwave infrared region (LWIR) around 10 µm (= 10,000 nm) and almost none in the visible. To use the LWIR radiation for imaging, sophisticated infrared cameras have been developed with exotic lens materials and special detectors that must be cooled down to the temperature of liquid air. The technical details of these devices are presented separately in a "Short ABCs of Thermal Imaging".

## 1.2 Who Can See Thermal Radiation?

After all, the humans have managed to surpass nature in this field with much high-technology – at least is appears to be so! But is that true indeed?

Already the busy bees at the blossoms of an apple tree in spring demonstrate that the evolution had not exhausted all possibilities in the development of the human eye. Specifically, the eyes of the bees comprise receptors in the ultraviolet (UV) spectral range. This is the region of shorter wavelengths directly beneath the blue light. This capability enables the bees to find blossoms easier, which typically reflect UV-rays extremely strong, and thus appear outstandingly bright to the bees.[7]

But even more surprising devices nature can offer in the field of thermal imaging technology.

In the cretaceous age, about 100 million years before our time, the development of the species of serpents began. Over the length of time, more and more specialized forms evolved (approx. 3000 today). In the Miocene, about 20 million years ago, when the numbers of warm-blooded mammals were steadily increasing, the family of the vipers (viperidae) emerged as one of the highest developed families. A subfamily of the vipers is the pit viper (crotalinae)[8] which was equipped by nature especially for hunting small mammals in twilight or during the night. These vipers did not get their name, because they perhaps were digging pits, but because each of the existing 169 species known today have two characteristic pit organs. These are arranged at each side of the head between the eye and the nostril; therefore, some species in Latin America are named "cuatro natrices" (four nostrils)[9]. This probably old designation is based on the mere appearance but not on the function of these organs. Even the first anatomical descriptions, being drafted many decades ago, did not catch the function of the pits. Only towards the end of the 1930s, when the scientists already had learned a lot about infrared technology, and the first thermal imagers had been built (see chap. 3), the mystery of the pit organs could be resolved.

In **1937**, two Americans, G. Kingsley NOBLE and A. SCHMIDT, working at the "AMERICAN MUSEUM

FOR NATURAL HISTORY" in New York, could proof, that the rattle snakes (crotalus) could detect heat with their pit organs. "The animals could even distinct a warm from a cold light bulb when the eyes were taped and the remaining sensory organs were disabled. As long as the pits remained clear, the snakes struck at the warm bulb; only after the pit organs were blocked too, they did not care about the light bulb anymore."[10]

In the **1950s**, Theodore A. BULLOCK and his collaborators at the University of California in Los Angeles proved that it is indeed the thermal radiation that causes neural stimuli in the brain of the snakes. "The pulses caused by a warm body ..., were equally strong in the light or in darkness. However, no response came when the same object was offered in a cooled state. The activity of the nerve fibers also vanished, when BULLOCK put a heat-absorbing glass filter between object and snake, although virtually all visible light passed the glass. The other way round, the response was changed only a little by a filter, that transmitted only infrared radiation."[11]

Thus, the pits are highly developed sensor organs for seeing thermal radiation, which enable the snakes to detect very small temperature differences of less than 1/100 degree Celsius. Moreover, both pits of a rattle snake are directed forward, and enable the snake to obtain stereoscopic infrared images. Therewith, a viper can judge position and size of a warm object in the dark with similar precision as a human can do with his both eyes at daylight.

In experiments, a blind rattle snake had struck its prey in 98 % of the attempts; when the pit organs were covered, the strike rate fell to only 27 %.[12]

Of course, the additional thermal imaging technology is a strategic advantage for the rattle snakes, which preferably hunt at twilight for lizards, birds and small mammals, because the warm-blooded mammals, and

9 – ***Head of a Texas rattle snake (crotalus atrox)***

*On the left side of the head between eye and nose the pit organ is clearly to be seen (yellow arrow) which helps the snakes to see thermal radiation emitted by their warm-blooded prey.*

*(Picture: Rainer Altenkamp)*

particularly the birds (body temperature about 40°C), are easily noticeable due to the high body temperature – as an observer with a modern thermal imager can easily see himself. Even 'coldblooded' lizards are not really cold, but typically show a small temperature difference relative to the surrounding due to their metabolic activities.

How can the pit organs obtain a thermal image?

A general problem of building a thermal imager is the availability of optical materials with enough transmission in the infrared (IR) region. Up to now, no organic material is known suitable for IR optics. Even artificial organic materials are sufficiently transparent only if they are very thin (about 0.1 mm). Moreover, these materials are not environmentally resistant and therewith not suited for practical application. The evolution has circumvented this issue elegantly by using no infrared optics at all. The pit organs are formed as a pinhole camera or "camera obscura" (Latin for dark chamber). The radiation coming from the scene enters a dark cavity through a small opening and generates a photographic image of the scene on the rear wall. When the opening is made smaller, the image is becoming sharper. However, this is at the cost of brightness of the image which even decreases with the square of the hole diameter.

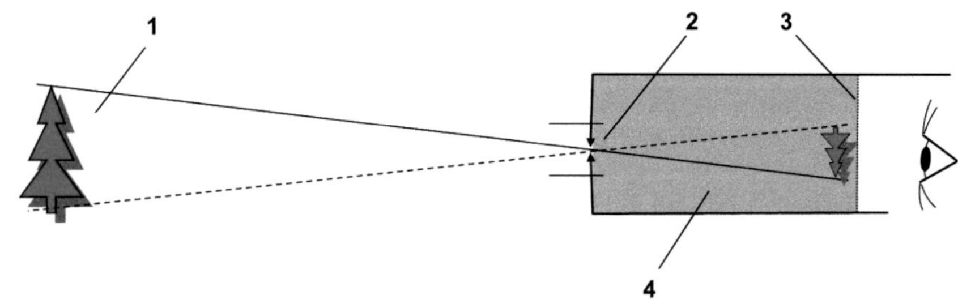


10 – *Principle of a pinhole camera (camera obscura)*
*The light coming from the scenery (1) passes through the pinhole (2) into a dark chamber (4) and produces a sharp image on the screen (3) that is upside down and laterally inverted. The image quality is almost perfect but the brightness is quite week compared to a conventional camera with a lens.*

Mankind is familiar with such pinhole cameras since ancient times.

In **330 BC**, the Greek philosopher and scholar ARISTOTLE described a "camera obscura" for the first time in his book "Problemata physica", and therewith effectively founded the science of optics. In the Middle Ages, artists fancied such cameras as a drawing aid.

The second big issue of thermal imaging is creating a detector highly sensitive to thermal radiation. For the pit organs, evolution has invented a type of detector that in today's thermal imaging technology is known as bolometer (from Greek bole = throw, beam).

In **1878**, millions of years after the rattle snakes emerged with the pit organs, such a bolometer was invented by the American astronomer, physicist and aviation pioneer Samuel Langley. In his bolometer, the radiation to be measured impinges on a strongly absorbing, thin metal sheet. Due to the small mass of the sheet, which is thermally isolated against its surroundings, is heated up significantly even by weak irradiation. By measuring the change of the electrical resistance LANGLEY could measure the intensity of the incoming radiation (see also chap. 2).

In the pit organs, the thin sheet corresponds to an only 15 μm thick membrane which is isolated by an air cushion against the surrounding tissue. The water content of the membrane, which is supplied with blood very well, ensures a high absorption of thermal radiation. The absorption causes a noticeable heating of the thin membrane. Therewith, the numerous heat sensitive receptors (nerve endings with many mitochondria) in the membrane are excited to emit neural stimuli, which are led directly into the brain of the snake by two branches of the trigeminal nerve (nervus trigeminus).[13]

This bolometer principle, and the manner how it is realized in the pit organs, makes the heat sensor of the pit vipers the most sensitive of the whole animal kingdom. Quantitative investigations have shown that the pit organs are a million times more sensitive than all other sensors one can find at mammals including human beings.[14]

It was proved that already a tiny temperature difference of only 0.003 °C induces the receptors in the pit organs to fire sensory stimuli. This value is on the level of the best technical thermal imagers available today.

The attached sketch of a pit organ may lead us to expect that the spatial resolution cannot be very high, due to the relatively large diameter of the opening being roughly one third of the pit diameter. Nevertheless, a rattle snake with fully covered eyes misses a target seldom by more than 5°.[15] Part of a possible explanation is perhaps, that the vipers have a sophisticated neural image processing which still is an object of scientific investigations today.[16]

Generally, both imaging sensory organs of the pit vipers (the visual system and the thermal imaging system) are not used isolated; the evolution has given to the snakes a capability on a neuronal level, that we would name image fusion by today's word usage. The researchers have found out that in the pit vipers' brain the thermal and optical image information is stored close together in a part called "tectum opticum", which is dedicated to vision. Therefore, subsequent information can be derived very quick by logical operations.[17] The results of these are again stored in the brain in parallel layers, and thus are available together with the original information, e.g. for a quick assessment of a potential prey during a nocturnal hunt (Is it a mouse, a frog or just a warm pebble?).

For this purpose, e.g. nerve cells exist providing a response that corresponds to the logical 'AND' and 'OR' functions, respectively.[18] Furthermore, other cells exhibit functions which can be outlined as 'light signal suppressed by infrared' or 'thermal signal suppressed by light'. By

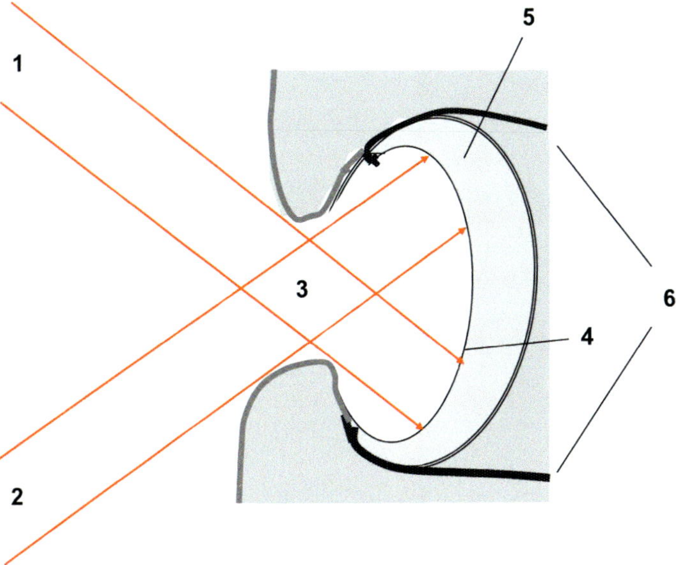


11 – *Principle design of a pit organ*

*The bundles of thermal radiation coming from different objects in the scenery (1 & 2) pass through the pinhole (3) and hit a membrane (4) only 15 µm thick in which thermally sensitive receptors are located. These are directly connected to branches of the trigeminal nerve (6) which conduct the neuronal stimuli directly into the brain of the snake. The membrane is thermally insulated against the surrounding tissue by an airspace (5). The very same principle is used in today's technical bolometer detectors.*

combination of several signals the pit vipers can create a neuronal software best fitting for their living conditions. For example, a combination of AND- and OR-function may result in a useful mouse-detector (mouse = warm, moving target).

And finally, the thermal imaging system is a major part of the 'fire control system' of the snake for the final strike which also triggers the bite reflex. How important the pit organs are for the cognitive and the locomotor system is demonstrated by an extreme experience: even after the death of a pit viper, a warm object, e.g. a warm hand, can still trigger the bite reflex. This can happen even an hour after the snake's death, and according to the statistics, about 15% of all snakebites treated medically in the United States were caused by dead snakes.[19]

**Technical Data of Rattle-snake Pit Organ**

**Detector**

Bolometer based on 15 μm membrane with air insulation and nerve-endings as receptors

| | |
|---|---|
| Wavelength range | approx. 1 - 1000 μm |
| Cooling | none |
| Receptor size | 60 μm |
| Number of image points | approx. 7000 |
| Infrared optics | pinhole camera, no lens |
| Pit diameter | 1 to 5 mm |
| Aperture | 1/3 of pit diameter, F/3 |
| Field of view | 130° x 105° |
| horizontal plane | -105° x +25° |
| vertical plane | -45° x +60° |
| Image reproduction | neuronal |
| Resolution | |
| angular | 5° |
| thermal | 0.003 K |
| Range (against mice) | up to 1.2 m |

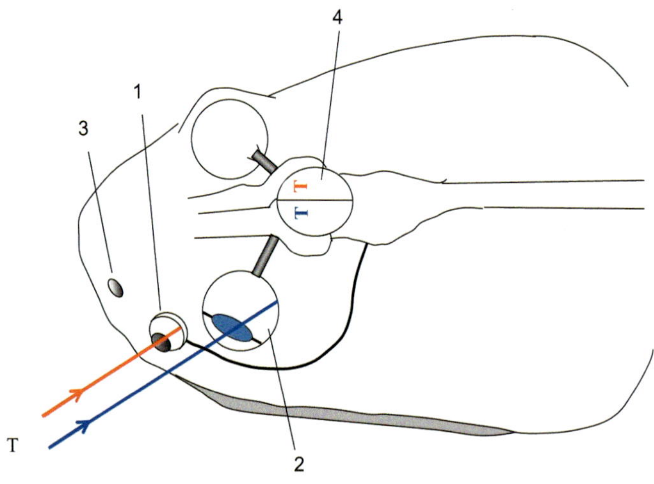


12

12 – *Processing of thermal and visual images (schematic)*
*The thermal radiation (red) coming from the object (T) is converted into neuronal stimuli by the pit organ (1) in a way that is analogous to what happens in the eye (2) with the incoming light (blue). Both images are stored in the same part of the brain which is known as tectum opticum (4) with an unambiguous assignment. Pit organs are located at both sides of the head between nostril (3) and eye (2), respectively.*

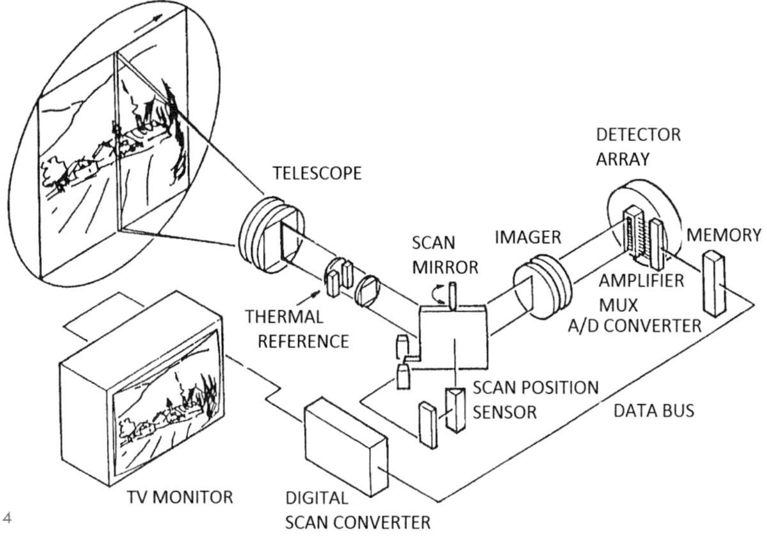


4

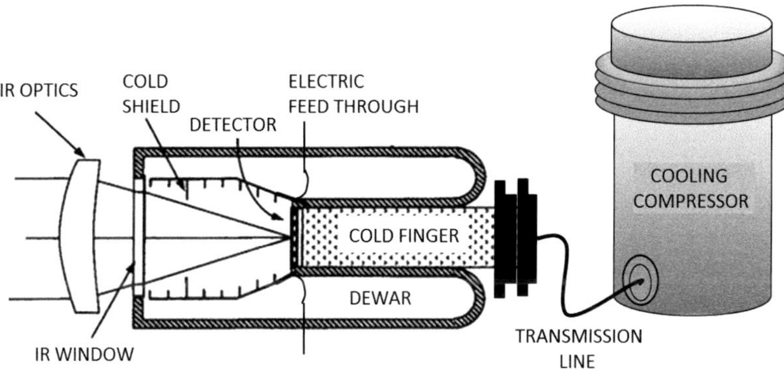


5

4 – *Principle of a typical thermal imager with scanner*

*The thermal radiation emitted by the scenery is imaged onto the surface of the detector chip via an afocal telescope (creating the specified field of view), the oscillating scan mirror and the imager. Because of the one-dimensional detector array (column detector) one column of the scenery is imaged at a time and the signals are stored in a memory after being amplified and digitized. By means of the rotation of the scan mirror, all columns of the scenery are captured and stored one after the other. From the contents of the memory a digital scan converter generates a standard video image to be displayed on a commercial monitor.*

5 – *Principle of a cooled infrared detector*

*The incoming thermal radiation is imaged by the IR optics on the surface of a detector chip that is mounted directly on a cold finger that has a temperature of approx. -200°C. To achieve an insulation as good as possible, the detector chip is encapsulated by a vacuum vessel known as Dewar which comprises an IR transparent window towards the optic. The Dewar also effectively prevents fogging of the detector. In front of the detector a carefully designed cold shield is mounted in order to block unwanted thermal radiation, originating from the interior of the housing, which may cause disturbing image artefacts. Today the electrical cooling machine mostly is a linear motor which is connected to the cold finger via a flexible transmission pipe through which the circular process in the cold finger is driven by exactly timed pressure waves. By this process, thermal heat is pumped from the detector side to the warm side of the system. Virtually all of today's cooling systems are using the Stirling cycle.*

staring thermal imagers had a relatively small number of elements (e.g. 256 x 256), today also megapixel detectors with 1280 x 1024 elements or more are available which allow to obtain infrared images according to the high-definition television HDTV standard.

The required cooling of the detector to approximately -200°C is a further costly necessity of thermal imagers to reduce the thermal noise to a tolerable amount. For this purpose, the detector chip must be isolated very well by a Dewar and sealed hermetically against condensation.

*6 – Recording of a thermal image*

*Due to the different temperatures, all objects of the scenery emit thermal radiation of varying intensity that is invisible to the human eye. On the way to the thermal imager the radiation must pass through the atmosphere and may be weakened due to absorption and scattering. The incoming radiation is imaged by the infrared optics onto the surface of the special infrared detector, where it is converted into electrical signals that can be displayed on a monitor and so viewed by the eye.*

This makes an elaborate optical-mechanical-electrical structure necessary. The deep temperature is typically generated by means of a cooling machine which works according to the Stirling-cycle, and uses the noble gas helium as a working medium.

For less sophisticated applications, also uncooled detectors are available today, using the so-called bolometer principle: The thermal radiation, absorbed by a blackened detector surface, heats the detector element up. The temperature change results in a change of the electrical resistivity, which is converted into an electrical signal that is proportional to the incoming radiation power. As cooler and Dewar can be omitted, these detectors are more cost-effective by far compared to cooled detectors. Their specific sensitivity, however, is roughly one order of magnitude lower.

## Capturing thermal images

If all objects of a scene to be imaged would have the same temperature, they all would appear equally grey in the thermal image, and would not differ from each other. Fortunately, practical experience shows that in a natural environment the dynamic interaction between solar radiation, wind and different heat absorption of the objects always generates temperature differences. Only in rainy weather it can happen that these temperature differences are as low as a few tenth of a degree. Modern detectors, however, are sensitive enough to generate sufficient contrast for a meaningful thermal image.

It is a happy coincidence for thermal imaging that the atmosphere at all is transparent for thermal radiation; otherwise long-range observation would not be

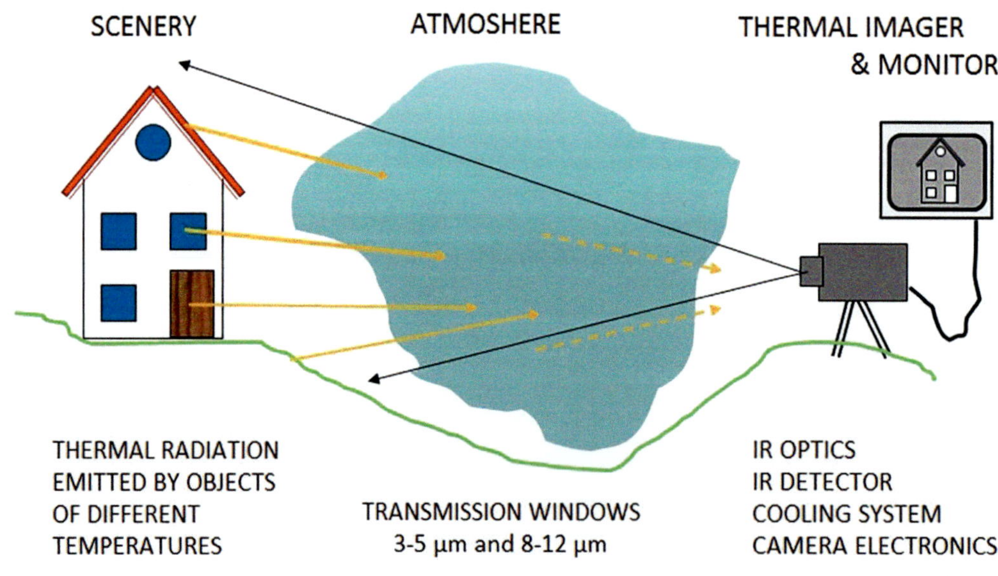

possible. For obtaining thermal images, two so-called infrared windows are of special importance: In the mid infrared region from 3 - 5 μm (MWIR) and particularly in the long wave infrared window at 8 - 12 μm (LWIR) the atmosphere transmits well under normal conditions. Luckily, both windows are in the region of maximum emission of thermal radiation by our surrounding. Outside these windows, transmission is strongly reduced, mostly due to absorption by the omnipresent water vapor.

Because the thermal radiation of a body is ever present, thermal images are available at daytime as well as at night. Therefore, thermal imagers have a night vision capability that is independent from any illumination (e.g. sun or searchlight).

During the day, direct vision or color TV cameras are generally favorable for surveillance and reconnaissance because of the higher spatial resolution and the color information. However, thermal imagers can deliver valuable information also at daytime, whenever large temperature differences between heat generating objects ("hot spots") and the natural background can be used. The waste heat of engine-driven land, naval and aerial crafts are examples as well as a hot gun barrel or a missile heated by friction. In these cases, the thermal image often is superior to a visible image because it can unmask the objects.

A normal atmosphere with sufficient transmission was assumed so far. To be dependent on adequate normal visibility, however, is a fundamental problem of optronics. In heavy fog, snow or a sand storm in the line of sight, usually no observation is possible. But specially in conditions of poor visibility, the thermal imaging equipment offer still benefits. Because the infrared radiation has an up to twenty times longer wavelength than visible light, it is much less scattered by airborne particles in smoke, haze or dust. Therefore, thermal imaging devices can better penetrate these disturbing phenomena of nature than visual observation means. Also, fire and backlight on a battle field have almost no dazzling effect, because they emit a comparatively low radiation power in the longwave infrared range.

## Validation of thermal imagers

The noise equivalent temperature difference (NETD) is widely used to describe the mere temperature sensitivity quantitatively and to measure it reproducibly. It indicates the temperature difference that two objects must have so that their signal difference is greater than the noise of the thermal image, and that the corresponding thermal images differ significantly in brightness. The NETD, usually given in the temperature unit Kelvin (K), is influenced by detector sensitivity, detector noise, transmission and the aperture of the imaging infrared optics but not by geometric resolution.

To account for all parameters, relevant for the performance of a thermal imager, generally the so-called "minimum resolvable temperature difference" (MRTD) is used as a figure of merit. The MRTD specifies the temperature difference between the two lines of a line pair in a scene that is necessary, so that the line pair is visible in the thermal image. The necessary temperature difference generally increases when the width b of the line pair becomes smaller. Therefore the MRTD usually is given as a function of MRTD ($\omega$) of the so-called spatial frequency $\omega$, defined as $\omega = 1/b$.

*7 – Principles of range calculation*

*The effective temperature difference curve ΔT(r) shows how the temperature difference seen by a thermal imager decreases with increasing distance between object and thermal imager. The MRTD curve of the thermal imager is at first hand a function of the spatial frequency and increases towards smaller object structures (= higher spatial frequency). Using the specified object size the spatial frequency can be converted into range. The intersection point of both curves marks the maximum observation range possible: For this distance, the effective temperature difference equals the minimum temperature difference that is needed by the thermal imager to resolve the structure of the object.*

The MRTD contains all relevant parameters that are decisive for the performance of a thermal imager, and is therefore understandably classified for many military systems. Vice versa, one can also say that the manufacturers of thermal imaging devices are responsible only for the MRTD; they really cannot influence the temperature of the observed objects and the quality of the atmosphere in practice.

Nevertheless, the users who employ such devices to the observation and reconnaissance over long distances, of course, like to have quantitative information about the performance of their device. To this, the specification of the observation range of the imager is useful.

Besides the MRTD, many external parameters enter into the range calculation of a thermal imager. Range specifications are therefore only useful and usable, if the underlying parameters are also specified. Thus, the following information should be listed always:

- Size of the object/target to be resolved (such as NATO standard target)
- Temperature difference between target and background
- Transmission of the atmosphere (weather)
- Reconnaissance task: detection, recognition or identification of an object

Even with the utmost effort, range evaluation basically remains a simulation, since in addition to the appropriate atmospheric models also assumptions must be made of how an observer perceives the target object.

What will happen in practice is a different story.

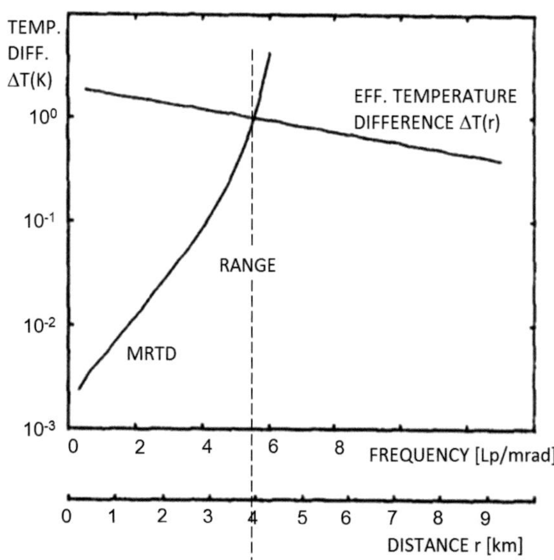

# 2

# FUNDAMENTAL RESEARCH UNTIL 1900: WORTH A NOBEL PRIZE

"There are rays coming from the sun … invested with a high power of heating bodies, but with none of illuminating objects. […]

The maximum of the heating power is vested among the invisible rays …

It may be pardonable if I digress for a moment and remark that the foregoing researches ought to lead us on to others …"

*Sir William HERSCHEL (1800)*[20]

„Geburtstag der Quantenphysik"

"Birthday of Quantum Physics"

*Max von LAUE (December 14, 1900) after the presentation of the new radiation law by Max PLANCK*

Whoever is searching for the origins of science not seldom will find something in the classical antiquity. This is also the case for optical imaging, as was already shown briefly in Chap. 1 by the example of the "camera obscura"[22]. That this is also true for thermal imaging technology may be surprising. Today's manifestation of thermal imaging can hardly be imagined without (digital) electronics, however, the idea of a pictorial presentation of an interesting heat distribution is very old indeed. Already in ancient Egypt it was common practice that a physician felt for temperature differences on the body of a patient by sweeping a finger over the skin of the affected parts of the body.[23] As we can confirm scientifically today, this was not unskillful at all, because the nerves in a human finger are even reacting to temperature changes which are smaller than 1/100 °C.

## 2.1 Antique Experiments on Temperature Distribution

The merit to have obtained the first pictorial representation of a heat distribution belongs to the Greek physician HIPPOCRATES whose name is familiar today even to the medical layman by the "Hippocratic Oath" that is named after him.[24]

By about **430 B.C.** HIPPOCRATES of Kos[25] was on the height of his work, and had accomplished already most of his pioneering achievements. Contrary to the conventions of his time, he proclaimed the separation of mystical and medical approach to a disease. The systematic observation of the progress of disease belonged to the main emphasis of his work (today's "clinical observation"), and writing of meticulous reports which allowed learning from experience. As the result of his studies, he could categorize the diseases in classes (acute, chronic, endemic and epidemic) which are in use until today, and he described a lot of typical disease patterns for the first time, among others also pulmonary inflammation and tuberculosis. During the observation of lung diseases, which he performed with special interest, perhaps he got the idea, to improve the diagnosis by a more precise investigation of the heat distribution of the chest. For this purpose, he put a moist, clay soaked tissue on the upper body of the patient and observed the drying process. Because the tissue dried faster at the patches where it was in contact with the warmer parts of the skin, these patches appeared brighter quite soon. After a few minutes, the tissue in total showed a pattern that displayed the heat distribution of the underlying skin. Such a presentation is commonly known today as "Thermogram", and HIPPOCRATES was the first to use this imaging method according to the deliverances from the past.[26]

We today could possibly guess that this narrative is only one of many legends which are told about HIPPOCRATES' work, and which in part are even contradictory.

In 1997 Japanese researchers[27], however, demonstrated in an experiment that the method is really of practical use, and that the image information obtained by this method, is comparable to a thermogram obtained with today's means. This positive result makes the narrative a lot more credible.

After this pioneering feat, however, several centuries had to go by until during the "Scientific Revolution"[28] in the 17th century new advances were made in the optical sciences.

## 2.2 Experiments on Thermal Effects of Radiation

As early as in **1660**, experiments were reported by the ACCADEMIA DEL CIMENTO in Florence[29] in which the heat of a hot body was focused onto a remote spot by collecting mirrors. The same was achieved also with the coldness of a cold body. One was talking about "heat radiation" and "cold radiation", and could demonstrate that both types of radiation could be reflected in the same way as light.[30] Other researchers (e.g. Edme MARIOTTE and Robert HOOKE) as well performed various experiments with thermal radiation from diverse sources. MARIOTTE, for example, observed in 1682 that the heating in the focal point of a collecting mirror was reduced by one fifth when he covered it with a glass plate.[31] As these researchers only had the human hand as a sensor,[32] the evidence of the experiments was rather limited, if not contradictory observations were made at all.

In **1672**, the English mathematician, astronomer, theologian and physicist (described in his time as a "natural philosopher") Sir Isaac NEWTON[33] published his important work "New Theory about Light and Colours". At the end of his studies at the TRINITY COLLEGE in Cambridge in 1665, the Great Plague (1664-65) had hampered further activities at the college. When the quarantine ended in 1667, he returned and

**13 – *Principle of colour dispersion with a glass prism***

*The white light coming from the left (1) is refracted by the inclined surfaces of the prism (2) and split into spectral colours. In optical science, this process is known as dispersion.*

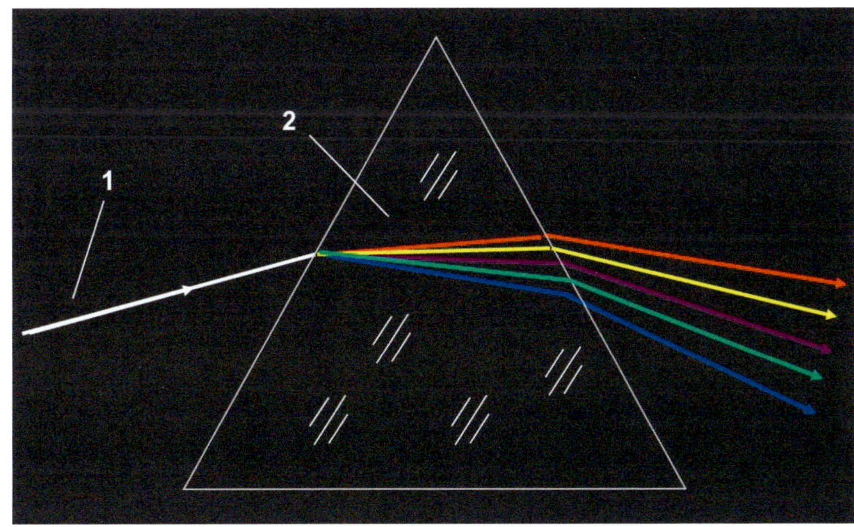

initially was elected as a fellow of the TRINITY, and became Lucasian professor for mathematics in 1669. Right at the start of his work, he gave lectures in optics and in parallel began his scientific work on the same matter: "I procured me a triangular glass prism to try therewith the celebrated phenomena of colours."[34]

With slit and prism, he performed comprehensive investigations of the color dispersion of light in the following period. His experiments led him to the conclusion that white light was composed from different-colored fractions which could be split by a glass prism. Therewith, he proved the antique idea wrong, that light was white by nature and the colors were added by the prism. For the first time, the pioneering idea was worded that there are different forms of light. In the publication mentioned above, NEWTON also was the first to use the term "spectrum" (Latin for appearance or apparition) for the colors of the decomposed light.[35] How much especially this knowledge of NEWTON had influenced the imagination not only of the scientists, can be seen from the marble statue in the chapel of the TRINITY COLLEGE, which was praised by England's best known romantic poet William WORDSWORTH in 1805 with the words:[36]

"The Antechapel where the Statue stood
Of Newton, with his prism and silent face"[37]

The prism became a symbol for the high appreciation that, in his times, was given to NEWTON's contribution to the clarification of the nature of light. Besides this, even his many other pioneering contributions, e.g. to mechanics, gravitation or infinitesimal calculus did fade out. In the following, generations of researchers have made many experiments to discover the entity of radiation and the interaction of radiation and matter, a work that ultimately took two centuries. NEWTON himself had forged his results into a general theory of optics.

In 1704, his principal optical work "Opticks or a treatise of reflections, refractions, inflections and colours of light" was published in which NEWTON presented the fundamental elements of his theory:

- Light comprises extreme small particles
- Light travels with very high (infinite) velocity
- Colors are formed by different particle size

With these assumptions NEWTON could explain all contemporary observations without problems, e.g. the origin of the colors of a rainbow. His particle theory, apparently inspired by his mechanical ideas, brought NEWTON in a strong opposition to the theory of the Dutch Christiaan HUYGENS[38] and his supporters who fancied light to be a wave (wave theory). At the time when the interference of ray bundles and double refraction in crystals was better investigated, it turned out that these optical effects could be explained easier by the wave theory. Hence, it seemed to the next generations of researchers as if NEWTON had been wrong with his particle theory. However, at the end of this chapter we shall see, that for a comprehensive explanation of the optical phenomena light particles are indeed useful – they will come back in the guise of the photons of quantum mechanics at the end of the 19th century.

The next scientific contribution about thermal radiation was astonishingly made by a female French mathematician and physicist. Even considering her aristocratic ancestry, it was quite unusual that a woman was active in science during the Age of Enlightenment in France.

In **1737**, Gabrielle Émilie Le Tonnelier de Breteuil, Marquise du Châtelet - best known as Émilie du Châtelet[39] - published a paper "Dissertation sur la nature et la propagation du feu" (Essay about the Nature and the Propagation of Fire). Therein she reported her research about fire, and, besides general considerations about the nature of light, produced the forecast that thermal radiation must exist.

Her further scientific contributions comprised, among others, the first correct calculation of the kinetic energy.[40] Her crowning achievement was the translation and commentary of NEWTON's main work "Philosophiae Naturalis Principia Mathematica" into French language. The competent commentary made the approach to this enormous piece of work much easier to her compatriots, and up to today her translation is considered as the standard work in French language.

Outside of science, Émilie du Châtelet was graced with various talents. She spoke Latin, Italian, Greek, and German fluently, made translations of Latin and Greek texts, and wrote several books about social and biblical issues. She also was not shy to employ her mathematical capabilities in successful gambling strategies. Moreover, she even performed cleverly worked out financial arrangements, which we today probably would call a 'derivative'.[41]

VOLTAIRE was fond of her capabilities, and declared in a letter to his royal friend King Frederick the Great of Prussia that she was "a great man, whose only fault was being a woman."[42] In person, he obviously was not sorry about that – for some time he was a lover of the marquise who lived in an arranged marriage with an older man.

In **1752**, the Scottish natural philosopher Thomas MELVILL[43] was the first to conduct experiments with flames which he made to shine in variant colors by injecting salts. For his observations, he also used a prism, and could detect the emission of infrared radiation by a sodium-flame.[44]

In **1777**, the German-Swedish apothecary and chemist Carl Wilhelm SCHEELE[45] published his "Chemische Abhandlung von der Luft und dem Feuer" (Chemical Treatise on Air and Fire) in which he presented the results of his measurements of thermal radiation. He was capable to discriminate between convection and emission of heat. For the latter, he coined the term "radiant heat" to lay emphasis on the discovered straight propagation of the radiation.

Also in **1777**, the Italian chemist, physicist and meteorologist Marsilio LANDRIANI[46] repeated NEWTON's experiment to disperse white sun light with a glass prism, and made quantitative measurements of the irradiation of the visible spectral colors with a thermometer.[47].

In **1791**, the Swiss scientist Marc-Auguste PICTET[48] in Geneva tried to measure the propagation velocity of thermal radiation. For this purpose, he built a measurement path between two collimating mirrors. He put a hot iron ball into the focus of the first mirror as a heat source. In the focus of the second mirror a thermometer served as a sensor. After unblocking of the optical path by removing the stop, PICTET observed that the reading of the thermometer raised without noticeable delay, from which he deduced that the velocity must be high. With respect to the limited technical means of his time, PICTET's approach was nevertheless noteworthy.

## 2.3 Research on Light and Radiation

Until the end of the 18th century, already some thermal effects of light and infrared radiation had been investigated empirically. "All the work performed prior to 1800 dealt with the study of the joint thermal effect of spectrally non-decomposed light and infra-red radiation. Although the experiments carried out in those days were actually handling infra-red radiation, there was no notion as yet of its existence, or of the differentiation of light from heat, the two latter concepts being merged into a vague single concept of "radiant heat".[49] In 1800, the well-known German-British astronomer William HERSCHEL (since 1816 Sir William) came to the topic by chance rather than by design. At least since his discovery of the Uranus in 1781, HERSCHEL was internationally established, and honored in England as "The King's Astronomer". Up to this, his life had been everything else than easy and straight forward. The 14 years old Friedrich Wilhelm HERSCHEL[50] enlisted for the Hanoverian foot-guards as a military musician (oboist). After French troops had occupied his hometown Hanover[51] only a few years later, he had to escape to England. There he could earn a living initially as a music teacher and organist. In 1766, he became music director in Bath, and first success came as a composer (in total he composed 24 symphonies). Moreover, his financial situation stabilized, so that he could send for his 12 years younger sister Caroline HERSCHEL[52] in 1772. Besides his musical activities in Bath, he had also the chance to follow his astronomical ambitions, inherited from his father, and he performed sky observations with telescopes, developed and built by himself. By the acquaintance with the English "Astronomer Royal" Nevil MASKELYNE his engagement in astronomy was encouraged, and he began to build and sell mirror telescopes, with the resourceful support of his sister Caroline. The high quality of his telescopes allowed HERSCHEL to make many new observations and was highly esteemed also by other astronomers. Within his working life, he had built, including his famous 40-foot telescope, more than 400 (!) astronomical instruments in total.

On **11 February 1800,** HERSCHEL tested optical filters which he intended to use in the observation of solar spots. The observation of the sun is not without peril. Due to the high irradiation of the sun, irreversible eye damages can happen very quickly if the radiation load is not dimmed by filters to a bearable level. HERSCHEL used in principle the same experimental setup with a dispersion prism as NEWTON had done more than hundred years earlier.

The incoming sunlight was decomposed by the prim into its spectral colors, the irradiance of which was measured quantitatively by a thermometer. So far, the experiment was not new. HERSCHEL, however, discovered the infrared radiation, when he arranged a second thermometer, which he used to monitor room temperature, beyond the red end of the visible spectrum. To his astonishment, this thermometer displayed a significantly higher temperature than that in the visible spectrum. Further systematic investigations revealed a distinct maximum of the heating effect that was far in the region of the invisible rays.

HERSCHEL named this invisible part of the light "the invisible rays", "the thermometrical spectrum", "the rays that occasion heat" or simply "dark heat". The term "infrared" used today was coined much later. By means of the same apparatus, HERSCHEL also investigated the radiation of fire, candles and a kitchen stove.

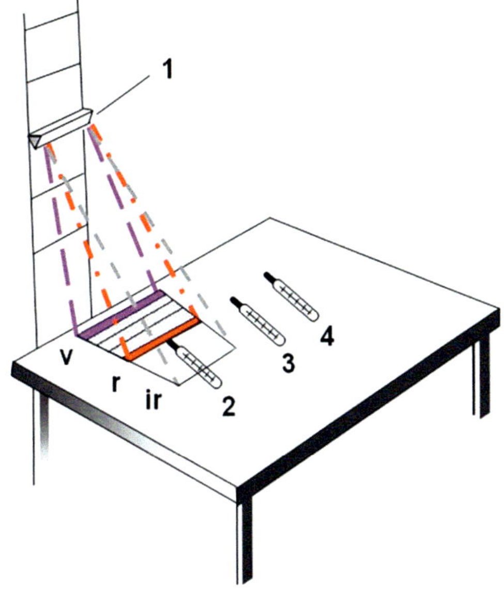


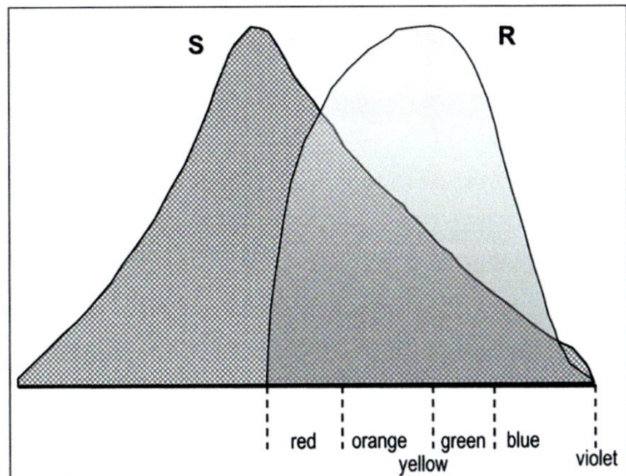


**14 – *Sir William Herschel's measurement of solar radiation***

*The sunlight coming in from the left is dispersed into its spectral colors by a prism (1) so that a colored spectrum is to be seen on the table plate. The visible part of the spectrum from violet (v) to red (r) is deviated more than the infrared (ir) part. The radiation is detected with blackened thermometers. In the sketch one thermometer (2) is positioned in the visible part. Two more (3, 4) are used to monitor the room temperature because the recording of a sequence of measurements takes a quite long time. To reach an accuracy of approx. 0.5°C with the blackened mercury thermometers of that time each reading needs a measuring time of up to 16 minutes. Later researchers have developed smaller and faster thermometers the reading of which was improved by a microscope so that they could achieve 0.1° temperature resolution. Herschel's curve (S), measured with the thermometer, was significantly different from the intensity distribution (R) observed visually.*

Moreover, he compared the temperature change with and without filter in front of the thermometer, which was a first rough measurement of the infrared transmission curve of the filter.

During his search for the best sun protection filter he measured in total more than 50 samples made of various materials like stained glass, natural crystals, water, gin, spirits, paper or textiles. In the end, he chose diluted ink in a glass cuvette as a sun filter.[53]

Despite some previous work of other researchers, one can say that HERSCHEL opened a new scientific field with his results, obtained within a length of time in 1800. He wrote that much also in his publication:

"There are rays coming from the sun…invested with a high power of heating bodies, but with none of illuminating objects. … The maximum of the heating power is vested among the invisible rays…. It may be pardonable if I digress for a moment and remark that the foregoing researches ought to lead us on to others."[54]

The most important contribution, however, was the good base he bequeathed to the subsequent researchers:

- He shows that useful sensors were available for further investigations.
- He asked provocative questions about the similarity of light and thermal radiation.
- He demonstrated that significant differences existed in the way how light and thermal radiation was transmitted or absorbed by various materials.

It is surprising that after the publication HERSCHEL virtually ignored his discovery, and only returned to it once in 1801. Despite of today's significance of infrared technology, it is thus almost logical that his contribution goes unappreciated by many an encyclopedia, and quite often is not mentioned at all. Not so in Sir William HERSCHEL's longtime residence Slough (Buckinghamshire, England) where the memory of his discovery of the infrared radiation is still alive. In 2011 a new bus station was built in the center of Slough, the architecture of which was inspired strongly by HERSCHEL's infrared experiment.

Shortly after **1800**, the German-Baltic physicist Thomas Johann SEEBECK[55] was one of the first who came into HERSCHEL's scientific inheritance, and investigated also the thermal effects of the variant colors of sunlight. Actually, he had studied medicine but then transferred to physical research and worked in Jena, Bayreuth and Nuremberg as private scholar, and finally at the Academy of the Sciences in Berlin. He worked, e.g. together with Johann Wolfgang von GOETHE on a theory of colors, and discovered the color sensitivity of moist silver oxide (in 1810) as well as the rotation of the plane of polarization by sugar solutions (in 1818).

At about the same time, the German physicist and optician Joseph FRAUNHOFER[56], coming from a poor background, had already acquired a reputation by manufacturing optical lenses and instruments.

In **1814**, FRAUNHOFER discovered the absorption lines in the sunlight, named "Fraunhofer lines" up to the present day, when he was testing his newly developed spectroscope. These dark lines in the sun spectrum had been observed before in 1802 by the English chemist William Hyde WOLLASTON[57], what perhaps was unknown to FRAUNHOFER. He thoroughly investigated the absorption lines and established a catalog with 570 lines in the visible spectrum, which later could be related to absorption bands in the gas envelope of the sun. Furthermore, he also observed the infrared lines caused by the absorption of the earth's atmosphere, and more than 700 of which were believed to be discovered by him.[58]

In **1821**, SEEBECK rendered a special service to the optical science by discovering the thermo-electrical effect, called "SEEBECK effect" even today. Right before, SEEBECK, who had already discovered the magnetism of nickel and cobalt, had come back to the Berlin university to work on the electrical magnetization of iron. When he had once formed a simple circuit with wires of distinct materials (one wire was from antimony, the other from copper[59]), he noticed a small electrical voltage as soon as a temperature difference existed between the contact points.

Therewith, he had found a connection between the thermal and the electrical effects, which could be used in many ways, e.g. to build electrical thermometers, called thermocouples, which up to now are state-of-the-art in many devices and machines.

In **1829**, the Italian artillery officer, veteran of the Napoleonic Wars, physicist and builder of measuring instruments Leopoldo NOBILI[60] was the first to make practical use of the "SEEBECK effect". He used the effect to build an electrical thermometer that was much more sensitive and faster than all mercury thermometers employed up to then. Due to the smaller mass of the sensor element, also the drawback of a temperature measurement on the original experiment could be significantly reduced. If the tip of the thermocouple (i.e. one of the contact points of the wires) was blackened as HERSCHEL and many others had usually done with the mercury thermometers, the new thermometer became an excellent electrical instrument for measuring radiation that indicated the power of the impinging radiation as an electrical voltage.

Already in **1833**, NOBILI and his Italian compatriot Macedonio MELLONI[61] had improved the measurement principle, and had developed and built a novel instrument that they called "thermomultiplier" or sometimes "electric thermoscope". This was a combination of a thermopile and a very sensitive galvanometer. The thermopile was a cascade of 10 thermocouples, formed from antimony and bismuth,[62] which led to an improved effect because the thermo-voltages of the single stages added to a total value being correspondingly higher. For his quantitative measurements NOBILI coupled the thermopile with his galvanometer, invented already in 1825 and compensated against the influence of the earth's magnetic field. With this highly sensitive electro-magnetic instrument he could convert the still relatively small thermos-voltages into a dependable deflection of the indicator.

It turned out soon, that the new instrument was well-made, and at least 40 times more sensitive than the best thermometer known so far. Employed as a heat sensor, it could measure the heat of a human at 10 m distance.[63]

15 – ***Principle of the thermoelectric effect (Seebeck effect)***

*Two wires A and B made from different materials together with a voltmeter (V) form a simple electrical circuit. In case both contact points have different temperatures $T_W$ and $T_K$ (e.g. $T_W$ is warmer than $T_K$) the voltmeter will indicate a small voltage. Typical thermoelectric voltages are in the range of 10µV per degree temperature difference.*

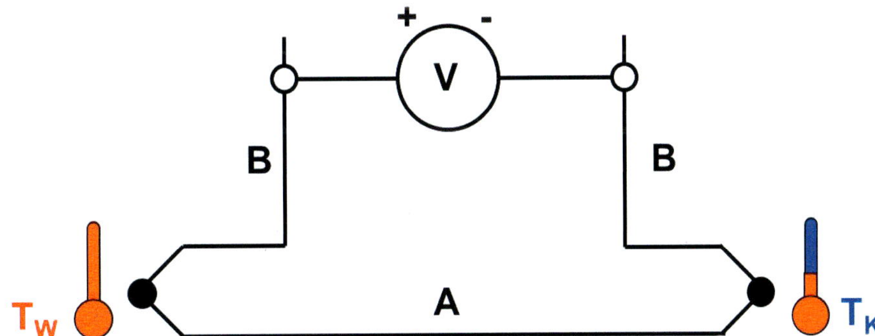

16 – *Principle of the thermopile of Nobili and Melloni*

*The incoming thermal radiation (2) hits a stack of thermo couples (1) and heats up the front side while the backside (3) is cooled or kept at constant temperature at least. Due to the temperature difference, each thermal element generates a small voltage based on the Seebeck effect. The thermal voltages of all elements are adding up and can be measured with a sensitive galvanometer (V).*

Especially MELLONI, who was professor and director of the Institute of Physics at the University of Parma, conducted comprehensive measurements of transmission, reflection, refraction and polarization of thermal radiation with the new instrument, and found out that thermal radiation and light behaved exactly in the same way. Nevertheless, this issue remained still in the scientific discussion for a long time, perhaps because the scientists of this age still had problems to measure the wavelength quantitatively. Even after Thomas YOUNG[64] in 1804 could measure the wavelength of visible light for the first time with his famous double-slit experiment, the advances to do the measurement for the infrared wavelengths still were very slow.

As the attached timeline shows, it needed nearly 100 years and the efforts of many scientists until the infrared spectral region was fully measured.[65]

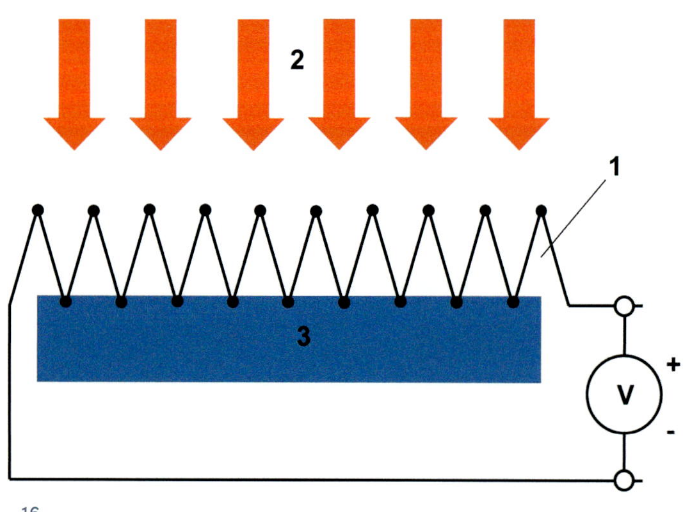

16

**International Contributions to Wavelength Determination**

| | |
|---|---|
| **1804** | Thomas **YOUNG** (England) is the first to measure the wavelength of light. |
| **1835** | André-Marie **AMPÈRE** (France) proves the similarity between light & thermal radiation. |
| **1847** | Léon **FOUCAULT** & Hippolyte **FIZEAU** (France) are able to determine wavelengths up to 1.5 µm. |
| **1847** | Hermann **KNOBLAUCH** (Germany) proves once more the equivalence between light & thermal radiation. |
| **1880** | Paul-Quentin **DESAINS** & Pierre **CURIE** (Frankreich) are able to determine wavelengths up to 7 µm |
| **1897** | Heinrich **RUBENS** (Germany) can determine wavelength up to 20 µm and measure the spectral distribution |
| **1898** | Heinrich **RUBENS** & Ernest Fox **NICHOLS** (USA) are able to determine wavelengths up to 150 µm. |

MELLONI also conducted thorough investigations of materials which he published in 1883 in a sequence of papers. He found out, for example, that the transmission of glass is strongly limited in the infrared region. Rock salt[66] however, showed a very good transmission of thermal radiation. This result made him to manufacture prisms and lenses from rock salt. With these components, MELLONI was the first to build an infrared spectrometer, and therewith he founded infrared spectroscopy which until today is an essential analytical tool in the fields of chemistry and pharmacy.

In **1835,** the British inventor and pioneer of photography Henry Fox TALBOT[67] came to the insight that "light traversing a transparent medium is able to excite motion among its particles."[68] Therewith he formulated the basic principle of resonance absorption of thermal radiation. Even if his statement might not be fully correct, he showed the way to a better understanding of the interaction between radiation and matter.

In **1839,** MELLONI used his new infrared spectrometer to investigate the infrared region of the solar spectrum and found absorption bands that he correctly attributed to the water vapor in the atmosphere.

Therewith the two Italian physicists not only achieved a reasonable advance of the measurement equipment. They were also the first to begin with the investigation of the atmospheric transmission properties.[69] For long-range observations with thermal imagers, the atmosphere is of crucial importance because long observation ranges can, of course, only be achieved at high atmospheric transmission.

The knowledge that there are special wavelength bands particularly in the infrared which are distinguished by a high transmission, we owe to John Frederick William HERSCHEL, the son of Sir William HERSCHEL. As son of a famous father, at the beginning of his career he had trouble to live up to the expectations connected to his person also publicly. Even when he was awarded the coveted Copley Medal, he was urged by Sir Humphry DAVY, President of the Royal Society, to follow "the example of your illustrious father, who full of years and honours, must view your exertions with infinite pleasure; and who , in the hopes that his own imperishable name will permanently connected with yours in the annals of philosophy, must look forward to a double immortality."[70] With the moral and also active support of his aunt Caroline HERSCHEL, John eventually found his own way in astronomy, and moreover in science generally. He began his astronomical work in the 1820s by validating and reworking the measurements which his father had made at many double-stars. Aunt Caroline, who was the only one being capable to do so, had created the prerequisites by compiling a catalog with over 2500 measurements from the confusing accumulation of her brother Sir William. At large, the work was very successful and brought, as a personal approval, the appointment of a knight of the "Royal Guelphic Order". At least since his 5-years, independently organized research journey to South Africa, Sir John HERSCHEL stepped out of his father's shadow. By means of a self-designed 21-feet telescope, in the years 1833-38 he observed and cataloged stars, nebulae and other objects of the southern sky. Sort of a sideline, together with his wife Margaret he created ambitious botanic illustrations of the South African vegetation, using a pinhole camera (camera lucida) to depict the exact outlines.[71]

In **May 1838,** Sir John HERSCHEL returned to England and right away applied himself at the generation of images in various wavelength regions. In this field, he had worked already before he had left to South Africa. For obtaining images with visible light, i.e. photography in the actual sense, he achieved remarkable contributions to photochemistry reaching to the beginning of color photography. He also coined the term "Photography", [72] and used in this context the adjectives "positive" and "negative" which are still common-place today. His pioneering

**17 – *Sir John Herschel's method to display a thermal image***

*By means of a concave mirror (2) the thermal radiation (1) coming from the warm object (T) is imaged onto a thin metal plate (3). The front surface of the plate is blackened with soot to achieve an almost total absorption of the radiation. Thereby the metal plate is heated locally and a thin oil or alcohol film (5) applied to the backside is evaporating unevenly. After a short length of time, a visible image of the temperature distribution of the observed object is developing on the plate – a thermal image or thermogram.*

results were presented in March 1839 and in January 1840 at the Royal Society in London.

In **1840,** Sir John HERSCHEL could present two excellent contributions to the field of thermal imaging. At first, he realized that there are certain wavelength regions in the atmospheric transmission that are distinguished by especially high transmittance values. The regions in between are unsuited for image capturing due to the high absorption, which in turn causes a strong attenuation of the thermal radiation needed for imaging. Altogether, he could identify three so-called "atmospheric windows"[73], among these also the standard windows of today's thermal imaging technology: The Mid Wave Infra-Red (MWIR) window at 3µm to 5µm and the Long Wave Infra-Red (LWIR) window at 8µm to 12µm. For these discoveries, he probably employed a measuring device for the direct measurement of the heating power of sun rays, that he had invented already in 1825 and named "Actinometer". It is known that John HERSCHEL used this device successfully in his photochemical experiments[74].

The second result in this year was even more spectacular. Since the ancient times of the Greek physician HIPPOCRATES, John HERSCHEL was the first who was again able to obtain a visible image of the temperature distribution of a scene. For the first time, his method used the imaging of thermal radiation for reproduction. His apparatus was based on a very thin metal plate blackened on one side with a material having a high infrared absorptance. The other side of the plate was coated with a thin film of an oil that evaporated at a rate proportional to its temperature. By a collimating mirror an infrared image was formed on the absorbing plate that selectively heated the tree-layer arrangement and caused a corresponding variation in the rate at which the oil film evaporated. The thermal image, which had been converted into a variable thickness oil

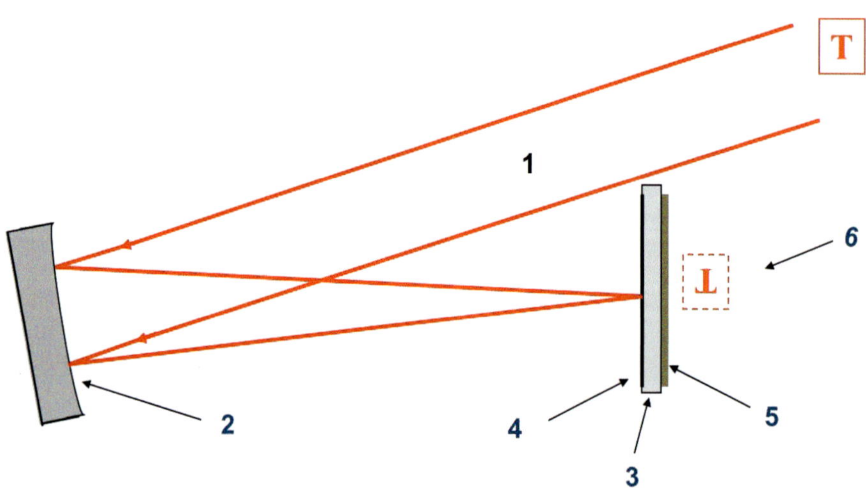


17

film, could be seen by reflected light where the interference effects in the film make the image visible to the human eye.[75] By this method, John HERSCHEL could generate a real thermal image on the plate, that he called a "heat picture".[76]

The interference effects are similar those that create opalescent colors when an oil stain is swimming on a puddle, and where the hue also is depending on the thickness of the oil film.

Under favorable conditions temperature differences as small as 1°C should have been detectable with this method, although it might have taken minutes for a picture to be formed[77].

Sir John HERSCHEL also managed to obtain a primitive record of the thermal image on paper, which he called "thermograph."[78] For this purpose, he imaged the thermal radiation "on the surface of thin paper covered with a layer of carbon black, and moistened with alcohol. From areas impinged upon by infra-red radiation alcohol evaporated sooner than from areas upon which no radiation or only radiation of lower intensity was incident. The dry spots on the paper differed from the remaining area which had remained moist."[79] With this method, Sir John could also obtain a visible picture of the infrared part of the solar spectrum. He concluded rightly that the dark areas were due to the selective absorption of the solar infrared radiation by the earth's atmosphere.

As is presented in Chap. 3, this image generation method was resumed and further developed in the 1930s by M. CZERNY in Berlin, and named "Evaporography". In the 1950s commercial "Evaporographs" and other non-electronic thermal imagers were built and put to market. The transient success of these non-electronic devices was ended by the great advances of the photoelectrical sensors in the 1960s.

18 – *Infrared spectrum of the sun after J. F. W. Herschel (1840)*

## 2.4 Laying the Theoretical Foundations

The investigations in the field of thermal radiation described so far were of a more experimental nature, and the knowledge was phenomenological. The German physicist Gustav KIRCHHOFF[80] was the first to discover deeper relationships.

In **1858,** he formulated a law that is named after him until today "Kirchhoff's Law of Thermal Radiation". It established a correlation between emission and absorption of radiation for all wavelengths:

"The radiation power P, emitted by an area, is proportional to its absorptance a and the radiation power $P_s$ of an ideally black area."[81]

The ideally black area is characterized by absorptance 1, corresponding to 100%; it absorbs all impinging radiation.

In **1860,** KIRCHHOFF found evidence that the radiation in a closed cavity is identical to that of an ideally black surface. This made an experimental realization of this important reference quantity possible.

In **1862,** KIRCHHOFF also coined the world-wide used term "Black Body" for an ideally black object.

In the late 1850s the English experimental physicist John TYNDALL[82], who is known to us today chiefly by the scattering effect in colloidal fluids named "TYNDALL effect" after him, began to investigate systematically the effect of thermal radiation on the constituents of the air.

TYNDALL for the first time measured quantitatively the infrared absorption bands of nitrogen ($N_2$), oxygen ($O_2$), carbon dioxide ($CO_2$), ozone ($O_3$), methane ($CH_4$) and water vapor ($H_2O$). Particularly he demonstrated the importance of water vapor for the atmospheric processes, and is commonly considered as a

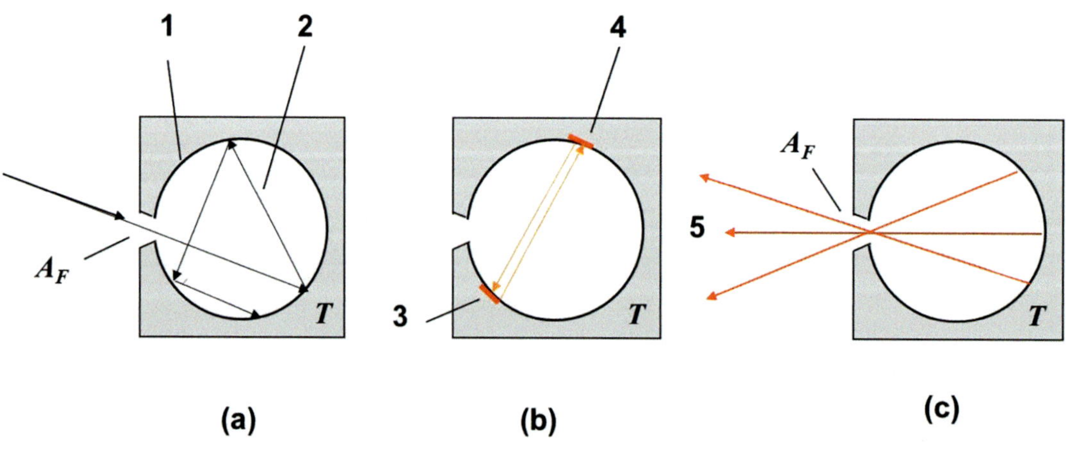


19 – *Principle of a cavity radiator as ideal black-body*
*The ideal black-body source comprises a cavity (1), for example a sphere with blackened internal surface. The radiant flux (2) coming in through a small aperture ($A_F$) is completely absorbed in multiple contacts with the internal surface. Compliant to the requirements for an ideal black-body the cavity swallows all radiation and virtually exhibits no reflection (Fig. a). Any given pair of surface elements (3, 4) inside the cavity is exchanging thermal energy by emission and absorption so that at thermal equilibrium the temperature is the same at all places of the internal surface (Fig. b). Therefore, the radiant power and spectral distribution of the thermal radiation leaving the small aperture ($A_F$) only depends on the absolute temperature (T) of the cavity.*

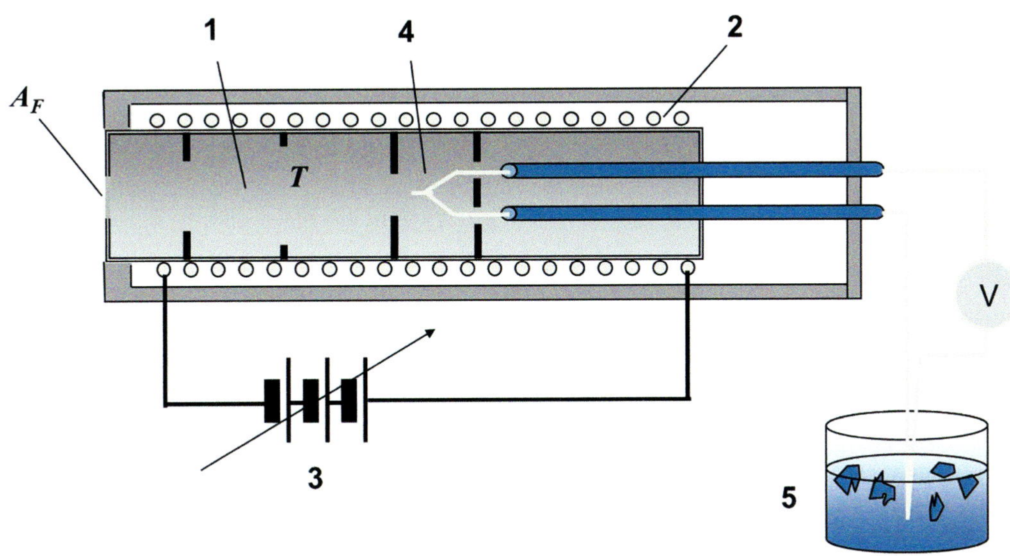


20 – *Technical design of a black-body radiator (schematic)*

*A cylindrical cavity (1) with blackened inner walls is tempered by a heating coil (2) which is supplied by a regulated voltage (3). The temperature (T) is measured accurately relative to a thermal reference (5), for example an iced water bath, by means of a thermocouple (4) and a connected voltmeter (V). The small aperture (AF) of the cavity emits thermal radiation which is dependant from the chosen absolute temperature T only and known as black-body radiation or black radiation.*

pioneer of infrared spectroscopy. The attached sketch of his experimental setup gives a vivid impression of the state of experimental physics in his time.

In **1872**, TYNDALL published the manifold results of his experimental work in the 450-page book "Contributions to Molecular Physics in the Domain of Radiant Heat".

On **20 February 1873**, the English electrical engineer Willoughby SMITH[83] published in the journal "Nature" a discovery, he had made almost by accident, which should have far-reaching consequences, however. SMITH was working with underwater cables for quite some time, and developed a test apparatus in this year, that allowed to test the cables during cable laying. For this, he needed a semiconductor material with high electrical resistance. Initially, rods from selenium (Se) appeared suitable, until they produced problems in practical use due to inconsistent results. Thorough investigations led to the surprising result that strong irradiation by light decreased the electrical resistance notably. Therewith, SMITH had found the first photoelectrical material, that was employed in the following decades in photocells of all kinds, and is sometimes even used today. It was discovered soon, that selenium is sensitive for radiation of the infrared part of the electromagnetic spectrum too. Experiments aimed in that direction, however, did not led to applicable results and were discontinued.[84]

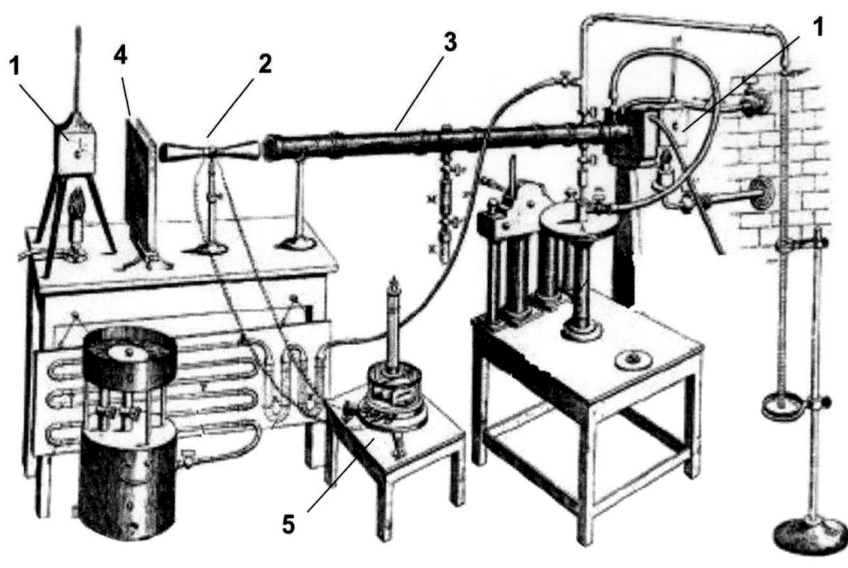


21 – *Tyndall's apparatus for measuring infrared absorption of gases*
*Black-body radiators (1) heated by gas burners are used as sources of thermal radiation. The cuvette (3) is a brass tube closed on both sides with salt (NaCl) windows. To detect the thermal radiation a thermopile (2) is used the voltage of which is indicated by sensitive mirror galvanometer (5). Furthermore, the sketch shows a heat shield (4) and various devices for gas mixing and cleaning.*

In **1874**, the multi-talented Max PLANCK was among the best who passed the German 'Abitur' in this year. Because he was very interested in studying physics, he went to the University of Munich and contacted a professor of Physics, Phillip von JOLLY, for orientation. To his great astonishment he became to hear, that "in this science virtually everything is already explored, and it remained only to close some nonsignificant gaps."[85]. The young man, however, was not to be discouraged. This turned out to be a piece of good fortune for science, as we shall see at the end of this chapter.

Besides an obviously failed career counseling, JOLLY's statement also unveiled a certain frustration and perplexity that seemingly had afflicted at least a part of the contemporary physicists. From today's view, replicable reasons for this negative perception are not visible. Up to then, new discoveries had been flowing like an ongoing stream, and there was no reason to expect that this stream would dry up soon. Only a few years later a new actor from the United States brought exciting advance to the scientific community.

In **1878**, the American astronomer, physicist, inventor and aviation pioneer Samuel Pierpont LANGLEY[86] presented his new sensor for thermal radiation. Its sensitivity was many times higher than that of all existing detectors. The functional principle of the new detector, which became known under the name "bolometer" (from Greek bole = throw, beam), was a blackened

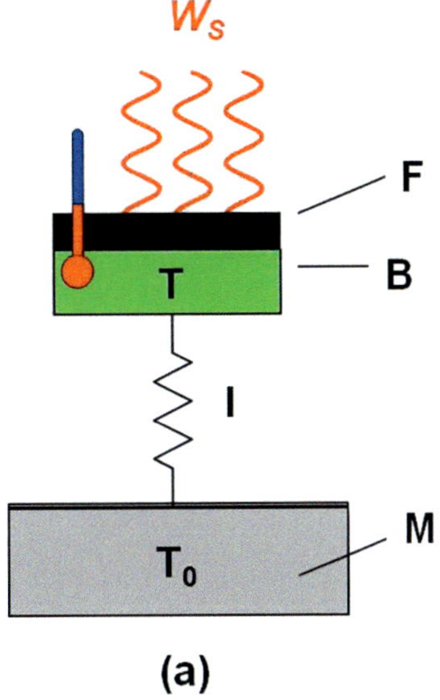


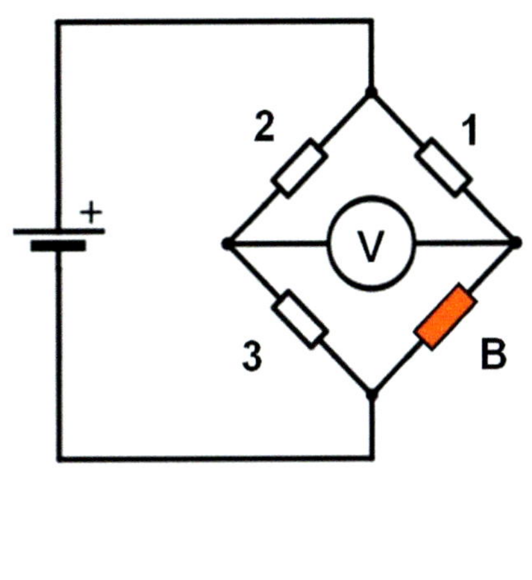


**22 – *Functional principle of a bolometer and a Wheatstone bridge***

*Fig. (a): The incoming radiation ($W_S$) is absorbed by the blackened surface (F) of the bolometer detector (B) and transformed into heat. Thus, the temperature (T) of the sensor is changed and its electrical resistance too. For a fast reaction, the bolometer needs a good isolation against the mass (M) of the rest of the housing. However, to limit the temperature increase during long lasting irradiation periods, the isolation (I) must allow for a controlled heat flow to the housing which is held at a constant temperature ($T_0$).*

*Fig. (b): To measure the electrical resistance the Bolometer detector is operated as a part of the so-called Wheatstone bridge. When the bridge is balanced, i.e. all resistance values (1, 2, 3 and Bolometer B) have the same value, the voltmeter (V) will read 0 volts. As soon as the Bolometer detector (B) is irradiated and its resistance is changed the voltmeter will read a voltage that is a measure for the power of the radiation.*

absorber of small mass which was heated by thermal radiation very fast. The temperature change in turn changed the electrical resistance which was then measured. As the absorption of radiation at the blackened surface was independent from the substrate material, LANGLEY could select the optimum material with respect to the variation of the temperature dependent electrical resistance.

For highly sensitive measurements of the electrical resistance, LANGLEY used a so-called "Wheatstone Bridge"[87], which was introduced by Sir Charles WHEATSTONE[88] in 1843. This circuit is ideally suited to measure very small variations of the electrical resistance.

In **1880**, LANGLEY presented an improved version of his bolometer in which he installed two strips from platinum with 0.05 - 0.2mm thickness as detectors. The electrical resistance of about 4 Ohm had to be matched with only 1% tolerance. The second bolometer detector substituted the resistor 3 shown in the circuit, and was used as reference detector to compensate the general temperature drift and the influence of the ambient irradiation. A battery served as a stable voltage supply, and a galvanometer was installed by LANGLEY as a sensitive voltmeter, as had been done already by Leopoldo NOBILI for his thermopile.

With the new bolometer an excellent temperature resolution of 0.00001°C was possible,[89] and the thermal sensitivity was sufficient to detect "a cow a quarter of a mile away."[90]

In the following years, LANGLEY could perform many interesting measurements in which the excellent sensitivity of his new bolometer detector was of benefit. Besides comprehensive material investigations he conducted accurate thermal measurements of the solar spectrum including a precise measurement of the Fraunhofer lines in the infrared spectral region. He used the bolometer also to detect infrared radiation from other astronomical objects, and made first efforts to determine the surface temperature of the moon.

His precise measurements of the interaction of thermal radiation with carbon dioxide ($CO_2$) in the earth's atmosphere even were the origin of first speculations about a possible greenhouse effect. So, in 1896 the Swedish physicist Svante ARRHENIUS used the values measured by LANGLEY as a base for his first evaluations of how the climate would change at a future doubling of the carbon dioxide content of the atmosphere.

In general, it was a valuable contribution to the scientific advance that LANGLEY with his bolometer provided a tool to all researchers so that they could measure the distribution of thermal radiation of all kinds of grey and black bodies much more accurate than ever before.

Alas, the international scientific community was still struggling to fully understand the nature of the invisible radiation.

In **1879**, the Austrian physicist and mathematician Joseph STEFAN[91] presented his law for the emitted thermal radiation as a function of temperature in a paper "Über die Beziehung zwischen der Wärmestrahlung und der Temperatur"[92] (About the Relation between Thermal Radiation and Temperature). He had found, that the radiation power emitted by a black body is proportional to the 4th power of the absolute temperature.[93] This means in practice, that each body with a temperature above the absolute temperature zero (0 K corresponding to -273,15°C) emits radiation, and that the power of this radiation rises much more than proportional to the temperature. If we compare the emission of a body with room temperature of about 300 K with the emission of a lamp filament of 3000 K, then the filament is only a factor of 10 hotter but the radiation power is increased by an incredible factor 10 000.

STEFAN had derived his law empirically, based chiefly on the experimental data of TYNDALL.

In **1884,** Ludwig BOLTZMANN[94], an Austrian physicist and former student of STEFAN, could theoretically derive the same law from thermodynamics. Therefore, we can find this law today in physics textbooks under the name "Stefan – Boltzmann Law".

Since **1880**, the term "infrared" (from Latin infra = beneath) spread more and more in the scientific literature as a general term for the region of invisible radiation beneath the visible spectrum. In course of time, the term was transferred into many languages as is shown by the following short table:

| | |
|---|---|
| German | infrarot |
| English | infra-red (since 1880 in Great Britain) |
| American | infrared (since 1920 in America)[95] |
| French | infrarouge |
| Russian | infrakrasnye |

Altough in Germany the term "ultrarot" was used in parallel for some time, it was a useful standardization.

In the time before the terms had been very individual, as may be seen from a few examples:

| | |
|---|---|
| Academia del Cimento | "heat radiation", "cold radiation" |
| Carl Wilh. SCHEELE | "radiated heat" |
| William HERSCHEL | "the invisible rays", "the thermometrical spectrum", "the rays that occasion heat" "dark heat" |
| John TYNDALL | "radiant heat" "ultra-red wave-motion" |

Despite all international efforts, the origin of the term "infrarot" could not be clarified yet.[96] Perhaps the time was simply ripe for a catchy term.

In **1893**, the German physicist Wilhelm WIEN[97] derived a law about the wavelength of the maximum emission of thermal radiation as a function of temperature that we know today as "Wien's Displacement Law". Based on thermodynamic considerations, WIEN found that the wavelength λmax of maximum radiation is inversely proportional to the absolute temperature T, or written as formula with the coefficient b: $\lambda max = b / T$

By using Wien's displacement constant b = 2898 K· µm and putting in the temperature in [K], one can calculate the wavelength in [µm] of the radiation maximum for various objects.

Already in **1896**, WIEN included empirical data in his considerations and could establish a full description of the wavelength dependence of the radiation power emitted by a black body. The law, named "Wien's Radiation Law"[98] after him, correctly gives the existence of a radiation maximum and its displacement with temperature, and was consistent with his displacement law. In 1911, WIEN was awarded the Nobel Prize for his work about thermal radiation.

Altogether, WIEN could describe great parts of the spectral distribution of radiation correctly. Noticeable deviations were only found at long wavelength; due to the limited accuracy of the measurements this could have been also measurement errors. Moreover, the limit of the radiation power for high temperatures postulated by the law, was not plausible for many scientists, but could hardly be ascertained by the contemporary technical means.

With a totally different approach, based on classical electrodynamics, the English physicist John William STRUTT, 3rd Baron Rayleigh (better known today as Lord RAYLEIGH)[99] tried to derive a new radiation law. Perhaps he hoped to resolve the (supposed) weaknesses of Wien's law.

**Wavelength of Maximum Thermal Radiation**

| | |
|---|---|
| Sun, colour temperature 5500 K | 0.53 µm |
| Filament, colour temperature 3000 K | 1.0 µm |
| Object with temperature 800 K = 527°C (red-hot) | 3.6 µm |
| Cooking water, temperature 373 K = 100°C | 7.8 µm |
| Human body, temperature 310 K = 37°C | 9.3 µm |
| Surrounding, room temperature 300 K = 27°C | 9.7 µm |
| Ice, temperature 273 K = 0°C | 10.6 µm |
| Liquid air, temperature 90 K = -183°C | 32.0 µm |

*23 – Graphical presentation of Wien's displacement law*

*By way of example, curves are given for a high temperature (3000 K, incandescent lamp) and room temperature (300 K). The different spectral locations of the maxima of radiation are clearly to be distinguished. This is compliant with Wien's displacement law. The curve for 300 K is enlarged by a factor of 10000 to make it visible in the diagram. This also shows the strong influence of temperature on the total emitted power of thermal radiation as is predicted by the Stefan-Boltzmann law.*

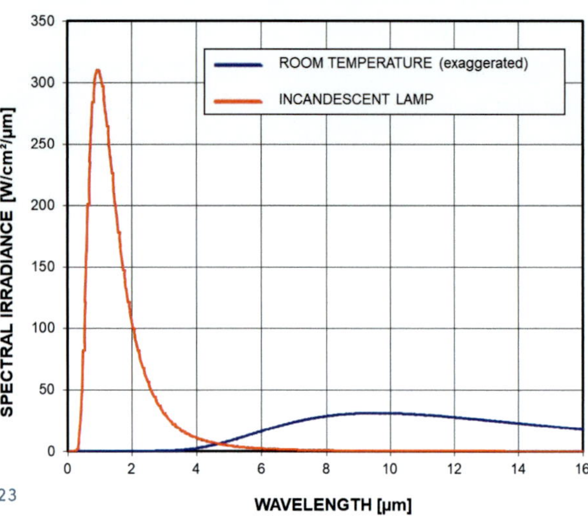


23

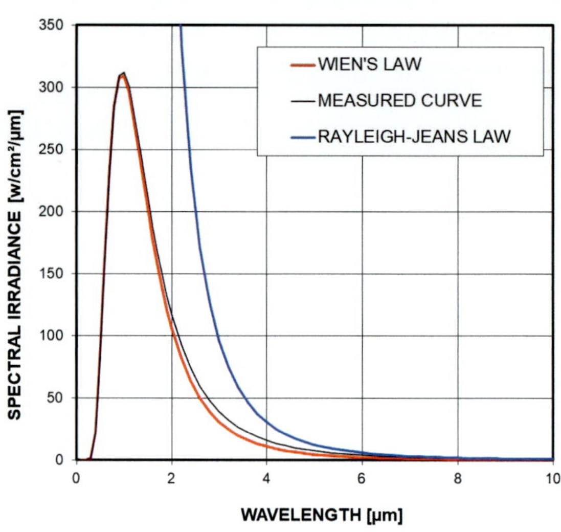


24

In 1900, RAYLEIGH presented his new radiation law in the publication "Remarks upon the Law of Complete Radiation". Initially the law contained a wrong prefactor, which was corrected in 1905 by another Englishman, the physicist, astronomer and mathematician James JEANS[100]. Therefore, the law is commonly known today as "Rayleigh-Jeans-Law"[101]. As is depicted by the diagrams, the law describes the behavior of the radiation very well at long wavelengths. At medium wavelengths, however, the predicted radiation powers are much too high, and are further away from the measured curves than WIEN's values. The German physicist Paul EHRENFEST in 1911 coined the term "ultraviolet catastrophe"[102] for this fact. The wording described drastically the large deviations that reached 'catastrophic' values in the ultraviolet spectral region – with respect to the significance of the law.

Apparently, a significant cognitive act was still missing to achieve a satisfying description of reality. In the same year 1900, the German physicist Max PLANCK[103] was capable to make a significant contribution which at the end should open the door to quantum mechanics. About PLANCK's entry into a scientific career we have already heard at the beginning of this chapter. After a study of mathematics and science in Munich and Berlin, PLANCK could finish his PhD-thesis with 'summa cum laude' in 1879. One year later, he was habilitated and with merely 22 years (unpaid) Privat Lecturer at the Munich University. In 1885, he became Professor of Theoretical Physics in Kiel, and therewith was one of only two theoretical professors in Germany, what casts an indicative light on the state of physics in Germany at that time. In 1889, he transferred to Berlin, where he got again the post of a full professor of theoretical physics from 1892 on. Although his PhD thesis and his habilitation treatise about thermal theory had caused only low response, he continued his research in the field of thermodynamics and investigated thoroughly the nature of energy and entropy. Since mid of the 1890s, PLANCK worked directly about the equilibrium of radiation, and tried to apply his thermodynamic knowledge to the radiation laws. He published, among others, 5 papers "About Nonreversible Radiation Processes", and he managed to derive "Wien's Radiation Law" from the radiation characteristic of a cavity radiator.

In the summer **1900**, precise measurements by the German physicist Heinrich RUBENS proved that the existing deviations of Wien's radiation law were errors of the law, and not due to plain measurement errors, as was guessed or hoped up to this point. In October, RUBENS told his friend PLANCK about the now experimentally proven facts that

- In the region of short wavelengths WIEN's law was accurate and the RAYLEIGH-JEANS law could be regarded as false
- In the domain of long wavelengths, it was vice versa.

Shortly after this conversation, PLANCK found "a luckily guessed interpolation formula"[104], by which he could correctly describe the entire distribution of radiation. He substituted the exponential term in WIEN's law by a more complex exponential function, and got a formula from which both known laws could be derived as limiting case for long and short wavelength, respectively.[105]

On **19 October 1900**, PLANCK officially presented this first result to the Academy after RUBENS had been able to experimentally confirm the formula once more in the days before. For the theoretical interpretation, however, PLANCK needed some more weeks because the formula did not match with conventional approaches. Only when PLANCK in an "act of despair" assumed that the radiation emitting oscillators could take discrete energy levels only, the theoretical derivation could be achieved. He introduced a constant h for the relation between energy and radiation frequency, which later turned out to be an essential natural constant, and which is known today as "Planck's constant". The two other constants in his formula could be related to three natural constants (speed of light, Boltzmann constant, Planck's constant)[106].

The day of **14 December 1900** can be viewed as the "birthday of quantum mechanics" according to Max von LAUE[107], because PLANCK officially presented his radiation law in a session of the Physical Society on this day. His result, however, was initially only as a very welcome formula, which characterized the radiation distribution correctly at last. The enormous consequences of the new law and of PLANCK's constant to the physical conception of the world became obvious only in the following years. After Albert EINSTEIN's[108] hypothesis of light quanta (1905) and his validation of the radiation theory together with Paul EHRENFEST[109] it was clear that the ideas of classical physics had to be extended. The upcoming discussion of light quanta (photons) and discrete energy levels marked the beginning of the theory of quantum mechanics - a new way of thinking which was hard to accept by many of the contemporary scientists. Even PLANCK himself had difficulties to accept his own results, and tried to find a second theory in 1912 and a third in 1914, both without any quanta. Alas, there was no way back, as we know today, and with hindsight, PLANCK came to a somewhat

24 – *Graphical presentation of Wien's law and the Rayleigh-Jeans law*

*For both laws curves of the spectral irradiance are given for a temperature of 3000 K. For comparison, the measured distribution of the radiation is also given. While Wien's law predicts the maximum of radiation correctly the values calculated with the Rayleigh-Jeans law are far from reality. Only for large wavelengths the values postulated by Rayleigh-Jeans are closer to the measured curve than those of Wien.*

melancholic and perhaps self-aware statement: "A new scientific truth is typically not gaining acceptance in the way that its opponents will be convinced and declare themselves enlightened, but in the way that the opponents become slowly extinct and the following generation is familiar with the truth from the outset." (Planck 1948)[110]

Beginning with a more accidental observation and simple experiments with the newly discovered invisible radiation, in about 100 years of scientific research not only the laws of infrared technology were formulated. The results led directly to the development of quantum mechanics, one of the greatest revolutions the scientific world had ever seen. Many excellent scientists have contributed to the results, several of these were awarded a Nobel Prize. At least at three award ceremonies, it was referred directly to the pioneering work in this field (WIEN in 1911, PLANCK in 1918, EINSTEIN in 1921).

**Planck's Law**

The law defines the radiated power per area and per unit of wavelength as a function of wavelength λ and absolute temperature T [K], also called »spectral radiant emittance«.

$$P(\lambda, T) = \frac{2\pi \cdot c^2}{\lambda^5} \cdot \frac{1}{e^{hc/\lambda kT} - 1}$$

The result depends only on three constants of nature:

| | |
|---|---|
| Speed of light in vacuum | $c = 2.99792458 \cdot 10^{8}$ m/s |
| Planck constant | $h = 6.626176 \cdot 10^{-34}$ J·s |
| Boltzmann constant | $k = 1.380662 \cdot 10^{-23}$ J/K |

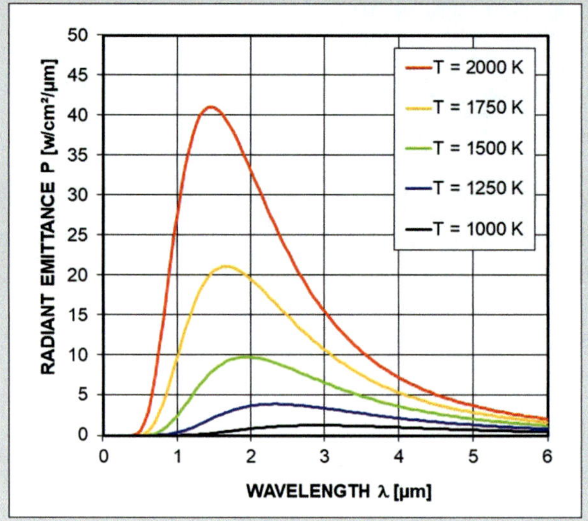


25 – *Mathematical formulation of Planck's radiation law*

3

# THERMAL IMAGING UNTIL 1945: DIFFICULT START

"The bolometer was refined enough to detect thermal radiation from a cow a quarter of a mile away."

*Samuel P. Langley, USA (1880)*

"The Germans lost the war ...
... but they clearly won the battle of the infrared."

*Richard D. Hudson, USA (1969)* [111]

As presented in the preceding chapter, in the 19th century a whole lot of extraordinary scientific accomplishments were performed resulting in the clarification of the nature of thermal radiation and leading to the new physical discipline of quantum mechanics.

However, this did not mean that all technical problems were solved, and that obtaining a thermal image was a trifle, for example. There were indeed a few useful detector elements (bolometer, selenium cell) available which could transform thermal radiation into electrical signals. To obtain a two-dimensional thermal image, however, was still not possible with acceptable quality and in reasonable time because a single detector element was only capable to measure one image point at a time.

Even though the thermal sensitivity of a single detector might have been sufficient "to detect thermal radiation from a cow a quarter of a mile away"[112], the recording of a two-dimensional image was much more complex. For that to happen, a suitable apparatus was necessary to scan the scene point after point in horizontal and vertical direction, and only the raster of the image points would create an image useful to the observer.

The technical means to perform the individual steps of this imaging chain were not sufficiently available at the beginning of the 20th century. Therefore, the researchers concentrated on improving the technical state of the heat sensors to the point that the detection of a local heat source in a homogeneous scenery became possible. In other words, for the moment one was content with a thermal image comprising one image point.

With this objection, development work on thermal pointers for military application was driven forward worldwide. The passive receivers should be used to detect ships on the cold sea or aircraft in the even colder sky by merely measuring the emitted thermal radiation.

Compared to the conventional observation methods the following benefits were expected:

- Because only the radiation emitted by an object is detected observation should be equally effective at day and night.
- Because the detection device does not emit any radiation itself, there is no danger to reveal the own position.
- Due to the long wavelengths of the thermal radiation one expected to achieve range advantages compared to visual observation especially under adverse weather conditions (rain, fog, dust).

There had already been very early international contributions to this kind of technology. Already in 1833, the Italian physicists Leopoldo NOBILI and Macedonio MELLONI had built their "thermomultiplier", the first detector that could generate an electrical signal proportional to the intensity of the impinging thermal radiation. While the "thermomultiplier" was not very sensitive and rather slow, the invention of the bolometer had marked a significant step towards to a practically useful thermal sensor. In 1880 Samuel LANGLEY had presented an improved version of his bolometer, which was capable of an excellent temperature resolution of 0.00001°C[113], and used it for astronomic measurements and spectroscopic investigations of the atmosphere of the earth. The innovative functional principle of the LANGLEY bolometer was to be the base of many thermal pointers until the mid of the 20th century.

"Many workers were intrigued by the possibility of heat seekers and were proposing a wide variety of infrared search devices... This era also gave birth to the myth that persists even today that infrared radiation has magical ability to penetrate fog, rain and clouds."[114] The myth tended to exceed all scientific knowledge and was fueled by each spectacular event.

In **1912,** for example, the sinking of the Titanic provoked a lot of inventions of nighttime iceberg detectors based on infrared sensors.[115]

During the development process of thermal detectors there had been great progress since the time of Sir William Herschel[116]. However, the scientific community was well-aware of the conceptual disadvantage of even the best of bolometers: the thermal radiation could be transformed into an electrical signal only indirectly by a temperature change of the detector element. Therefore, researchers kept on searching for appropriate detector materials that were capable of a direct photo-electric conversion. In the decade before the Great War especially lead salts were investigated for the ability to detect infrared radiation.[117] The objective was, of course, the development of more sensitive and faster detectors.

In the years **1916-18,** the American inventor Theodore W. CASE[118], who became well-known later for the development of a commercially successful sound film system ("Movietone sound-on-film"), was searching systematically for photo-conductive materials. Effectively, he continued the work of Willoughby SMITH who had discovered the photoconductivity of selenium in 1873. After having tested more than 200 lead salts including various mixed crystals he concluded that sulfides had the best prospects of success. Initially, he had the best success with bismuth sulfide and granular lead-antimony-sulfide.[119]

In **1917,** CASE developed the so-called "thallofide cell", a sensor comprising a thin film of the substances thallium, oxygen and sulfur formed on a quartz plate. "The

**Improvement of Detector Sensitivity**

| Year | Inventor | Type of sensor | Factor |
|---|---|---|---|
| 1800 | W. Herschel | Mercury thermometer | 1x |
| 1829 | Nobili | Thermocouple | 5x |
| 1833 | Nobili & Melloni | Thermo-multiplier | 40x |
| 1901 | Langley | Bolometer | 1200x |

cell had a spectral response through the visible region out to about 1.45 μ, with peak responsivity at a wavelength of 1 μ."[120] Based on this sensor CASE developed a top-secret classified infrared signaling system for the U.S. Navy. The prototype of the system comprised a 60-inch searchlight as radiation source and a "thallofide cell" in the focus of a 24-inch parabolic mirror. [121] A communication from ship to ship was successfully tested with this system off the coast of New Jersey in the same year. No less a figure as Thomas Alva EDISON, the most eminent inventor of all, was engaged by the U.S. Navy as an expert. [122] He could confirm that the infrared signal system, being invisible to the human eye, "sent messages 8 miles through what was described as 'smoky atmosphere'."[123]

However, the new detectors had two grave shortcomings:

- They could be damaged by irradiation with short wavelength very easily.
- The production of the cells was non-reproducible and the sensitivity varied badly from cell to cell. It was not unusual that the electrical signal neither was proportional to the intensity of the incoming radiation nor to the height of the applied voltage.[124] Eventually the functional mechanism and the exact composition of the material were not sufficiently known (the material was also termed incorrectly as "thallium-oxysulfide", a substance that does not exist).

In **1918**, the work on the "thallofide cell" was discontinued.

In **1917**, a first "Infra-red Search and Track" (IRST) device might have been developed in Great Britain according to several literature sources. It is reported only that the system "could detect aircraft at a distance of 1 mile and people at a distance of 1000 ft." [125] In any case, one can guess that "unlike their American counterparts, the interest of British military centered on the detection of small aircraft by the heat they emitted." As early as 1916 a corresponding proposal had been submitted by Frederick A. LINDEMANN (the later Lord CHERWELL)[126] but was not pursued. Moreover, an investigation conducted by the Admiralty in 1926 was also not promising.[127]

In **1919** a system similar in ambition and performance was published and patented by the Californian S. O. HOFFMAN.[128] The system comprised crossed thermopiles and a galvanometer to display the signal.

In the years **1914-18**, during the time of the Great War, researchers at the German company CARL ZEISS made their first experiments on thermal pointing. Rumors say that also Emil MECHAU[129] should have been a part of that but chronological data in literature appear to be inconsistent, at least to the author. MECHAU got much appreciation for his flicker-free projector when he was with the company LEITZ in Wetzlar later.

In **1926**, an active system for detecting airplanes was built in the United States. In the country of the inventors of bolometer and "thallofide cell" various active and passive infrared devices for reconnaissance of ships and airplanes were developed in the 1920s.[130] In parallel, commercial applications were implemented in the chemical industry. Infrared spectroscopy based on the identical detector technology was an analytical tool generally accepted in the laboratories of the American chemical and petro-chemical industry around 1935.[131]

## 3.1 Thermal Images without Electricity

All these applications used a single detector element (thermocouple, bolometer) for sequential measurements. It was not possible to obtain a two-dimensional thermal image with acceptable quality and in reasonable time with this technology. Except the ancient Greek physician HIPPOCRATES and the British astronomer and photographer Sir John HERSCHEL no one had been able to obtain an visible image based on thermal radiation. Both used thermo-optical techniques for generating an image which did not need any electrical signal processing.

The principle of John HERSCHEL's method was that a fluid film was heated locally by thermal radiation. Because the fluid evaporated faster at the warmer areas than at the colder the film became thinner where it was warmer. Therefore, the difference in the intensity of the thermal radiation was transformed into a difference in the thickness of the fluid film. If the fluid film was illuminated with white light, the light was reflected on the front and on the backside, and interference effects occurred between the two bundles, known as NEWTON fringes. Depending on the thickness of the fluid film it appears in variant colors. By this method the thickness difference was made visible to the human eye and could also be photographed. Due to the limited technical means of his time, John HERSCHEL was not able to realize his brilliant idea perfectly in 1840.

In **1929**, Prof. Marianus CZERNY[132] turned to this method. He was measuring vibrational and rotational spectra of molecules in the infrared (in Germany called "ultrarot") wavelength region at the University of Berlin. His objection was to develop a spatially resolving sensor for infrared spectroscopy based on this method.[133] CZERNY's sensor essentially comprised a glass tube that was closed with a plug on the top side and a thin celluloid membrane on the bottom to form a closed volume. The membrane was made from Zapon lacquer (Z-116 from AGFA-Wolfen) by CZERNY himself. The exterior side was blackened with soot to achieve a high absorption of thermal radiation. On the internal surface a thin, white layer of camphor or naphthalene was formed by evaporating the material in an electrically heated vessel.

We now can fancy the sequence of obtaining an image to be as follows:

1. By brief heating, naphthalene (for example) was evaporated out of the glass vessel and condensed on the inner surface of the membrane as white layer.
2. Where thermal radiation was impinging on the blackened side, the membrane was heated locally.
3. At the warmer patches, naphthalene evaporated and the white layer became thinner and therewith darker or even vanished totally.
4. When illuminated with white light, the thinner areas of the naphthalene layer looked darker than the unchanged areas.
5. On the membrane appeared a thermal image in the way that warm parts appeared dark and cold parts bright.

CZERNY used the new sensor in a spectrometer with a calcium fluorite ($CaF_2$) prism that is transparent for thermal radiation of wavelength up to almost 10 µm. By measuring absorption lines between 1.5 µm and 6 µm with this apparatus he could also prove that the new

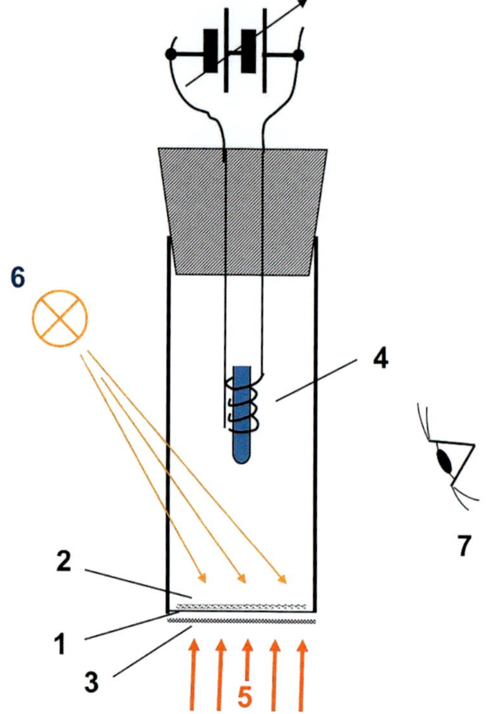


26 – *Czerny's apparatus for capturing thermal images*

*The material (camphor or naphthalene) which is evaporated out of an electrically heated tube (4) is condensing as a thin white layer (2) on the membrane (1) which is blackened with soot on the outside (3). When thermal radiation (5) impinges onto the membrane the white coating is evaporating at this place. Thus, the warmer spots appear darker under visual illumination (6). The spatial intensity variation of the thermal radiation by this process is transformed into correspondingly varying grey levels that can be observed with the eye (7).*

detector was sensitive for thermal radiation in the longwave region. After this initial success CZERNY's method was more and more improved over the following years.

In **1929**, "the Hungarian physicist Kálmán TIHANYI[134] invented the first infrared-sensitive electronic TV camera for air defense in Britain".[135] This is reported by several sources of the Anglo-Saxon literature. The first infrared TV camera was classified and eventually published as late as 1956, then called "Evaporograph". Therefore, the invention could not have influenced the developments of the following year. Nevertheless, it makes sense to have a closer look at the possible substance.

After his studies in Bratislava and Budapest, TIHANYI developed and patented a fully electronical television system in the 1920s. In 1928, he went to Berlin where he worked with TELEFUNKEN and SIEMENS. From 1929 on, he worked on the military application of remote control by means of video images. For the BRITISH AIR MINISTRY in London TIHANYI built a radio-controlled airplane.[136] As the invention of the electronic IR-camera fell in this busy year, one may guess that it was a mere combination of the available TV-camera with an IR-image sensor based on CZERNY's method. For an experienced inventor like TIHANYI this was an obvious conclusion, the more so as he had lived in Berlin the year before where CZERNY had developed his thermal sensor. To what extend the invention was put into practice is not known to the author. Some sources publish pictures of the IR-camera of the BAIRD company in this context. As the IR-optical system of the BAIRD camera is made from germanium this is for sure not a realistic description of the contemporary state-of-the-art, especially for the simple reason

that the production of germanium in the necessary purity was possible only after 1949.[137]
In **1932** an improved version of the active direction finder of 1926 was available in the United States. It could track an airship up to an altitude of 6300 ft. However, the development was discontinued due to the lack of powerful radiation sources. In the same year "the possibilities of 'seeing' by heat waves were being explored by a UNITED STATES ARMY SIGNAL CORPS team led by Marcel GOLAY."[138]

Also in **1932**, Herman WILLENBERG could present significant technical improvements of CZERNY's method and a more practical apparatus in his PhD-thesis and a further paper.[139] In place of the naphthalene a thin fluid film was formed on the membrane in the manner of John HERSCHEL's experiment of 1840. The visible image was not created by difference in remission any more but by means of interference effects in the light reflected by the film. The optical effect is the same that is generating a shimmering play of colors caused by a drop of oil swimming on a puddle. In both cases, the color hue is defined by the thickness of the oil film.

As a trade-off of all relevant criteria a mixture of fluid hydrocarbons ($C_nH_{2n+2}$) was chosen. The boiling range was restricted to 260°-280°C to hold the issues of different rates of evaporation in bay. The space above the fluid film was not closed any more but could be evacuated to adjust a quite repeatable vapor pressure. The membrane of approximately 40 mm diameter was again made from Zapon lacquer. To keep out disturbing environmental influence as far as possible, the membrane was protected by a housing. On the infrared side, it was closed by an IR-transmitting salt (NaCl) window while on the opposite side a glass window allows for visual observation.

We now can imagine the image formation to be very similar to CZERNY's original method:

1. By heating briefly, enough oil was vaporized to precipitate a thin but closed oil film on the inner surface of the membrane. The surface looked uniformly bright when illuminated with monochromatic light.
2. Where thermal radiation was impinging on the blackened surface of the membrane, it was heated locally.
3. At the warmer areas, more oil was vaporized than at the cooler, so that the oil film became thinner there.
4. Illuminated with monochromatic light (e.g. green), the areas where the oil film was thinner showed a different brightness due to the change in interference conditions.
5. In the reflected light, a thermal image in different grey levels could be viewed.

**27 – *Generation of a visible image by interference***

*The incoming thermal radiation (1) is absorbed on the blackened side (2) and heats up the membrane (3) locally. The oil film (4) on the backside evaporates faster at this place and becomes thinner. When the oil film is illuminated light is reflected at its front and back side. The two reflected bundles will experience interference whereby the bundles may extinguish or amplify each other depending on the path difference. As the path difference depends directly on the film thickness the interference parameters of bundle (6, 6') at a thinner spot differ from that at the unchanged parts (5, 5') of the membrane. Therefore, the observer will see a difference in brightness.*

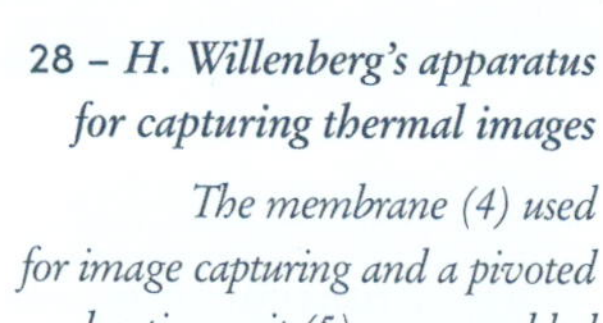

**28 – *H. Willenberg's apparatus for capturing thermal images***

*The membrane (4) used for image capturing and a pivoted heating unit (5) are assembled in a brass housing (1) which is closed with a common salt window (2) on the front side and a glass window (3) on the backside. The housing has a connector (7) to a vacuum pump, and the pressure can be monitored with a manometer (6). The thermal radiation (12) coming in through the salt window changes the temperature distribution of the membrane. This results in variations of the thickness of the fluid film on the backside of the membrane, which in turn leads to changes in brightness or interference colours. The thermal image is made visible by illumination with a light bulb (8), and can also be captured with a customary photo camera (11). By means of a spectral filter (9) and a heat protection filter (10) the image quality can be improved.*

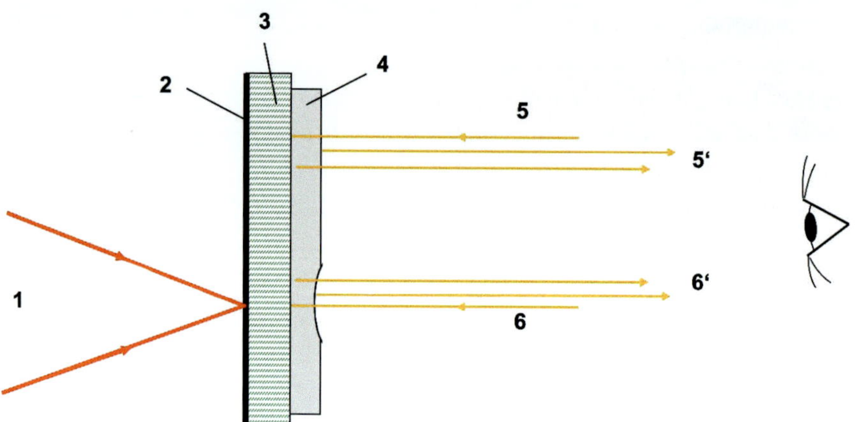


27

28

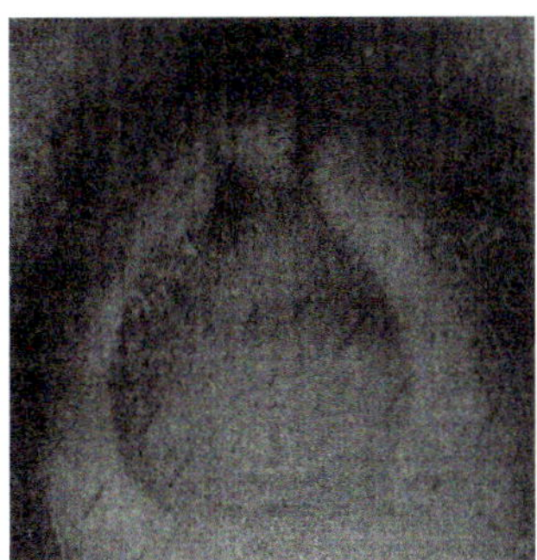

29 – ***Thermal image and photographic picture of a warm object***

*The thermal picture (on the left) of a blackened glass flask filled with hot water of 90°C was captured solely using the natural thermal radiation. The time for recording was approx. 30 seconds. The picture on the right shows a photograph of the scene.*

## 3.2 Thermal Pointers with Bolometer Detectors

In Germany, the international advances in the field of thermal imaging were carefully watched at the beginning of the 1930s. The political class took the experiments in the United States very seriously and particularly the military staff was downright alarmed. The foreign office presumed that the western allies had attached superior strategical importance to the thermal imaging technology. They concluded that it was high time for own activities and thus great emphasis was given to the development of German infrared technology until the very end of World War II.

As we know today, the priority of allied sensor development was radar technology where operational equipment could be realized faster with the contemporary technical means.

Early in the **1930s** a development landscape was formed in Germany which was structured to cover the whole infrared region from 0.8 μm to roughly 15 μm.

In this book, mainly the developments in "pure" thermal imaging shall be followed, based on the thermal radiation emitted by objects near room temperature. In the 1930s, the accessible wavelength regions were in the mid infrared MWIR (2 – 5 μm) and in the longwave region LWIR (8 - 15 μm).

The declared objective was, to develop passive pointing devices which fully relied on the thermal radiation emitted by the targets, without any need of additional illumination. Besides the companies ELAC in Kiel, Carl Zeiss in Jena, Zeiss-Ikon in Dresden and parts of the AEG also the physical institutes of the Universities of Berlin, Erlangen and Prague were working in this field. Although only a few thermal imagers were deployed until the end of the war in 1945, the participating institutions provided pioneering scientific and technical contributions as is shown in the following.

A second direction of development represented by the companies AEG in Berlin and LEITZ in Wetzlar as well as the FORSCHUNGSANSTALT DER DEUTSCHEN REICHSPOST (Research Establishment of the German Imperial Post) worked in the field of active target detection in the near infrared region from approximately 0.9 μm to 1.4 μm. These devices worked outside the human sense of vision that is limited to the wavelength range from 0.38 μm to 0.78 μm. Because the natural thermal radiation in this spectral region is much too weak to be used for direct detection, the devices were depending on natural or artificial illumination of the scene by the sun or by searchlights, respectively. For night observation special near-infrared searchlights had to be developed additionally to the image converter tubes. The successful work produced some target sights

**30 – *German research on thermal imaging between 1930 and 1945***

*Valuable contributions to the development of instruments in the field of thermal imaging came from the universities in Berlin, Erlangen, Frankfurt and Prague as well as from several companies like Carl Zeiss in Jena, Electroacustic (ELAC) in Kiel and Zeiss-Ikon in Dresden.*

(e.g. Spanner I & IV, Panther, FG 1250, Uhu, Falke, Vampir) which were deployed in considerable numbers at army and air force before the end of the war. The applied technologies were quite different from thermal imaging and their interesting history deserves an adequate treatment at another place.

As early as in the 1920s, semiconductor materials and the photoconduction in solid state was investigated in Germany. At the beginning of the 1930s, the improvement of infrared detectors became on object of interest.

In **1933** Edgar W. KUTZSCHER[140] at the professorship of physics of the University Berlin, "discovered

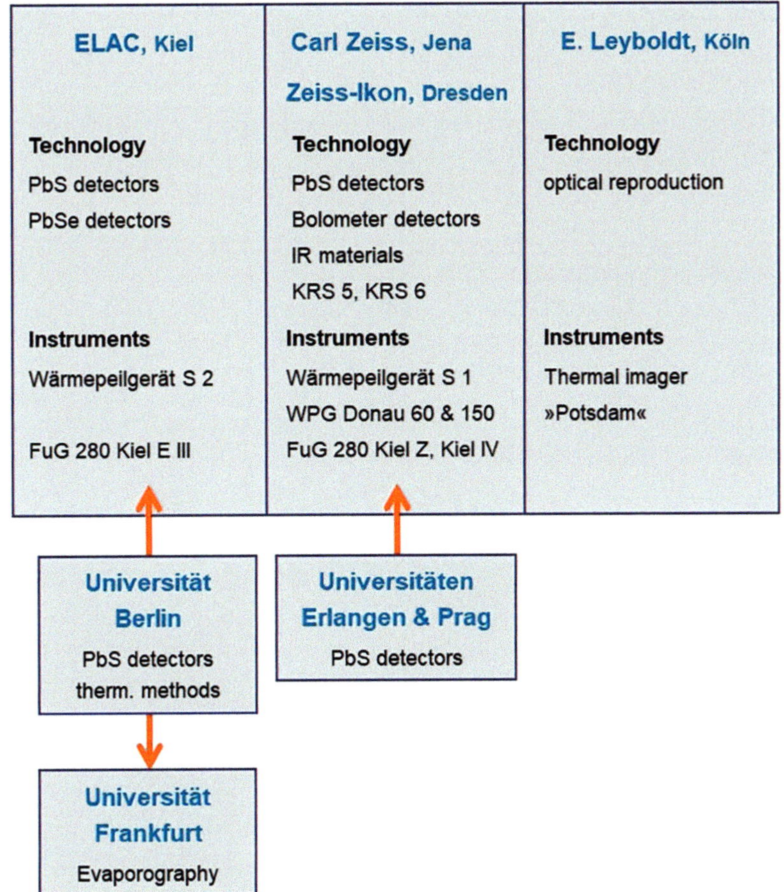


31 – *Functional links between German research institutions in the 1930s*

that lead sulphide (from natural galena found in Sardinia) was photoconductive and had response to about 3 µm."[141] His lead sulfide crystals (PbS) were much more sensitive in the near infrared than all crystal detectors known by then. By simple cascading of several crystals KUTZSCHER could build a PbS-cell which achieved similar performance as the bolometers used at that time.[142] "Like his British and American contemporaries, within a year he obtained military sponsorship – from the German army, in this case." A few years later, KUTZSCHER transferred to ELECTROACUSTIC (ELAC) in Kiel. During the war he managed the department for infrared research with a core of seven collaborators "with whom he had studied the basic physics of detectors, materials and the atmosphere as well as production techniques and applied systems engineering for infrared detection."[143] He assisted to build up a series production of PbS-cells and pioneered "methods of depositing IR-sensitive surfaces by chemical precipitation"[144]. The production of photoconductive PbS-cells began around 1943. Towards the end of the war the detectors were produced in high numbers of up to 1000 per month.[145]

In parallel to these activities Bernhard GUDDEN[146], Professor for Experimental Physics at the FRIEDRICH-ALEXANDER UNIVERSITY in Erlangen and from 1939 on at the KARL-FERDINANDS UNIVERSITY in Prague, investigated photo-electric effects especially in semiconductors. At this time already, he was ranked "together with his teacher and mentor Robert Wichard POHL among the grey eminences of experimental solid state physics in Germany" and had published "fundamental work about light-electric effects in solids and about ionic conduction in crystals."[147] As an alternative to the production method of ELAC, GUDDEN developed an evaporation method to form poly-crystalline PbS films of a thickness of approximately 1 µm.

GUDDEN's method was taken over by Paul GÖRLICH[148], who had transferred to ZEISS-IKON AG[149] in Dresden in 1932, shortly after his doctorate about light-electric effects in fluids at the TECHNISCHE HOCHSCHULE (technical university) in Dresden. "There he founded the laboratory for vacuum and photo cells and developed photo-cells for the beginning sound film technology as well as image converters for military night vision, image sensors for the beginning television and receivers for missile guiding systems. At the same time, he began the development of photo-cells based on the internal photo-effect as elements for automatization and for military IR technology."[150] Based on the fundamental development of GUDDEN and own experience with photo-cells, PbS-cells were produced at ZEISS-IKON in reasonable numbers.[151]

In **1935** Dr. GÄRTNER, representing the OBERKOMMANDO HEER OKH (High Command of the German Army), took the initiative and "the work on the issues of heat pointing on a thermal basis was re-started at the company CARL ZEISS together with Wa Prüf 8 at the suggestion of Dr. Gärtner, OKH/ Wa Prüf 8."[152] This was the starting shot for a long term development of thermal pointers with bolometer detectors at CARL ZEISS in Jena, about which a comprehensive report exists , filed by H. PLESSE mid of 1944.[153] Some of the essential parts of this report are cited in the following. "The developments and investigations were conducted on behalf of the Heereswaffenamt (Army Weapon Office)."[154] The first objection was quite sober. "The task initially presented by the office was: Creating of a barrier device, i.e. a device by which certain terrain sites like road crossings, mountain passes, hollows, harbor entrances, river fords etc. could be watched with respect to the transition of targets." Thus, a device was requested that we today would call an infrared motion sensor. Those devices for controlling the outside lighting can be bought today in every DIY market for a few Euros. However, the ranges needed for the mentioned surveillance tasks were in the order of kilometers and not meters, as is sufficient for a house entrance. After the first considerations it was already decided to use a bolometer as a detector and a measuring bridge "set up in the manner of a WHEATSTONE bridge" to build a "robust instrument". "At that time not the question of the highest achievable sensitivity was considered because at first the basics of thermal pointing in a terrain had to be explored."

In **1935**, a passive device with a bolometer detector for the detection of ships was tested in the United States. For this purpose, some of the largest ships of that time were chosen as targets (perhaps to be on the safe side or to attract more attention)[155]. The ships could be detected at a distance as listed below:

| | |
|---|---|
| SS MAURETANIA | at 13 miles |
| SS NORMANDIE | at 17 miles |
| SS AQUITANIA | at 10 miles (befogged) |

It turned out that the sensor had enough sensitivity to discriminate the dummy funnel of the MAURETANIA from the three funnels in use.[156]

"In **1935**, R V JONES[157] was developing infrared detectors at the CLARENDON laboratory at Oxford under LINDEMANN's guidance. He was occasionally diverted from his work … to produce detectors for a retired American Navy inventor, Commander Paul H. MACNEIL, who was promoting his own version of an infrared detector of aircraft. While the MACNEIL device was also unsuccessful, it reinforced interest at the AIR MINISTRY, which in January of that year had set up a Committee for the Scientific Survey of Air Defence. Jones and an NPL scientist, J S ANDERSON, performed their own trials late in 1935, again with poor results. Detecting the radiation from hot engine surfaces appeared difficult.

The Committee nevertheless asked JONES to continue with full-time development, even if it was recognized to be a peripheral line of investigation. Unlike the concurrent radar research, which "had a large team … devoted to it", JONES "for much of the time, had only himself". He devised equipment based on infrared detectors coupled to a small telescope, with signals amplified by a four-stage valve amplifier and indicated on a galvanometer – an arrangement employed tentatively in spectroscopy laboratories since the 1920s."[158] Various versions of the system were developed over two years.

In **1936**, JONES "developed a search system with an 11-cm aperture and a bolometer detector. It was ground tested … and detected an aircraft at distance of 1 mile during the day and 2 miles at night."

In **1937**, an improved system of JONES "was flight tested … and showed detection ranges of about one-third mile against aircraft. This may be the first time that one aircraft was detected in flight from another by use of infrared equipment" (in another source one-half mile is stated).[159] "Compared to radar, though, this radiometric equipment was incapable of detecting the range (distance) of aircraft. And experience belied the myth: the detection was not effective through clouds."

"In **March 1938**, the small project was ended in favour of radar. Infrared 'light' could be detected in some circumstances, to be sure – especially when emitted by cooperative targets – but appeared too weak to be measured for the military application."[160]

These American and British activities probably have confirmed the German officials in their assessment of the high strategic importance of infrared technology, especially if they had no detailed knowledge about the British conclusions.

On **1 October 1937**, CZERNY became head of an institute in Frankfurt. Together with P. MOLLET, he published further technical improvements of his method.[161] The front side of the membrane (diameter 55 mm) now was blackened with an opaque layer of bismuth or aluminum. For the fluid film 5-times distilled paraffin oil was used, the refractive index of which was measured with an Abbe refractometer made by ZEISS.

CZERNY and MOLLET also presented a thermal camera, which looked quite serviceable, and performed comprehensive investigations to compare it to near infrared photo cameras.

To make comparisons in short form possible, he named "our method "Evaporographie" in contrast to infrared photography with sensitized plates." With the new wording, which gave an indication of the functional principle of evaporation (from Latin evaporo = to vaporize, to transpire), CZERNY coined a name for his method, that became internationally used from that time on (engl. evaporography, franz. évaporographie).

For the comparative testing both cameras were illuminated by an identical, specially prepared light source. It used a monochromator with a quartz prism to isolate the chosen wavelength out of the white light of a so-called Nernst lamp[162] (special incandescent lamp). The cameras were illuminated with different wavelengths (= colors) in sequence.

The quantitative comparison with the AGFA 1050, the most sensitive contemporary infrared film, produced instructive results. For several wavelengths in the near infrared region between 0.9 μm and 2.5 μm the exposure times, necessary to obtain a useful image, were measured.

**Comparison: Exposure Time for a Good Picture**

| Wavelength | IR-Film AGFA 1050 | Evaporograph |
|---|---|---|
| 0.9 μm | 0.2 s | 5 s |
| 1.05 μm | 0.2 s | 6 s |
| 1.2 μm | 4.0 s | 5 s |
| 1.3 μm | 32.0 s | 4 s |
| 1.5 μm | —— | 7 s |
| 2.0 μm | —— | 8 s |
| 2.5 μm | —— | 16 s |

The values showed that the infrared film is useful only up to a wavelength of 1.3 μm at maximum, a value that represents the state of technology until the present day. The exposure times of the evaporograph were constant virtually over the whole wavelength range. The experiment could not be carried on beyond 2.5 μm because of the limiting infrared transmission of the employed quartz prism.

With a salt prism, it could be proven experimentally by measurement of infrared absorption spectra (e.g. of furfuryl alcohol $C_4H_3O \cdot CHO$) that the evaporograph was also sensitive in the longwave IR range (LWIR) beyond 10 μm. An amazing result also was that the well-known $CO_2$ absorption band at 4.3 μm and the broad vapor absorption band around 6.3 μm could be measured in the room air.

In the following years of the World War, the sources of information about evaporography, being the only available method to obtain thermal images at that time, were drying up. New information about the achieved technical state would 'surface' – in the literal sense of the word! – not until the end of the war (see below).

At the end of the **1930s**, the Research and Development Department of CARL ZEISS in Jena was organized in 9 General Laboratories for fundamental research, 8 Department Laboratories for development series devices and 5 Design Offices for structural development of the series devices.

The General Laboratories conducted fundamental research in virtually all scientific disciplines that were relevant for the Zeiss products. The scope of work reached from physics, chemistry, metallurgy and crystallography via surface technology to electronics, measuring technology and test engineering. In this structure, the new topic infrared technology was incorporated in the way that the development of new infrared sensors (photocells and bolometers) was the responsibility of the Electric Lab of Werner WEIHE. One of the two Crystal Labs, headed by Rudolph KOOPS, was commissioned with the search for new optical materials with sufficient transmission in the infrared spectral region.[163]

A few years later, the nuclear physicist Georg JOOS, who had a reputation as former professor of theoretical physics at the Universities in Jena (as from 1929) and in Gottingen (since 1935), could be appointed a director of the whole Research and Development Department including the mathematical offices. Initially on leave granted by the University Gottingen for only one year, on 1 April 1941 JOOS began his work at CARL ZEISS, and it became clear very soon that it should last much longer. Exactly two years later he became a member of the Geschäftsleitung (Board of Management), and in this function, he had to supervise the post-war relocation of the company to Württemberg, which was prosecuted by the American Allies, and the start-up of the new ZEISS OPTON company in Oberkochen.[164]

End of **1938**, systematic investigations of the ZEISS researchers led "to the necessity of new bolometer designs which were adapted to the special purpose of thermal pointing" as H. PLESSE has reported.[165] First tests with known bolometer types "with a 60 cm mirror had resulted in rages of about 800 m against a driving truck and the offset was severely drifting.“ Thus, a systematic improvement was inevitable. As "Sperrgerät" (blocking device) at first the development of a "Gleichlichtgerät" (continuous light device) with self-elaborated wire bolometers was launched. The installation of two bolometer surfaces, laying closely side-by-side was intended for minimizing the thermal influences and the effects of the thermal radiation from the surroundings. The principle, which was used already by LANGLEY in his improved bolometer of 1890, was termed "spatial difference principle" by Marianus CZERNY[166], a term that was also used at CARL ZEISS. The wire bolometer was operated in a Wheatstone bridge with AC voltage. Therewith, it was easy (i.e. manageable with the contemporary electric means) to amplify the generated measuring signals.

The wire bolometer itself, the actual sensor, was formed from the thinnest iron wire commercially available, which was coiled to a closed spiral. This was done to achieve a sufficiently high heat dissipation together with a low thermal capacity.

32 – *Organisation of research and development at Carl Zeiss after 1935*

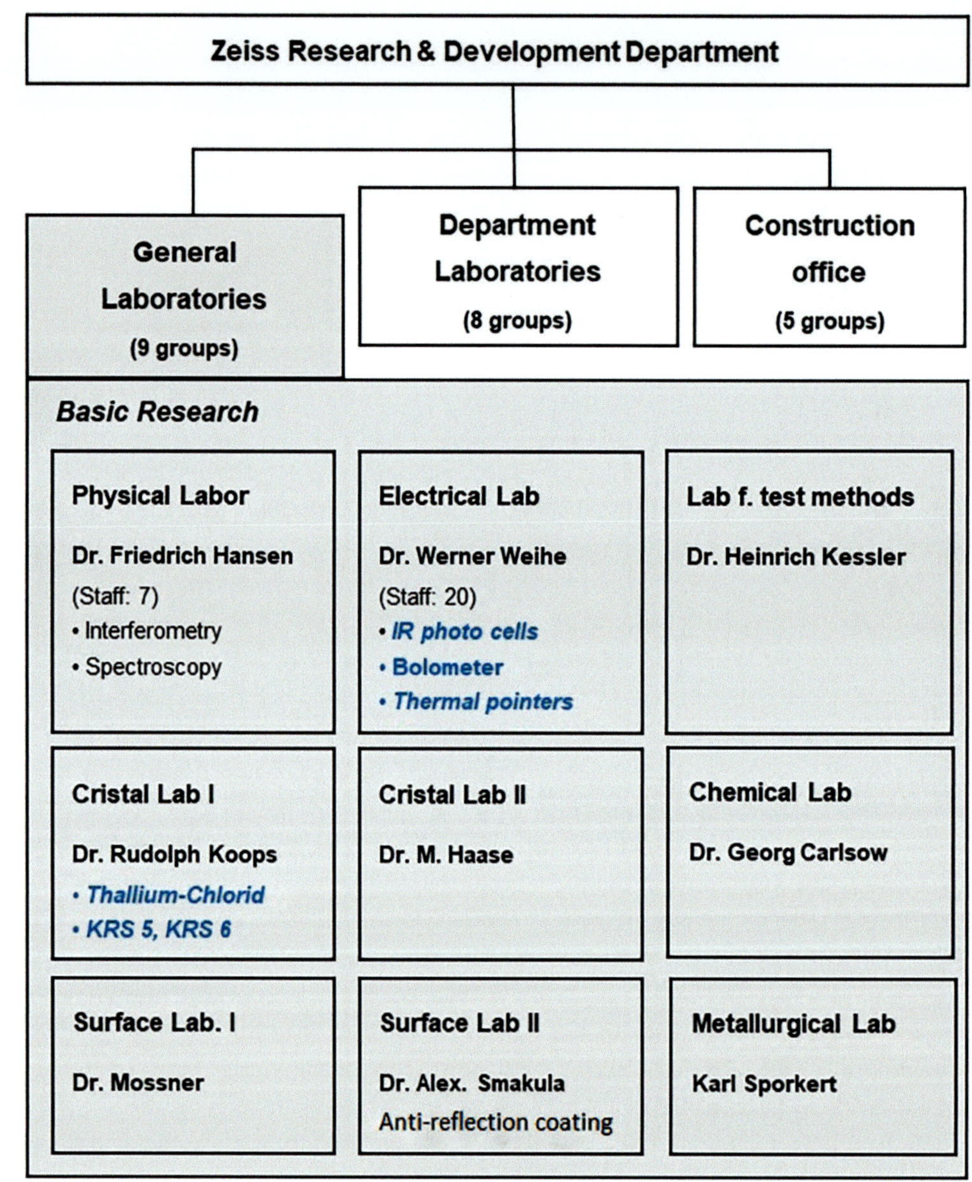

Due to the lack of suitable windows for closing a vacuum vessel the sensor had to be designed as an air bolometer. "At that time, for the longwave IR in the region from 8 to 12 μ not even means for normal air exclusion were available. It was frequently tried to protect rock salt by a coating of lacquer or plastic. If the layers were thick enough to withstand the weathering effects, the absorption was too high. If, on the other hand, the transmission was sufficient, the layers were not durable." To hold the thermal influences in bay, a block of copper was used as a housing. The wire bolometers were mounted in excavations in the copper block which in turn was fixed "in two further metallic hulls which enclosed it as symmetrically as possible".[167]

The properties of the new blocking device were measured at the Physical Lab of Friedrich HANSEN. As the encouraging results showed "the sensitivity was on the level of sensitive thermocouples, it was higher than that of all air bolometers known by then", whereby the ZEISS researchers in all honesty noted "that earlier data in literature are mostly incomplete".[168]

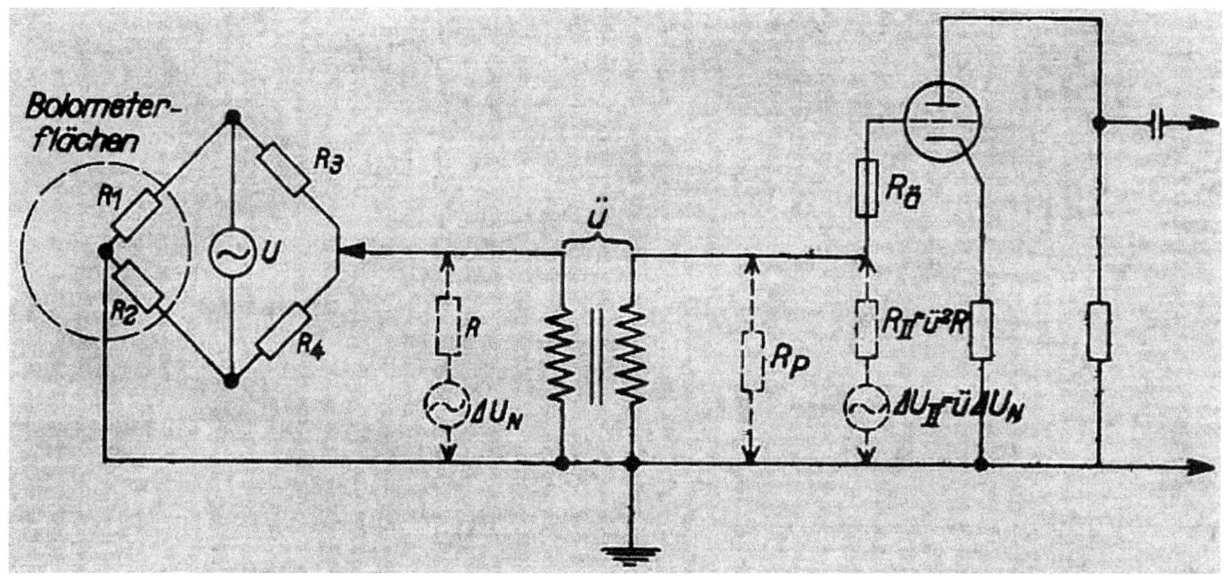


33 – *Electrical principle of the Zeiss bolometer sensor*

*Two wire bolometers ($R_1$ and $R_2$) are part of a Wheatstone bridge ($R_1$ to $R_4$) which is operated with alternating voltage (U). A transformer (ü) transmits the measured signal into the measuring electronics where it is amplified and rectified.*

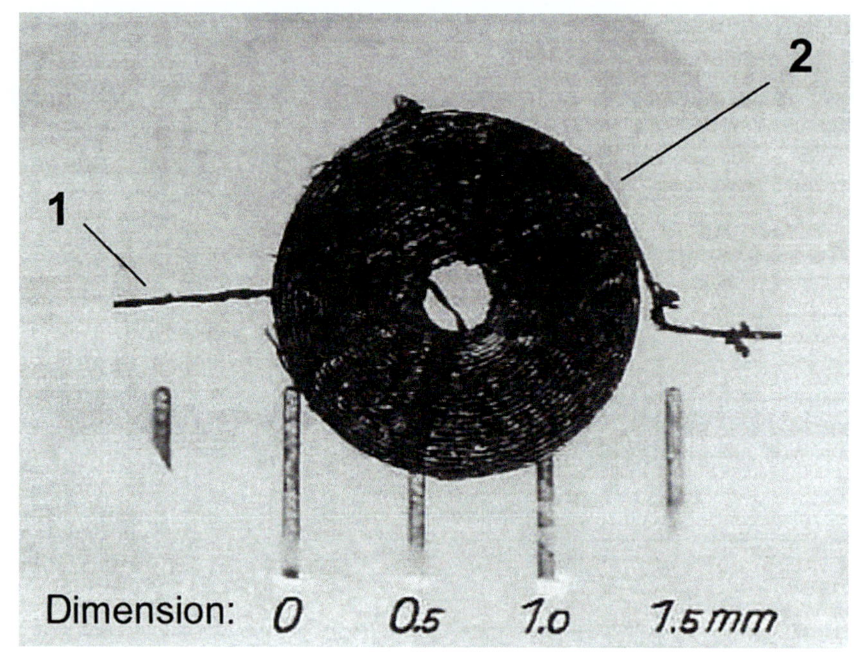


34 – *Assembly of the first wire bolometers of Carl Zeiss*

*A very thin iron wire (1) of 15 µm diameter is wound to a spiral (2) with 26 windings. The diameter of the spiral is approx. 1.3 mm as can be estimated using the scale at the bottom of the picture. The electrical resistance of the wire bolometer elements must be inside a tolerance margin of 5% to ensure the proper function of the electrical circuit.*

## Performance values and ranges

The detectable minimum temperature change was $1.5 \cdot 10^{-5}$ °C, "whereby the measured deflection was 3 x as large as the uncertainty." The evaluated sensitivity limit for thermal radiation was $3 \cdot 10^{-7}$ W/cm$^2$ at a setting time of 2 seconds. "This corresponds to the irradiation of a hand in a distance of about 4 m."[169]

"With a 60-cm mirror now instead of 800 m, after gradual improvement, up to 4 and 5 km range against driving cars were achieved." "The first detailed tests conducted together with the Oberkommando Marine OKM (High Command of the Navy) in Gotenhafen in 1940 … against ships with a 150-cm mirror led to ranges up to 30 km in an especially fortunate case."[170]

Apart from that, it was the same as usual: "After reaching the longer ranges with the "Gleichlichtgerät" (continuous light device) also the other requirements were increased." According to the report of H. PLESSE this was related mainly to the following shortcomings of the "Gleichlichtgerät":[171]

- "Too great lag.
  The setting time was approx. 0.7 seconds"
- "Drifting of the zero point.
  Despite the housing in copper sheet and careful shielding to the exterior, again and again a drifting occurred so strong that the detection of targets was often heavily hampered."

- "Surfaces too small, especially too short for many purposes.

  Bolometer surfaces in the depicted form cannot be made really large, because the mechanical stability is too low."

For these reasons, the ZEISS researchers were "thus forced to tread other paths. In consequence, they led to using a "Wechsellichtbolometer" (modulated light bolometer)."

As early as in **1939**, the experts concerned with bolometers at CARL ZEISS had contacted their colleagues of the material research and had discussed with the head of the Crystal Lab Rudolph KOOPS. A lab for crystal growing in a solution, later in a melt, was re-established only in 1930. After growing trials of calcium fluoride ($CaF_2$), initiated by Ernst ABBE himself, were discontinued without success in 1898,[172] it had apparently taken some time to overcome the failure.

In the discussions about the development of a suitable window material for the thermal imaging wavelength region (LWIR 8 - 12 µm) also the head of the Surface Lab Alexander SMAKULA was involved, who had developed the first practical anti-reflection coating in the middle of the 1930s.

The result was "that thallium chloride despite of some shortcomings like low optical and mechanical quality and low water-resistance would be sufficient as air seal for first applications." Thallium chloride (TlCl) is a poisoning, color-less crystal with a high transmission in the infrared wavelength region from about 2 to 20 µm. "All initial experiments were made with this substance."

The element thallium enters - besides chlorine - also into chemical compounds with other halogens like fluorine, iodine and bromine and even mixed crystals with several halogens are possible. Therefore, KOOPS continued his search for a better suited material that avoided the limitations of TlCl while having an equally high transmission.

In **1941,** KOOPS could finalize his development and produce a mixed crystal that was almost compliant to the raised requirements. Up to now, this material is known as KRS 5 and is still used for cuvettes and windows in infrared spectroscopy. The term KRS introduced by CARL ZEISS was derived from "**Kr**istalle aus dem **S**chmelzfluss" (crystals out of the melt), giving an indication of the production process.

Besides KRS 5 KOOPS developed and produced as a further mixed crystal also thallium-bromine-chloride $TlBr_{0.3}Cl_{0.7}$. This material was named KRS 6, the properties were less favorable, however.

According to PLESSE therewith "a suitable material was available to infrared technology. With KRS 5 vacuum vessels can be closed, the plate can be polished so that also lenses are manufacturable. The resistance against air moisture is good."

Also new optical glasses with excellent near infrared and mechanical properties reached the state of industrial production at that time[173]. An example for that was Duran glass[174] made by SCHOTT GLASWERKE in Jena. Due to the high resistivity against temperature changes and the mechanical stability it was readily employed not only for optical-electrical parts (e.g. cathode ray tubes and radio tubes) but also as substrate material for telescope mirrors. Another example was the very pure quartz glass produced by HERAEUS QUARZGLAS in

**Properties of the IR-Glass KRS 5**

| | |
|---|---|
| Chemical Composition | 42% Thalliumbromide, 58% Thalliumiodide |
| Colour | red |
| Transmission range | 0.7 - 30 µm |
| Best transmission | 1 - 20 µm |
| Refractive index | n = 2.4 |
| Chem. Stability | not soluble, not hygroscopic, sensitive to organic solvents, alkali and Phosphate |
| Mech. properties | mechanical not totally rigid, tendency for creeping |
| Methods of working | Diamond-sawing, grinding, polishing |
| Health risk | toxic |

Griesheim[175], that is distinguished by a very broad transmission region reaching up to a wavelength of 4 µm on the infrared side. This all is combined with very good mechanical, thermal and chemical properties.

Also in **1941**, according to PLESSE "the development of a "Wechsellichtgerät" with a low-persistent bolometer was begun."[176] At the "Gleichlichtgerät" "preferably equal surfaces were arranged very close side-by-side" in order to "eliminate the influence of temperature changes." As practical trials showed the application of this "spatial difference principle" was not sufficient to achieve a satisfying stability of the zero point. The researchers hoped for significant improvements by designing a new instrument at which additionally the "temporal difference principle" was applied also. "The receiver surfaces of the bolometer are formed as thin foils with such a small thermal lag that under illumination the end temperature is reached in less than 1/10 sec. Due to modulation of the radiation with a chopper blade with a frequency of 10 Hz the output at the bridge is an alternating voltage with equal frequency, the height of which is dependent on the impinging radiation only. Local temperature changes, that typically are slow, are eliminated."

This measurement principle that is widely used and termed "Lock-In-Amplification" today, was not invented at ZEISS but was known from the scientific literature in the 1930s already.[177] In the subsequent time it found entrance into many other ZEISS instruments like spectrometers and polarimeters.

After the "first not very promising experiments "had shown that "before all other … it is practically impossible to eliminate the vibration-sensitivity of the amplifier for low frequencies" it was clear that a direct measurement of the 10 Hz-modulation of the light was not sensible. Therefore, it was decided to use "the bridge not with DC voltage but operated with a voltage of a frequency of 4000 Hz. We get primarily as a measurement voltage a modulated voltage at a frequency of 4000 ± 12 Hz at 12 Hz chopper frequency."[178]

Besides employing the modulated light principle, one hoped for some improvements by a better design of the bolometer surfaces. Conceptually two configurations were known from literature.[179]: metal foils or precipitated layers. As metal foils "could be made hardly thinner than approx. 1 µ", what would have resulted in "a thermal lag that would be of significant importance at a beam chopping frequency of 10 Hz", the decision

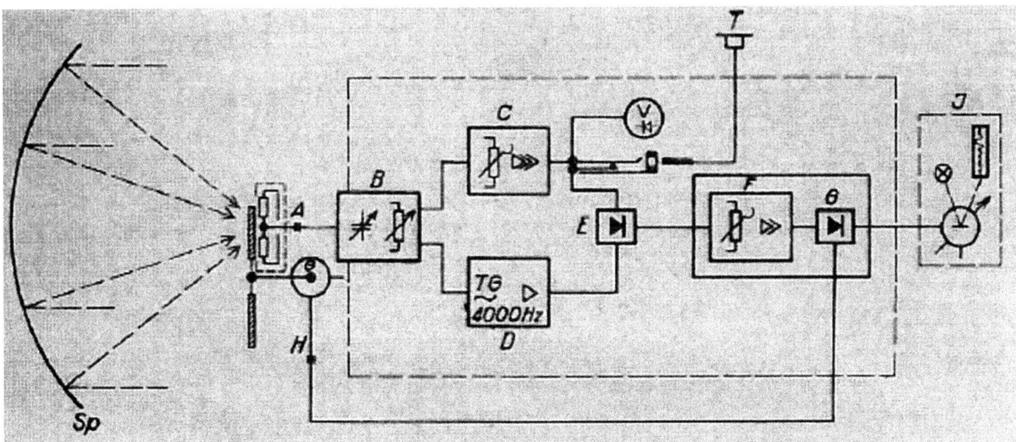


35 – *Block diagram of the bolometer with modulated light of Carl Zeiss*

*The incoming thermal radiation is imaged by a parabolic mirror (Sp) onto the bolometer surfaces (A). In front of the bolometer a rotating shield (chopper), driven by a motor (H), generates a modulation of the radiation with 12 Hz. The bolometer units are connected to a tuning device (B) via an electrical line that can be up to 15 m long. The bolometer units and the tuning device make up a Wheatstone bridge. The sound generator (D) generates 4000 Hz AC voltage for driving the bridge and the sound amplifier (C) amplifies the modulated 4000 Hz measuring signal. Behind the first rectifier (E) for the frequency 4000 Hz the measuring signal is still modulated with the frequency 12 Hz. After amplification by the low-frequency amplifier (F) and rectification by the second rectifier circuit (G) the measuring signal is feed in an indicating instrument (I). The signal strength at the exit of the sound amplifier can be monitored by an instrument (V) or a loud speaker (T).*

was made for "lacquer foils with precipitated layers". These "can be made even thinner without losing stability."

During the search for the best bolometer metals not only the temperature coefficients were considered but many other significant properties too like specific heat, density, heat conductivity, boiling point (with respect to production) and the specific electrical resistance. It turned out "that all these values were exceedingly favorable at antimony."

In the assembly of the new bolometer "a skin of Zapon lacquer was formed over a glass ring into which feed electrodes were melted" (see also Chap. 4, CZERNY 1929). On top of this the antimony (Sb) was precipitated. The antimony was not blackened (e.g. with bismut-black) because this led in practice to thermal and electrical shortcomings, and was not durable enough. Besides this "experiments with a blackening … yield an improvement of only 20 %."

These bolometer surfaces were installed in a glass flask which was closed airtight by a KRS 5 window. Therewith the bolometers no longer were exposed to the environmental influences as had been the case with the air-bolometers. Furthermore, the glass flask was "pumped out to a few torr", which was favorable for the following reason: "Around the bolometer surfaces heated air cushions form which perform smaller movements under vibration than the bolometer surfaces themselves, so that they are shifted to spaces with higher or lower temperature. Evacuation down to approx. 2 to 3 torr virtually eliminates this vibration sensitivity."[180]

According to PLESSE, the application of the new "Wechsellichtgerät" by the user was as follows: "When the target image runs through the field of view one gets a deflection to one side, it goes through zero when the image is exactly between the bolometer surfaces, and goes to the other side afterwards." If a target is perfectly adjusted the signal is the same - that is signal strength zero - as if there were no target at all in the field of view. The consequences of this odd constellation became quickly obvious.

For a thermal pointer, it is obvious that it will be used for the purpose of target tracking. Then the observer has a problem: It is "unfavorable that for tracking the deflection must be made zero to achieve an exact horizontal direction because then the operator has no measure of the radiation coming from the target and, especially at small or distant targets, has no indication whether the target is exactly in the line of sight or lost. As a remedy for this issue, the bolometer in the mirror is paned around the center position every second. Therewith, a slow target transition creates a shape of curve as depicted on the right of the picture." In particular, the modified "Wechsellichtgerät" now provided a maximum signal modulation on the display device once the target is exactly on the line of sight; on decreasing modulation amplitude it was easy to counteract and hold the target in the visor.

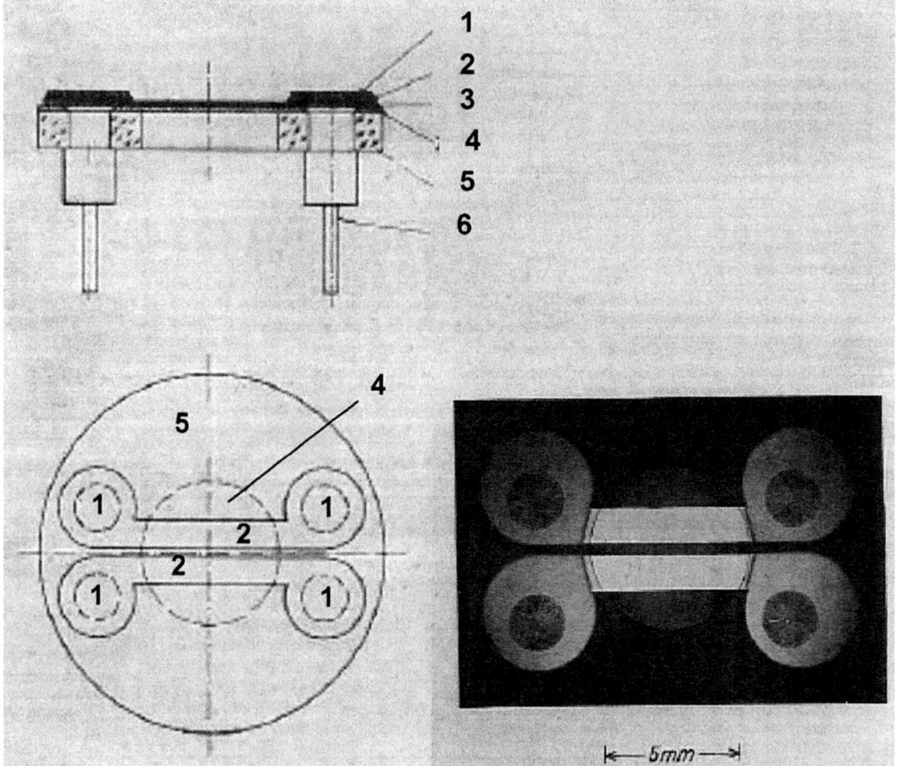


36 – *Design of the bolometer detector for modulated light of Carl Zeiss*

*On a glass ring (5) with connection wires (6) moulded into it, a thin foil made from Zapon lacquer (4) is applied leaving the patches for the electrode open. These are coated with copper (3) at first, and then with a layer of antimony (2). Eventually the patches (1) around the electrodes are coated with copper once more.*

Compared to the "Gleichlichtgerät", the expectations were indeed fulfilled. The temperature resolution, which was also not a problem of the "Gleichlichtgerät", was on the same level. A significant improvement was not possible because the "Wechsellichtgerät" had some loss mechanisms (only 50% absorption of the bolometer surfaces, 30% reflection loss at the KRS window, 20% loss in input circuit) which could not be avoided with existing means. The stability of the zero position and the general noise immunity was greatly improved, and that had been the main objective of the development[181].

About the achievable ranges one can find different values for various targets in the literature as is shown in the table below.[182] Of course, the range always depends on the temperature difference and the size of the target. Additionally, in field tests also the weather conditions are of fundamental importance. During tests of visual systems, it might be possible to tell something about the visibility conditions. For devices working in the region of thermal radiation, it is virtually impossible because the long wavelengths are not part of our visual perception.

To make the practical use easier, the "Wechsellichtgeräte" were equipped with some auxiliary instruments. For better adjustment of bolometer and mirror, but also to make pointing easier, a monocular sight 7x50 was installed as pointing telescope:

- Magnification 7x
- Diameter of objective 50 mm
- Pupil diameter on the eye-side 7.1 mm
- Field of view 7.3° (128 m on 1000 m)
- Illuminated crosshair in the shape of the bolometer surfaces

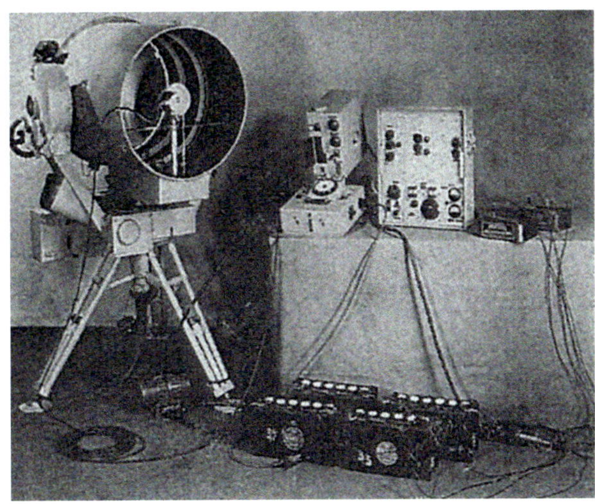

37

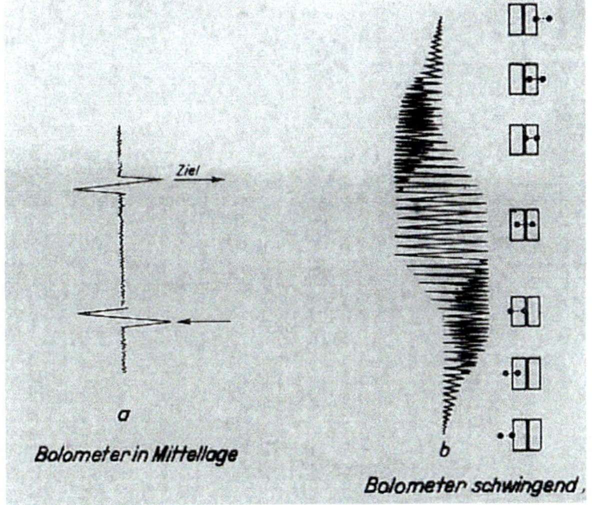


38

37 – *Thermal pointer Wärmepeilgerät WPG - Z*

38 – ***Target display on the monitor of a bolometer with modulated light***

*Signal curves are given for a target transition as shown at the right margin of the figure. Fig. a shows the signal curve of a bolometer that rests stationary in the central position. When the target is perfectly tuned to lie in the center of the detector, the signal is zero and therewith the same as if there were no target at all. Fig. b shows the same situation with a Bolometer that oscillates around the center position. When the line of sight is perfectly aligned to the target the signal shows maximum modulation.*

**Datasheet of Wechsellichtbolometer**

| | |
|---|---|
| **Optical System** | |
| Imaging optics | parabolic mirror |
| Material | glass substrate |
| Type of mirror | front surface mirror |
| Focal length | 641 mm |
| Mirror diameter | 60 cm (optional 150 cm) |
| Rotating slot aperture | 12 Hz modulation |
| Field of view, vertical | 10ˉ = 10/16° = 0.625° |
| Field of view, horizontal | 5,5ˉ ≈ 5/16° = 0.313° |
| **Bolometer** | |
| Detector surface | coating layer of antimony (Sb), not blackened |
| Wavelength range | approx. 8 µm - 13.5 µm |
| Number of surfaces | 2 identical surfaces |
| Detector area | 6 x 1.5 mm |
| Distance | 0.5 mm |
| Carrier material | membrane made of Zaponlack |
| Mounting | fixed on a glass ring |
| Housing, internal pressure | glass flask, 2 - 3 Torr |
| Window pane | KRS 5, cemented with Oppanol |
| Surrounding temperature | -30°C to room temperature |
| **Amplifier** | |
| Sound amplifier (4000 Hz) | 150 000x to 200 000x, 3-stage |
| Low-freq. amplifier (12 Hz) | 90x, 2-stage |
| **Performance data** | |
| Angular accuracy | 1ˉ to 2ˉ = 1/16° to 2/16° |
| Temperature resolution | 0.0001°C (temperature difference on Bolometer surface)<br>~ 10 nV voltage change |
| Range | up to 15 km (vs. torpedo boats) |

**Published Range Values**

| Source (Year) | Marine Dv 291 (1944) | Carl Zeiss (1954) | Lusar (1962) |
|---|---|---|---|
| Tank | — | — | 7 km |
| Speed boat | 12 km | — | — |
| Submarine | 16 km | — | — |
| Maritime target | — | 20 km | 20 km |
| Battleship | 40 km | — | — |

The visible display of the measuring results was not a trivial task in that time. To improve the general overview and to ease the operation a special display unit was developed, originating from a proposal of Prof. JOOS himself.[183] The display unit comprised a moving-coil instrument to which the signal voltage, generated by the bolometer and amplified several times, was applied. On the moving-coil a plane mirror was fixed by which the light of a small light source was deflected and made visible "on a slowly rotating drum with a phosphorescent layer." The drum was actuated by a clock work. Modulations of the signal voltage resulted in a rotation of the moving-coil instrument and therewith also of the mirror. By this the light point is deviated sideways and "the phosphorescent layer conserves the history of the deflection for several minutes to ease the overview over the proceedings to be monitored."[184] To enable an intuitive operation "the bolometer branches … (were) so connected and the phase … (was) so adjusted that targets

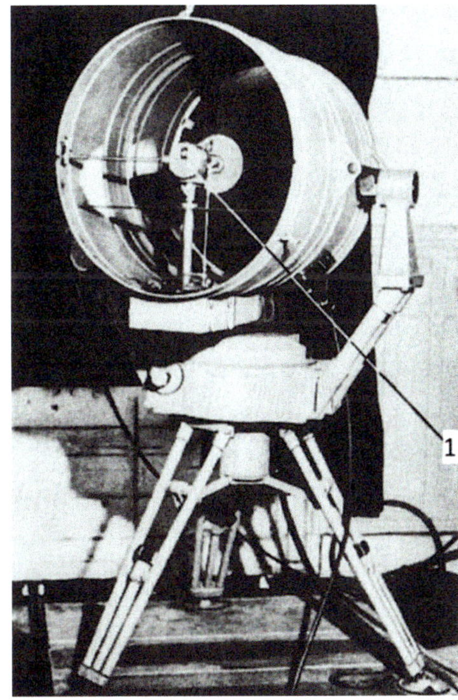


39

39 – *Zeiss Wärmepeilgerät S1 for the navy*

*In the picture, the arrangement of the bolometer (1) in the focus of the parabolic mirror is clearly to be seen.*

40 – *Control units for the Zeiss Wärmepeilgerät S1*

*The instrument S1 of the navy used the identical assemblies (display unit (2) and lateral steering unit (8) as the WPG-Z; only the amplifiers were packed differently. The picture shows the mirror instrument for indicating the bearing (7), the indicators for the bridge (4) and operating voltage (5), the low-frequency amplifier (6), amplifier (9), bridge balancing (10), sound generator (11) and measuring unit (12).*

moving to the right at first cause a deflection to the right and then, during the shifting of the target image to the other bolometer surface, a deflection to the left."

A plausible operation, for sure, was especially advantageous for the target tracking procedure. For tracking, the so-called lateral steering had to be operated "to reposition the mirror sideways after detection of the target." This could be done "mechanically with a hand-wheel … or electrically with a path-speed steering again with a hand-wheel or by means of an automatic path-speed steering operated by a console unit." In either case, the operator had to regulate the lateral panning of the parabolic mirror with an accuracy high enough to prevent that the target was lost out of the narrow field of view of 0.3° only. He could either turn the hand-wheel with the correct speed or he managed to adjust the electrical drive so that "the mirror movement was matched to the target movement". He could also do a combination of both – neither option sounds really comfortable to us today, even though "a good conformity of both movements … was achieved in general with a few corrections only."[185]

The lateral steering had special importance also for the search of undiscovered targets. To do this, it comprised "an apparatus for periodic panning between two adjustable values of the graduated dial by means of an electrical drive to find the targets." After activation of the search function "the mirror was periodically panned to and fro in a range limited by switch cams with an angular velocity of approximately 28⁻/sec (≈ 1,8°/sec)." This rotation perhaps appeared slow but "limited by the thermal persistence of the bolometer surfaces, it must not be greater than approximately 33⁻/sec (= 2°/sec)." Otherwise the target image might scurry over the bolometer without causing a significant deflection.

Taking together all described functions of the "Wechsellichtgerät" - and its limitations too - we may conceive a typical search-and-track mission in the following manner:

- Set bolometer to center position
- Adjust the lateral steering to the angular range to be surveilled
- Set necessary (fixed) elevation angle
- Start search procedure: a straight line without significant deflections appears in the display; two small lamps indicate the actual search direction
- Double significant deflection (see Fig. 38a): → target detected
- Stop search procedure, sway back the mirror a little
- Switch bolometer to 'Schwinglage' (modulation mode): a clearly modulated curve appears on the display (Fig. 38b)
- Switch lateral steering to 'Verfolgen' (tracking) and adjust, with hand-wheel or rocker switch, the path-speed steering so that the maximum modulation amplitude is kept constant permanently (the lateral swaying now is equal to the target motion).

In today's usage of language missions of this kind are commonly known as "Infra-Red Search and Track" IRST.

## 3.3 Thermal Pointers with Lead-salt Detectors

The advances in the infrared field in Europe were noted world-wide, and induced the United States to root the topic in the NATIONAL DEFENSE RESEARCH COMMITTEE NDRC founded in June 1940. In the department for "Detection, Controls and Instruments" the two sections "Instruments" and "Infrared Detection" were commissioned with infrared sensor research.[186]

In **1941**, the NDRC engaged the physicist Robert J. CASHMAN,[187] who was newly appointed as professor at the Northwestern University in Illinois, to conduct corresponding research work. Since 1935 CASHMAN had already resumed the work that had been discontinued by CASE in 1917, and had continued the research on the "thallofide cell". He discovered that a controlled oxidation of the thin thallium sulfide ($Tl_2S$) film was the key for permanent function of the cells as well as for reliable production. The improved detectors were employed in various devices by the United States during

World War II.[188] Also Bernhard GUDDEN conducted comparable investigations on this material at the KARL-FERDINAND-University in Prague.[189]

Whether or when the German side got notice of these American activities and how accurate this information had been, is not known to the author. However, "early in the war the Germans concluded that the Allied were using infrared equipment to detect Nazi U-boats and aircraft. Because of this erroneous conclusion, the Germans concentrated much of their research and development on infrared equipment and on means of counter measuring it. By contrast, the Allies concentrated their efforts on the development of radar. As a result, the Germans lost the war but they clearly won the battle of the infrared."[190]

In **1942** at about the same time, the photoconductivity of lead selenide (PbSe) was discovered in Great Britain and Germany.[191] It is the author's guess that the German work was done at the ELAC company in Kiel where also the series production of PbSe detectors was prepared even before the end of the war.

It was also discovered that not only the noise of photoelectric detectors could be reduced by cooling (e.g. with liquid nitrogen of -196 °C = 77 K) but that the spectral range could be enlarged also:

| Spectral range (cooled) | lead sulfide PbS | 1 – 3.5µm, |
|---|---|---|
| | lead selenide PbSe | 1 – 7µm |

For the first time, the cooled PbSe detector opened the possibility to realize a fast, photoelectrical detector which could measure the thermal radiation of objects at room temperature.

At CARL ZEISS, the development of the "Wechsellicht-bolometer" had not ended yet. In fact, all new developments in detector technology were analysed to find out whether these could be successfully used for further development of the own devices. It was backed by experimental investigations "that bolometers detect mainly in the longwave region between 8 and 13.5 µ, what is the main region of transparency of the atmosphere." In contrast, photoelectric sensors, like the PbS cell, were available only for the short wave infrared region, as "a detection above of 3 µ was not performed practically yet". PLESSE[192] derived the possible applications from the fact that "the total radiation energies in the 0 - 13 µ region on the one hand and the 0 - 3 µ region on the other hand … have a ratio of approximately 300:1 at a warm object of 100° C, and 3000:1 at an object of 50° C". This meant a conceptual advantage of the bolometers. On the other hand "an essential advantage of the [photoelectric] cells is, however, … the low persistence." From this the researchers at CARL ZEISS derived the conclusion "that for cool and slow moving targets the bolometer is the best detector, while for fast and hot targets the cells are better."

There were enough relevant military objects of both categories at that time, from surface ships (slow, cool) to aircraft (fast, hot exhaust pipes). Consequently, also thermal pointers with PbS cells were built at Carl Zeiss until the end of the war. The PbS detectors were provided by the own production facility ZEISS-IKON in Dresden.

In spring **1943**, the allied air forces operated with strongly increasing success against German submarines. Besides other measures the Allies had deployed newly developed radar devices in the S-band (10-cm band) which "not only brought immunity from detection by

[German] "Metox" but also gave increased ranges of detection on U-boats."[193] "With this upswing in Allied aircraft success, the Germans became convinced that Allied aircraft were using some new detection device and started frantic activity to identify and counter it. For a time, they occupied themselves with the idea that it was an infrared detector, since they had tried to develop one of their own, and experimented with special paints intended to give no infrared reflections. They also considered the possibility of a frequency-scanning radar and developed a scanning receiver with a cathode-ray tube presentation. This was of definite advantage to the operator, but still covered only the same meter-wave band. The sinking of U-boats continued."[194]

Latest in **1943**, ELECTROACUSTIC (ELAC) in Kiel[195] were building the Wärmepeilgerät (thermal pointer) "S2" for the German navy. Besides CARL ZEISS in Jena also ELAC was very active in the field of thermal imaging and, by hiring Edgar KUTZSCHER from the University Berlin in the second half of the 1930s, had gathered excellent know-how.

For the "S2" photocells made from photoconductive lead sulfide (PbS) were used for detection which had been developed by ELAC itself in the years before and were produced in-house then. The PbS-cells were cooled to -78°C (195 K) with dry ice or snow of frozen carbon dioxide to improve the sensitivity (better signal/noise-ratio). The impinging infrared radiation was focused by a parabolic mirror (field of view: 2° horizontal, 1° vertical) onto a rotating slit aperture which was fixed in front of the photocell.[196]. The radiation and therewith also the electrical signal of the cell were modulated with a frequency of 800 Hz. This alternating signal was amplified by pre- and main amplifier and

41 – *Receiver assembly of the ELAC Wärmepeilgerät S2 for the navy*

*(Photo: Fritz Trenkle)*

displayed in an instrument. Additionally, "on the shaft of the slit aperture a small generator was fixed which provided an alternating voltage with the same number of periods as slits were passing the photocell."

By "comparing this voltage … with the output voltage of the main amplifier in a compensator device" a phase-depending signal could be generated. In case "both voltages were in-phase … the target had to be located exactly on the optical axis", because "the generator voltage reaches its maximum value when a slit is exactly in front of the center of the photocell" and also exactly "at this moment the highest irradiation" was impinging on the cell. Did the irradiation happen earlier or later then "compared to the generator voltage a phase-shift occurred which could be used for a left-right-indication."[197]

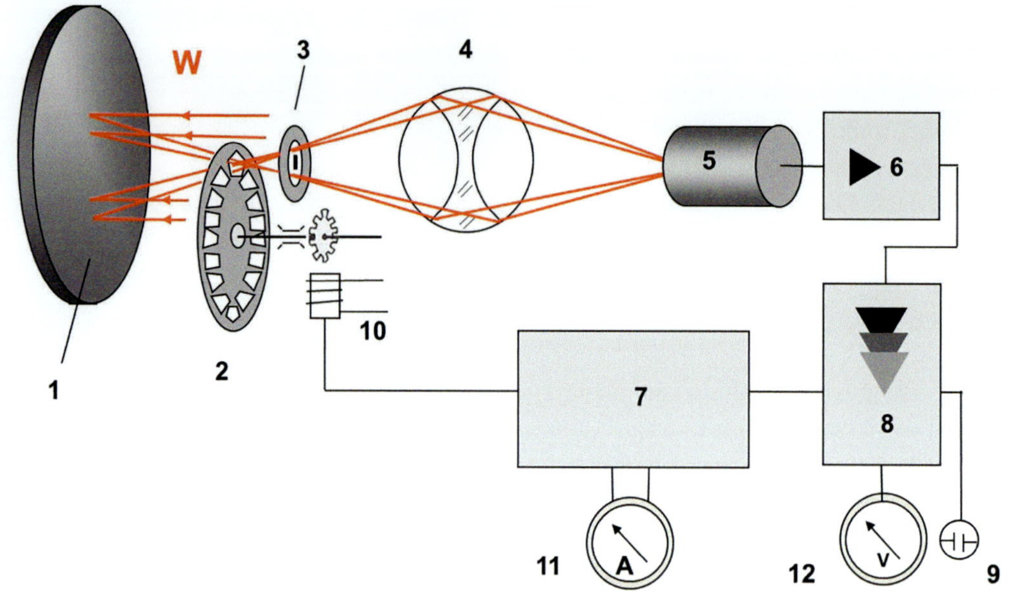


**42 – *Principle of ELAC Wärmepeilgerät S2***

*The incoming short-wave thermal radiation (W) is imaged by a parabolic mirror (1) onto a mechanical stop (3). The rotating slit aperture (2) modulates the radiation with 800 Hz. The radiation is focussed by a collecting optics (4) on a photocell (5). The 800 Hz signal generated by the photocell is transmitted via pre-amplifier (6) and main amplifier (8) to the indicator unit (12) and additionally to a glow lamp (9) for optical indication. The generator (10) generates an AC voltage synchronous to the modulation by the aperture, which is compared to the signal voltage in the phase indicator (7). A left-right-information is derived thereof and indicated with the pointing instrument (11).*

As achievable ranges 20 to 30 km are given in literature without any information about the kind of target. However, a temperature difference of 100°C between target and background was necessary.[198] This is a direct consequence of the fact that PbS-cells, even when cooled, were sensitive only in the shortwave infrared region from about 1 to 3.5 μm. Only objects with elevated temperatures emit sufficiently strong thermal radiation in this region to be measured. Towards the end of the war these detectors are believed to be produced in high quantities of a 1000 pieces per month despite the discussed limitations.[199]

In **1943**, further fundamental investigations at the "thallofide cell" were conducted by the MASSACHUSETTS INSTITUTE OF TECHNOLOGY (MIT).

In **1944**, the "NATIONAL DEFENSE RESEARCH COMMITTEE" "NDRC contracted GENERAL ELECTRIC in West Lynn, Massachusetts to manufacture the cells. Within 11 months some 6800 had been produced with a reported 90% yield ."[200]

By **1944**, the first lead selenide photoconductive cells were ready for production at ELAC.[201]

"Apparently dissatisfied with search receivers as a means of achieving immunity from radar detection, the U-boat Command turned to more drastic measures during the last year of the war. The development of the "Schnorchel" (snorkel)was carried out in the latter month of 1943."

Early in **1944** the first German submarines were equipped with Schnorchel. "Using Schnorchel for Diesel intake and exhaust, the U-boat could run at periscope depth with only a small Schnorchel head above the surface, yet accomplish the ventilation of the boat and recharge of batteries that previously had required surface operations. Before the submarines had to surface routinely to charge the batteries, now they could stay submerged and only show the small snorkel head with intake and exhaust pipe at the surface. Starting in spring 1944, the U-boats spent very little time on the surface, employing Schnorchel and electric propulsion almost exclusively. Original Schnorchel gave reduced radar echoes merely because of their small size. As a result, the average range of detection is reduced to about a third of that on a surfaced U-boat. In addition, the echo is frequently too small for detection until it has entered the area of sea returns which mask it quite effectively, … Even without camouflage, a Schnorchel is a difficult target to detect by radar. Not content with this state of affairs, the Germans developed non-reflective coatings for application to Schnorchel which still further reduced the echo from it."[202] For the period November 1944 to March 1945 an analysis was made of a region near the British Isles and showed that the "operational results have been even more discouraging". The effective sweep width of an aircraft equipped with radar system was only "about 1/10 mile, only about 1 per cent of the value on a surfaced U-boat."[203] "This made the submarine virtually undetectable by radar and, had it been available earlier, could have prevented Allied success in the Battle of the Atlantic."[204]

Until then "the British made no serious attempt to exploit infrared radiation because there were no suitable detectors", and perhaps also because they relied on the effectivity of their radar.

"As a last hope, the Admiralty asked for infrared trials on the detection of Schnorkels or exhaust emissions from U-boats." The solution of this task was not impossible as was demonstrated in that time by the German thermal pointer WPG 15 which could detect a snorkeling submarine in 1 to 2 km distance. "This work continued for several years and was to make an important contribution to the development of thermal imaging, but a much more interesting German equipment was captured at the end of the war."[205]

In winter **1944**, also Robert CASHMAN turned to the lead sulfide (PbS) cells. Very soon it was clear to him that this cell had much higher potential than all sulfide cells he had investigated before. Moreover, he discovered that the photo-sensitivity could be improved further by precipitating the PbS in the presence of oxygen. In December 1944, CASHMAN could form successfully thin PbS films with a chemical deposition process. At that time, he could also purchase some German PbS photocells which he judged better than his own due to the smaller size. Consequently, he also turned that way and soon could achieve comparably good performance values.[206] Thereby his cell, manufactured at the NORTHWESTERN UNIVERSITY became the first PbS-cells ever produced in the United States.[207]

Also in **1944**, a groundbreaking new thermal imager, the Wärmebildgerät Potsdam was presented in Cologne. The Potsdam of "the German firm E. LEYBOLD'S NACHFOLGER was the prototype from which many thermal imagers, both British and American, were later developed. This contained a double optical system in which an infrared detector was scanned across the scene while at the same time a spot of light was scanned in a corresponding pattern across a film.

The output from the detector was used to modulate the light spot, so that when the detector picked up infrared emission, the light scanning the film became more intense. As a result, a visible picture corresponding to the thermal scene was built up. At the end of 1944, when the PL3 version of this equipment was built, there were no photodetectors sensitive enough to pick up the radiation, with waves 10 μm or so in length, which is emitted by bodies at room temperature. The Germans, however, did have fast bolometers with which they were able to obtain a simple picture of a person."[208]

The simplest realization of the described concept is by using a plane scan mirror coated on both surfaces which can be tilted around two axes. The frontside is used to scan the thermal radiation while the backside deviates the visible light, used for reproducing an image, in an analogues way. This turned out to be a brilliant concept to build useful thermal imagers despite the limited possibilities of the contemporary analog electronics. Many companies (e.g. BARNES, TEXAS INSTRUMENTS, CARL ZEISS etc.) adapted it, it was used by several 10,000 Common Module imagers and was superseded by the 2$^{nd}$ generation thermal imagers with digital electronics not until mid of the 1990s.

Even today hundreds of devices based on this concept are still employed. A telling example is the German battle tank Leopard which is deployed by many NATO countries. The upgrade of the old scanning thermal imagers by modern staring device started only recently in 2016.

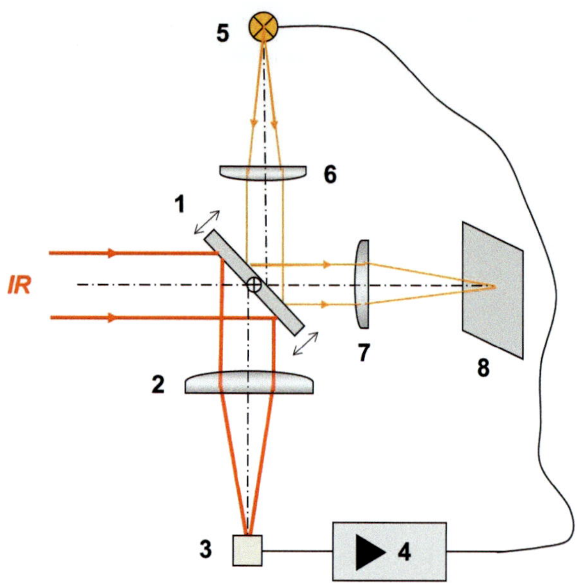


43 – *Optical image reproduction in the Potsdam thermal imager of E. Leybold*

*The incoming thermal radiation (IR) is deviated by the front side of the scan mirror (1) and is imaged by the IR imager (2) onto the IR detector (3). The detector signal is connected via an amplifier (4) to a small light source (5). The higher the thermal irradiation is on the detector, the brighter is the light of the lamp. The divergent light bundle emitted by the lamp (5) is collimated by the lens (6) and becomes a parallel bundle. This bundle is deviated by the backside of the scan mirror in a way analogous to the thermal radiation and imaged by the lens (7) on a photographic film (8) as a visible image.*

In spring **1945**, it became apparent also in Germany that the war in Europe would end with German defeat very soon. On the edges of a hotly debated event [209] it surprisingly happened that fresh information about the state of "Evaporography" 'surfaced' (literally!).

A submarine was the only means left to exchange information and strategic goods between Germany and its far away ally in Japan. Alas, also the 90-days travel to Japan was an extremely high risk, on average 75% of the submarines got lost due to allied air and naval dominance or plain technical defects.

Now a last German submarine was equipped and loaded for the long journey to Japan. U 234 was a submarine of the type X B, with nearly 90 m of length it was one of the biggest submarines of the Kriegsmarine

(German Navy). Surface running at 10 knots, it could travel a route of up to 18,450 sm.[210] The GERMANIA shipyard in Kiel had originally built U 234 in 1943 as a mine sweeper but converted it to a cargo submarine only a year later.

In the afternoon on **25 March 1945,** U 234, with 300 t freight and supplies for 6 to 8 months, put to sea towards Norway for the moment. The freight (8 t documents, 1 t diplomatic pouch, 210 t military goods) comprised weaponry, ammunition, innovative aeronautical equipment (e.g. a complete turbine fighter aircraft Me 262), radio & radar technology and nuclear technology (10 cases with uranium oxide!).[211] As for sure the highly complicated devices were not self-explaining to the Japanese addressees, also 12 passengers (German officers and civil engineers as well as two Japanese engineer-officers) were on board. The technicians and engineers, e.g. of MESSERSCHMITT in Augsburg, were to help with the integration of the most modern technology in Japan. One of the civil passengers was Marinebaurat Dr.-Ing. Heinz SCHLICKE (commissioner of naval equipment), a well-recognized German radar and infrared specialist.[212]

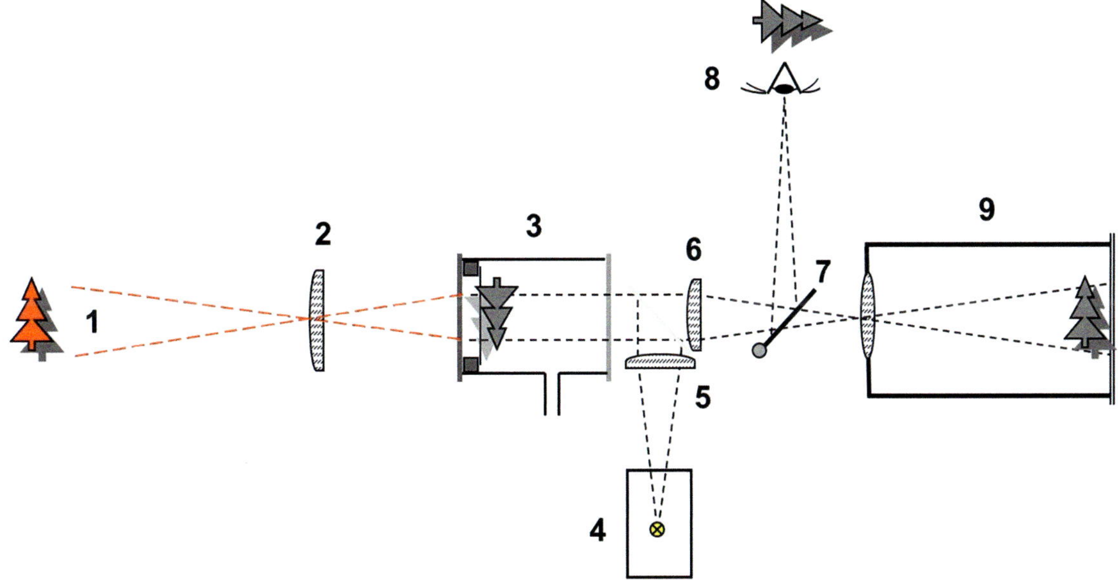


***44 – Principle of obtaining a thermal image with an evaporograph***

*The thermal radiation coming from the scenery (1) is imaged by an IR lens (2) onto the receiver membrane of the thermal sensor (3). The backside of the membrane is illuminated by a light bulb (4), a collimator lens (5) and a semi-transparent mirror. The resulting interference fringes in the fluid film are transmitted by a second collimator lens (6) either to the observer (8) or to a conventional photo camera (9). Switching between observer and camera is done by a swivelling mirror (7).*

The progress of the journey was tedious due to repeated technical troubles.

In **May 1945,** the journey came to a definite end due to the end of the war, and after some entanglements U 234 surrendered to the Americans. At the arrival at Portsmouth in New Hampshire the crew and the remaining passengers (the two Japanese officers had avoided the encounter with the Americans by suicide) were confronted with a heated-up and hostile atmosphere and had to endure a hard time.

Based on intelligence reports, the American OFFICE OF NAVAL INTELLIGENCE ONI had spotted SCHLICKE as important knowledge carrier and very soon transferred him to Fort Hunt in Virginia for interrogation. With his three PhD-degrees and experience from several executive positions (e.g. naval testing, news transmission) SCHLICKE could provide qualified information about virtually all aspects of radio and radar technology. Due to his last occupation as coordinator of the German naval research, especially in charge of the development of the infrared devices in the wavelength region from 0.7 µm to 200 µm, he was also familiar with the latest state of German thermal pointers and imagers. And he could give account about a device he called "Evaporigraf" that we can easily identify as CZERNY's Evaporograph. In detail SCHLICKE reported the following:[213]

- A parabolic mirror was used as infrared optical system
- The thermal radiation was focused on a blackened glass screen (obviously this referred to the membrane)
- The backside of the 'screen' was covered with a thin oil film with a thickness of a fraction of the wavelength
- The oil film was enclosed in a chamber in which the pressure was decreased to a value lower than the surrounding air pressure to make the image visible

According to SCHLICKE's report, one may imagine the performance as follows:

- After approx. 6 seconds a vague silhouette appeared
- After approx. 10 seconds a full picture was visible
- The "Evaporigraf" could measure up to 100 m against a person

SCHLICKE pointed out two special advantages of the "Evaporigraf":

- The device did not emit any radiation, i.e. it could not be located
- Due to the long wavelength, the sensitivity of the device was not affected by water vapor

As his knowledge had proved to be useful to the United States, SCHLICKE was admitted in the exploitation program to absorb scientific and technical knowledge called "Project Paperclip" and got a legal status for the residence in the United States. Until 1950 he participated in various classified projects at government research organizations and transferred to the American industry afterwards.

## 3.4 Thermal Pointers in Military Service

With the thermal pointer concepts of ELAC and CARL ZEISS basically all designs are discussed that were produced and employed in numbers before the end of the war. No other nation deployed genuine thermal pointers which passively detected the thermal radiation emitted by the target.

### Deployment at the Naval Artillery

Due to their technical prerequisite, the thermal pointers with bolometer were especially suited for moderate warm targets near room temperature that moved only slowly.

As was discussed at the beginning of the chapter, that is because these targets are emitting thermal radiation strongest in the atmospheric window from 8 to 12 µm in which the bolometers are very sensitive. The optionally available PbS detectors, however, were totally blind at these wavelengths. Due to their functional principle, the persistence of the bolometer allowed only the measurement of relatively slow intensity changes of the impinging radiation. The image of a fast-moving target passed the bolometer without being detected.

Due to its character, the ZEISS Wechsellichtgerät was predestined for deployment at the naval artillery. Based on the seafront their task was the protection of the coast and the surveillance - and in case combatting - of the shipping traffic off the coast. Seen from the coast, ships were exactly the kind of target for that the Wärmepeilgerät S1 was suited best: moderately warm and slow. Moreover, it was advantageous that the sea offered a uniform and relatively low radiation background. All this suited the new infrared technology, being still in its infancy, very well.

In 1939 the German naval artillery had to secure about 1500 km of coastline mostly at North Sea and Baltic Sea, whereas it was more than 15,000 km in 1941. The coastline then comprised the Atlantic coast from North Cape to the Pyrenees and parts of the coast of Yugoslavia and Greece; later also parts of the North-African coast came in addition.[214] Especially to protect the Belgian, Dutch and French coast the "Marine Artillerie Regiment" 21 to 26 were raised in 1940 which could be equipped only with captured artillery pieces in several cases.[215] During that time, the preparations for an invasion of the British Isles (code name Seelöwe (sea lion)) were brought forward which should start from the French channel coast. To keep away the superior British Fleet by long-ranging artillery, the "Marine Artillerie Abteilung" MAA 240, 242 and 244 were formed and deployed between Calais and Boulogne. For a secure deployment of the guns hardened shelters and gigantic bunkers with up to 4 m thick walls were built. The new batteries got the most modern fire control systems of the navy, e.g. the 10-meter stereoscopic range finder[216] of CARL ZEISS and the latest radar systems Freya and Würzburg. Early in 1941, some of the artillery departments got "Ultrarot-Wärmepeilgeräte" (infrared thermal pointers).[217]

In **April 1941,** positions of the MAA 240 in Zeebrugge, Oostende, Dunkirk, Blanc Nez und Cap d'Alprech were equipped with thermal pointers. Additionally, a depot for spare parts was built in Wimereux. In the subsequent time, also the equipment of the remainder departments of MAA 240, the MAA 242 near Cap Griz-

45 – *Thermal pointer of the MAA 242 at an artillery position near Blanc Nez*

*(Photo: Michael Schmeelke)*

Nez and the MAA 244 in the Calais area were provided with thermal pointers. The detection range of the infrared devices shall have amounted up to 18 km.[218]

In autumn 1941, when the invasion planning was given up, the naval batteries at the Channel coast got as a new task the bombardment of the English south coast with the harbors Ramsgate, Dover and Folkstone as well as the blocking of the Channel for enemy shipping. In the period of 15 month all batteries together fired 2,450 shots, of them 1,242 were aimed on enemy ships.[219]

In **1942**, at least parts of the "Heeresküsten-Artillerie-Regiments" HKAR 180 (army artillery regiment), that was deployed in Denmark, were also equipped with thermal pointers. The equipment of the Army Battery 8/180 in Fanø[220] comprised a thermal pointing control station and 4 additional thermal pointers. All bearings were given as a fire control order to the guns.[221] This equipment can be assumed typical for all coast batteries.

In **1943**, further Large Batteries along the southern coast of France were subsequently equipped with thermal equipment.

From **1941-44**, in total 50 thermal pointers of the so-called DONAU 60 type were said to be employed for coast defense by the MAAs.

In much smaller numbers, even larger thermal pointers of the WPG 15 and Donau 150 types were also employed. "The equipment had a 150cm collecting mirror and a 15 or 30 mm PbS cell" and "the Germans had claimed that their IR coast watching equipment WPG 15 could detect a schnorkeling U-boat at 1 to 2 km range."[222]

## MAA 244 Battery Oldenburg

To get a realistic impression of the conditions under which the thermal pointers were used in that time, we may follow the fate of the MAA 244 Battery Oldenburg as an example. Not that the Oldenburg battery had been anything special (biggest, nicest, best etc.) - it was typical middle class.

In **1939**, the battery had its position on the German isle of Borkum. It was equipped with 2 pieces of the 24-cm SK L/50 type, which were not at all straight out of the factory. These guns were of Russian origin and had been captured during the Great War in 1915 near Libau. They were upgraded by KRUPP in the 1930s, and adapted to the German caliber 238 mm. The length of the barrel was 11,900 mm and the projectile weight was 148,5 kg. After 270 shots (!) the measured muzzle velocity was still 900 m/sec, which made possible a maximum range of 28,000 m.

On **15 June 1940**, the Oldenburg battery had its position in the South Blackwood Forrest and supported the advance of 7. Army on the MAGINOT line.

End of **1940**, the battery transferred to the Channel coast near Calais, and was equipped with thermal pointers a year later as already mentioned.

From **1942** on, the Oldenburg battery also was engaged in fighting enemy ships on the Channel. Because the pointing accuracy was high enough the spotted enemy ships were fired at with the bearing values of the thermal pointer.[223] It is reported that the Oldenburg battery even managed to sink a ship during such a deployment of a thermal pointer. A similar success is also reported by the MAA 242 Batterie Großer Kurfürst positioned in Boulogne.[224]

On **30 September 1944,** the Oldenburg battery was the last to be conquered by the Canadian 79. Armoured Division by means of a Churchill Mk. IV tank, specially equipped for bunker fighting with a flame thrower canon.[225]

After **1945,** the Russian guns of the former Oldenburg battery were scrapped. Cutting the barrels in pieces needed a special gas mixture for the cutting torches which costed more than the scrap value of the guns.[226]

## Deployment on Surface Ships

The German navy tested the deployment of thermal pointers not only at the land-based navy artillery but also aboard of destroyers and cruisers.

Already in **1941,** the destroyer Z-31 was equipped with a thermal pointer, probably it was a "Gleichlichtgerät" of the type described above.

In **1943,** the cruiser Prinz Eugen, on which a thermal pointer was installed on trial, fired on the target ship Hessen from 18,000 m range.

In **1944,** the destroyer Z-31 used the bearings of the (possibly new) thermal imager to fire also on the target ship from 11,500 m range.

Despite of these first successes, the development was discontinued, perhaps by the following two reasons:[227]

For one, the field of view of 0.3° was too narrow, so that it was virtually impossible to hold the target in the crosshairs even at moderate movements of the ship. To achieve that, each thermal pointer had needed to be mounted on a stabilization system.

Stabilization systems were indeed available already at that time. They utilized the property of a freely mounted mechanical gyroscope to preserve its angular direction in space.[228] From these gyros, control signals were derived to operate electro-mechanical actuators which held the line of sight of the optical device constant, independent of the ship's motion. Such stabilization systems were installed on German battle ships in small numbers already since mid of the 1930s. They were built at the ANSCHÜTZ company in Kiel where they had also been invented.[229] Besides the stabilization of large search lights the ships were mainly equipped with stabilized range finders, nick-named "Wackelköpfe" (bobbleheads), for the anti-aircraft guns to hold their line of sight on the flying targets. With the contemporary means the technical effort was quite high, and so it is understandable that the navy spared the high expense for the uncertain benefits of the new thermal imagers.

Secondly, the rough conditions on board a ship led to functional problems so that temperature differences could not be detected reliably.

## Deployment aboard of Submarines

During World War II, the German submarines were increasingly under threat of by Allied aircraft and suffered directly or indirectly from very high losses. In relation to the total losses even "nearly half of these losses (43%) … were to ascribe to air forces."[230] Thus, the aircraft had become the greatest enemy of the submarine. In so far it is understandable that passive thermal pointers were developed to be employed on a submarine for detecting approaching aircraft.

About the technical design of thermal pointers named Kormoran und Marabu, which are repeatedly referred to in literature, little is known to the author.[231] Based on the knowledge of contemporary technical means, however, some solutions may be 're-constructed'. Because of the very agile aerial targets and the hot temperatures (exhaust system, engine covers, open radiators) these devices probably used PbS cells. Thinking of the limited means of a submarine on a long voyage, a cooling, e.g. with dry ice, does not appear probable. An issue was for sure, to install a pressure-resistant window with sufficiently high transmission in the infrared. Compared to the small windows of the contemporary periscope windows, this window had to at least 10 to 15 cm in diameter to so that enough thermal radiation could reach the detector. Furthermore, the window had to be relatively thick (30–40 mm) to withstand the pressure during diving. The window material should be chosen to make full use of the wavelength range of the PbS detector (about 1 to 3 µm) and have a high transmittance in this region. Fortunately, all this could be achieved by using quartz glass. Due to the necessary salt-water resistivity, exotic infrared materials or rock salt are no option.

In the end, the problem to create a pressure-tight feed-through into the pressure hull could not be solved satisfyingly, and it is not astonishing that the development was dropped.[232]

In the United States, similar concepts were followed towards the end of the war. In 1944 for example, a patent[233] was filed for an infrared device to be attached on submarines which should be used to detect ships. The device was claimed to be hermetically sealed and pressure-resistant. The window should be made from rock salt (NaCl), the transmittance of which should secure that enough thermal radiation would reach the thermopile detector. The rock salt window should be protected against sea water by a thin layer of chlorinated rubber.

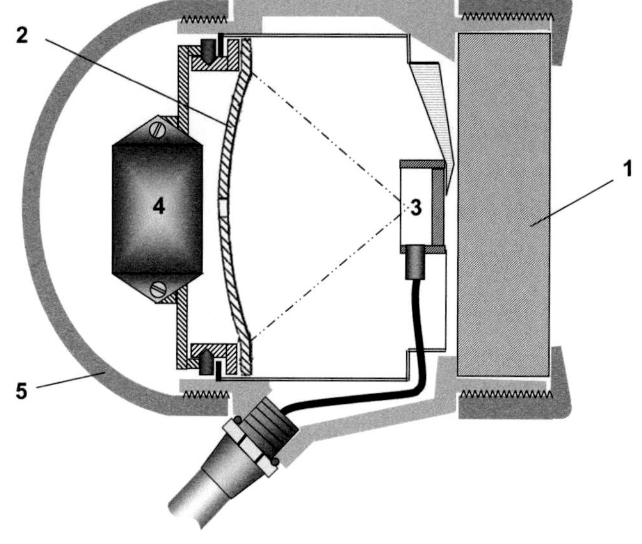


46 – *Principle of a pressure tight thermal imager for submarines*
*The pressure tight housing (5) is closed with a window (1) made from an appropriate optical material (most probably quartz glass). The thermal radiation comes in through the window and is imaged by the concave mirror (2) onto the infrared detector (3). By means of an electrical motor-gearbox unit (4) the concave mirror can be swivelled horizontally and vertically by a small angle.*

### Deployment against Aircraft

Early in the beginning of World War II, the first British bombing raids as well as some propaganda flights were undertaken over Germany at day light. When the flights more and more involved heavy losses for the British, they rescheduled their missions into the darkness of night. A consequence of this decision was that both parties bump into entirely novel issues. The British raiders had to deal with the issues of navigation and orientation at night what eventually led to a boost to radar development in Great Britain. The German defenders had the issue to spot the bombers in the dark night sky before they possibly could attack.

To this purpose the anti-aircraft guns were supported by large search lights which searched the sky with their light cones and tried to catch an enemy aircraft in the light spot. In case of success, the search light crew had to keep the target in the light as long as possible to give the anti-aircraft guns time for firing. The German night fighter, at the so-called "helle Nachtjagd" (bright night chase), had to rely on the (more or less) well-coordinated search batteries. If this was not to happen, the night fighters had no other chance than use the (from today's view horrible) "Mattscheibe" (matt screen) procedure[234], which was based on the effect that the bellies of the bombers were sufficiently bright illuminated by the reverberation of the burning cities. Naturally, each improvement of the situation was highly welcome.

In **1942**, based on the thermal devices with PbS cells a thermal pointer for the detection of bombers was developed. To collect as much thermal radiation as possible from the relatively small targets, the device was equipped with a mirror of 150 cm diameter. With this system bombers could be detected in 12 km range.[235]

### Deployment onboard of Aircraft

End of **1943**, the "Aerial Battle of Berlin" raged in the night sky above Germany. Night after night 500 to 900 four-engined, mostly British bombers streamed into the German air space, each carrying 3 to 5 tons of bombs. This bomber stream was countered by the German night interception with approximately 800 mostly two-engined aircraft.[236] From a primitive beginning at both sides in 1940, besides the real fighting, a battle for technical superiority had developed at which British radar technology stood against German "Funkmesstechnik" (radar technology). Beyond this, every other means was naturally welcome, provided it could bring at least a small advantage over the opponent.

As the fighting had spread over the whole territory of the German Reich, it was no longer sufficient to fight the bombers at a few neuralgic positions. Therewith, the German night fighters had the aggravated problem to find the British bombers in the dark night sky. Electronic radar systems, like the Lichtenstein system, were developed for this purpose, and became standard very soon, but were successfully interfered by the British quite often. Therefore, parallel to the electronic development, the German side designed and tested several infrared search devices. These were called Ultrarot (UR)-Zielsuchgerät (infrared seeker device) in Germany and were counted among the radio devices by the German air force for logistic reasons.[237] The first reasonably promising devices, known as Spanner, used an image converter tube in the near infrared region (wavelength approx. up to 1.3 µm). The invisible thermal radiation emitted by the target aircraft was imaged on the entrance surface of the tube which converted it into a visible image on the exit surface. The shortcoming of this concept was that only thermal radiation of very hot parts of the aircraft, i.e. the exhaust pipe, were bright enough in the near infrared to be detected by the Spanner devices. Unfortunately, as viewed from a night fighter these parts of an enemy aircraft often enough were concealed by wing and/or fuselage. This disadvantage could only be solved by using a thermal pointing device that was sensitive at longer wavelength and could 'see' thermal radiation also of other parts of the aircraft.

In **November 1944,** the UR-Suchgerät FuG 280 Kiel Z (Z stood for Zeiss) was installed in the nose of a Ju 88 G. The Ju 88 G, from which 2500 pieces were built, was the night fighter version of the reliable multi-role Junkers Ju 88 aircraft.

The basic principle of the device was the same as that of the thermal pointers with the faster PbS cells. The mirror optic had to be adapted to the limited space and reduced in diameter. From the photos, a diameter between 100 and 150 mm can be derived. To achieve a field of view sufficient for the application the mirror simultaneously performed a rotational and a wobbling motion. Therewith, a field of view of in total 20° was scanned in circular motions. For a synchronous display of the thermal signal, the electron ray of a cathode ray tube (with crosshairs) was deflected by two alternating voltages which were generated by two transducers coupled via appropriate toothed wheels. Normally the electron beam was blanked, but was switched bright as

47 – *UR-Suchgerät FuG 280 Kiel Z*

*The left picture shows a front view of an infrared Kiel device. The right picture shows how the instrument was installed in the nose of a night fighter aircraft Junkers Ju 88 G. The four antennas belong to a short range radar device FuG 220 Lichtenstein.*

*(Photo: Fritz Trenkle)*

48 – *Functional principle of Ultrarot-Ortungsgerät FuG 280 Kiel*

*The mirror (Sp) is rotated by an electro motor (M) with 15 revolutions per second around the mirror axis (SpA). This axis is slightly tilted relative to the optical axis (OA). As a consequence, the line of sight does a circular movement around the mirror axis. By means of an appropriate reduction gear unit the complete mirror assembly rotates around the axis of the sensor unit (EA) with 2 revolutions per second. The resulting trajectory, which is shown top left, is run through two times per second. Synchronized to the rotational movement two transducers generate voltages (with a frequency of 2 Hz and 15 Hz, respectively) for the deflection of the electron beam in a cathode ray tube. The signal of the PbS cell is amplified several times and induces a dark / bright keying of the electron beam as soon as thermal radiation from a warm target is impinging on the PbS cell. Additionally, the whole sensor unit can be swivelled relative to a vertical axis (SA).*

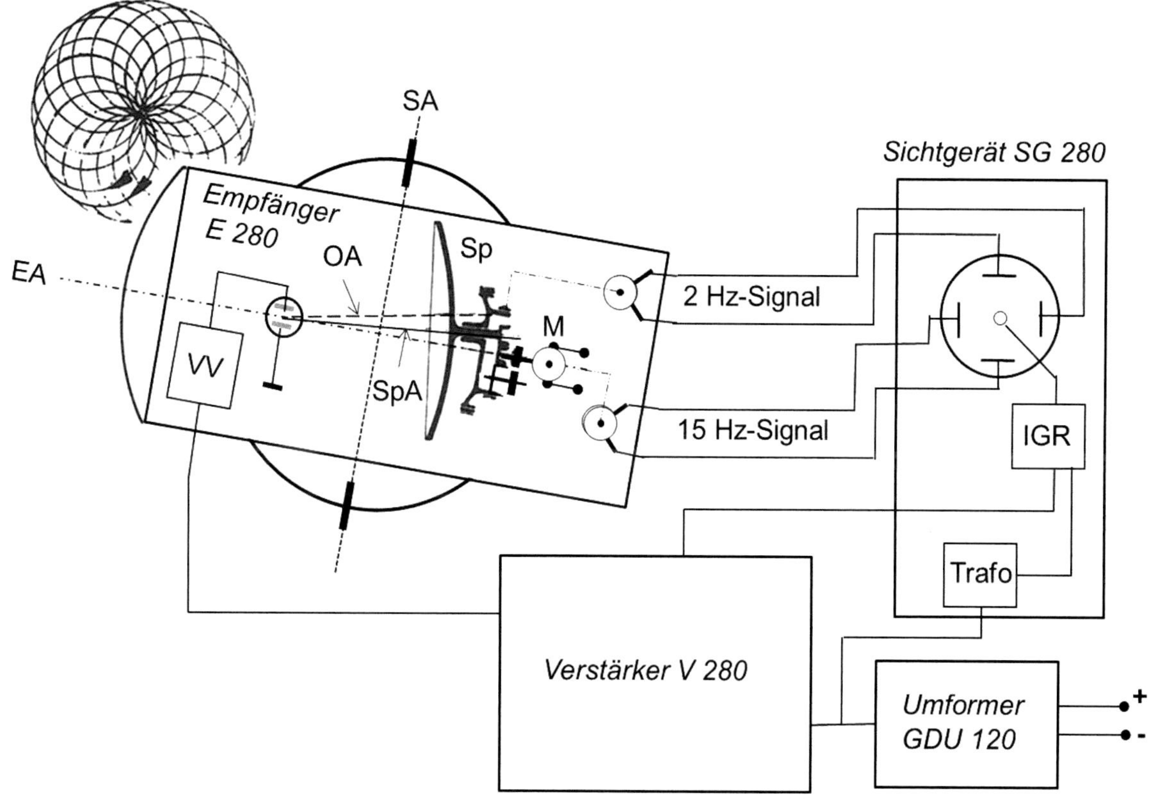

soon as thermal radiation was impinging on the PbS cell. Therewith, the screen on which the line of sight was marked with crosshairs, was lighted at a point corresponding the mirror position. Additionally, the whole mirror system could be swiveled ± 35° by hand. In airborne tests, the device could detect aircraft in 4 km range.[238] The weight of the Kiel Z was 42 kg, and only a small number was built for experimental purposes.[239]

In **January 1945,** "the production was accelerated after a bottleneck in the tube production had slowed down the production of the cm-wave radar devices."[240]

"In last weeks of the war the first series devices were deployed, and e.g. could not detect enemy aircraft, if an extensive fire was in the same direction." Such wild fires could be detected in over 100 km distance.

In **May 1944,** virtually at the same time of the FuG 280 Kiel Z development for night fighters, a version of the Kiel E III was finalized by ELAC for bombers to detect warm land and naval targets. During the successful testing "against larger industrial facilities ranges from 15 up to 40 km" could be proven. An ordered series of 200 devices, however, became obsolete with the cancellation of all bomber projects end of 1944.

In Allied sources, repeatedly a scanning search system for airborne deployment named KID is mentioned which used a mirror of 5 in diameter and should have had a range of 5 miles against unspecified targets. The correspondence of the data suggests that the FuG 280 is mend by these sources.

By **1944/45,** a last version of the UR-Suchgerät (IR search device) was made available. According to American infrared experts, "the most successful wartime development during war time was the Kiel IV, an airborne infrared system that – unlike Jones' English prototypes – had excellent range, and which was produced at Carl Zeiss in Jena under Werner K Weihe."[241] "This search set was able to detect a bomber at a range of 32 km."[242]

Further developments in this last phase of the war were dedicated to so-called "Temperaturbildgeräte"[243] (temperature imaging devices) for scanning of the terrain flown over. Two versions for different applications became known:

- scanning of a strip, e.g. for the V1 missile (an early cruise missile), to report the crossing of a coast line
- scanning of 10 strips to establish a heat map during the flight, e.g. to detect industrial facilities

Considering the external circumstances, we can safely assume that these projects were never brought to a documented completion.

## Deployment of Guided Missiles

The seeker head of a guided missile not necessarily has to provide a thermal image, an information about the direction to the target is sufficient, but is needed very quickly. To solve this requirement, since the beginning special technical solutions have been developed which are distinct from the devices presented so far. In detail, it was required to measure the angular deviation from a hot target to the line of sight of the sensor. "By rotating or swaying stops with special geometry (slits, sectors, spiral, chessboard etc.), the object, that was imaged as a hot spot in the image plane of the sensor optics, should generate a time coded signal in the detector that provided information about its position in the field of view of the sensor."[244]

A brief discussion of the infrared sensors for missile application appears to be sensible to the author, because they are based on the same laws of radiation physics. As early as in the 1930s, "the development of (passive) infrared sensors for missile application … was encouraged by the following motivation:

- By the availability of infrared-sensitive photo-cathodes and photoelements; and
- By the realization that military interesting objects contrast of their surrounding by higher temperature."[245]

Despite comprehensive investigations of dissipation properties of infrared radiation in the atmosphere and many field trials to measure "the thermal radiation of landscape, clouds, sky and military objects, … it was not possible to make an infrared sensor for guiding of a missile ready for deployment. Not satisfying before everything else was:

- The suppression of the disturbing background effects and
- The transformation of the measured signals of the target position into control commands."[246]

The problems with the background are understandable when we consider that as fast detectors only photocells were at hand (e.g. at the seeker head "Hamburg"), and the sensitivity of these was limited to 1.3 µm at maximum, "where the solar radiation is strong and the transparency of the atmosphere not much better than in the visible region."[247] Furthermore, only very hot objects with temperatures of more than 1000°C show a reasonable emission. Moreover, variations of the background and non-uniformities, which were hardly to be avoided in the production of the photo-cathodes and photo-resistors, could pretend objects "[248] which then erroneously would be taken as targets.

During the further course of the work, it was tried "at least to develop sensors for the hot exhaust pipes of the enemy aircraft, to be deployed at night." For this, the only available detector for thermal radiation was the PbS cell, which was sensitive nearly up to 4 µm, and also fast enough.

By **1943**, the seeker head Madrid was developed for air-to-air missiles[249] which used a PbS detector.

In **1944**, in "a report to the German Aviation Office, infrared search devices were praised as technically more promising for guiding missiles than radar or acoustic methods because of the simplicity and state of technical progress."[250]

However, a decade after the war, "KUTZSCHER clarified that the construction of an efficient system, designed for the application recognized as the most important, namely the guiding of missiles, was not realized at the end of the war."[251]

The missiles, which should have carried these sensors, were not ready either.[252]

## 3.5 State of Thermal Imaging in 1945

The author would like to devolve the summary and assessment of the state-of-the-art reached in the first half of the 20th century upon well-recognized American and British infrared experts and book writers. Henry L. HACKFORTH,[253] Richard D. HUDSON, Jr.[254] and Sean JOHNSTON have discussed exactly this question thoroughly in their books.

"The military organizations of the United States, Great Britain, and Germany began experimenting with infrared equipment soon after 1900. During World War I, experimental blinker signaling, voice communication, and search equipment were developed. Although neither side put any infrared equipment into production before the end of the war, the experience gained with the experimental equipment clearly demonstrated the military potential of infrared technology."[255]

Also, the progress at civil applications was limited for the moment. "The earliest applications of IR spectroscopic techniques to the chemical industry occurred in the 1920s. By 1935 the value of IR spectroscopy as an analytical tool in the laboratories of the chemical and petroleum industries was fully realized. However, prior to the outbreak of World War II, the few spectroscopes in existence were primarily hand-built models used in laboratories, and interest in IR techniques was chiefly academic. ... In Germany, the potential applications of IR were fully recognized, and a great deal of fundamental investigation in the basic fields of emission, atmospheric attenuation, and background radiation, as well as in the development of new IR detectors, optical materials, and scanning systems, was carried out in the 1930s. With the outbreak of war in 1939, the intensity of IR research in Germany was greatly accelerated. Accent was, of course, mainly on military applications, but such significant contributions were made that a brief discussion of German achievements in this field is considered essential to an understanding of the rapid growth of IR techniques."[256]

"Fundamental studies of atmospheric absorption effects were made. Laboratory studies of IR transmission through short paths of water vapor of various concentrations were carried out and verified in field tests. The dispersion and attenuation effects on IR in fog, rain, haze, and artificial smokes were investigated. The effects of background and camouflage were probed and special camouflage paints and coatings were developed.

Thorough investigations of detecting devices of all kinds resulted in the development of new and sensitive types of IR detectors. Lead sulfide photoconductive detector cells were developed and produced by the Electroacoustic Company (ELAC) in Kiel, where methods of depositing IR-sensitive surfaces by chemical precipitation were pioneered. ZEISS-IKON at Dresden produced lead sulfide cells by the GUDDEN evaporation method.

Techniques for improving the detectivity and wavelength response of these cells by introduction of impurities and by cooling with solid carbon dioxide or liquid air were developed. Lead sulfide cells ... were produced in quantity and used in practically all the IR devices employed during the war.

By 1944 the first lead selenide photoconductive cells were ready for production by ELAC. New types of highly sensitive thermocouples and bolometers were developed."[256]

As we can see from the previous citations, there is a consensus among the experts. "During World War II, workers in Germany made noteworthy contributions to the development of photoconductive detectors, and they were the first to demonstrate the increase in sensitivity that could be obtained by cooling detectors."[258]

In the words of the American experts, the state of the IR optical systems is characterized as follows:

"Great advances were made in the development of new materials and optical systems. Among the important new materials developed were KRS-5 and KRS-6 optics by ZEISS for IR transmission out to a wavelength of 40 μ or more. New optical glasses with excellent IR properties included aluminate glasses, Duran glass developed by the SCHOTT GLASS WORKS, and fused silica produced by HERAEUS.

Both reflecting and refracting optical systems were used, and methods of reducing spherical aberration and coma by using correcting mirrors and Schmidt correcting plates were invented. Optical systems for IR use were achieved with f numbers as low as 0.6. The Germans use mirror optics wherever possible for maximum efficiency in the IR region. Protective and antireflection coatings for front-surface mirrors were evaporated, deposited, coated on glass or aluminum. IR prisms of quartz, fused quartz, alkali halides, and various special glasses were developed by ZEISS, STEINHEIL, and I.G. FARBENINDUSTRIE. A great deal of work was devoted to special filters designed for IR detecting systems, IR photography, and for IR spectroscopic applications. Methods of reducing interference caused by internal effects in detector cells, by background radiation, and by fluctuations in atmospheric IR radiation were pioneered. … The evaporograph, developed by CZERNY, extended the useful range of IR photography from 1.3 μ, the limit of conventional emulsions, out into the far-IR region. …

Similar development and research work was carried out by the Allied nations during latter part of the war … However, German leadership was generally recognized, as was British leadership in radar development, which at that time received a much higher priority than IR research. The primary interest in the IR development that was carried out, as in Germany, was centered on military applications. Various IR detecting, direction finding, and ranging devices were produced, of which the sniperscope and snooperscope, developed in the United States, are probably the most familiar."[259]

HUDSON also states that "during World War II numerous infrared devices were proposed and investigated on both sides. However, a close study of this period shows that relatively little infrared equipment reached production status."[260]

Moreover HUDSON states that the "the development of the infrared image converter tube on the eve of World War II made it possible for Germany to be the first country to deploy infrared equipment in the field."[261]

As shown above, and perhaps unknown to HUDSON at his time, the first long wave thermal pointers were deployed at the German Navy Artillery as early as in 1941. Latest in 1944, the first infrared image converters were integrated in the fire control systems of several German tanks. "These systems, which were used on the eastern front in 1944, proved to be remarkably effective in nighttime battles. … The best known U.S. development, the sniperscope, consisted of an image converter and an illuminator mounted on a carbine. … The sniperscope was first used in combat in April 1945 during the invasion of Okinava."[262]

Besides these devices, no further nation did employ infrared devices. "Influenced by the German successes, the Japanese had plans for the production of similar active devices, but the war ended before they could be put into effect."[263]

"Despite post-war claims by NDRC that 'American scientists won by a wide margin in their race to be the first to make practical use of infrared light', German work clearly surpassed it in pursuing new technical directions and concepts"[264] the British infrared expert Sean JOHNSTON wrote in 2001.

Already in 1969, his American counterpart Richard HUDSON, Jr. had analyzed: "… the Germans concentrated much of their research and development on infrared equipment and on means of counter measuring it. By contrast, the Allies concentrated their efforts on the development of radar. As a result, the Germans lost the war but they clearly won the battle of the infrared."[265]

4

# NEW BEGINNING IN GERMANY IN 1945: THE ZERO HOUR

"This Zeiss plant symbolizes the fate of Germany like few other factories visible to the world."

„In wenigen Werken von Weltsichtbarkeit ist so das deutsche Schicksal markiert wie in diesem Zeiss-Werk."

*German Federal President Dr. Theodor HEUSS*
*(Speech given in Oberkochen on May 1, 1954)*[266]

The further development of infrared technology and thermal imaging after World War II, of course, was quite different for the protagonists so far in this field. Although the allied victorious powers had some problems to return to peacetime economy, the routed Germany was another world. In many respects, it was the often quoted "Zero Hour". Economy at rock-bottom, people had the severest difficulties just to master their every-day life. Military developments were out of a question, particularly as it was prohibited by the Control Council. It took 10 years until the two states on German soil were integrated in the military structures of the respective power blocks. It took another 10 years until German companies were able to make significant contributions to infrared technology again.

Therefore, it makes sense to follow the German development separately for a certain time. With the accession of the Federal Republic of Germany to NATO by joining the Paris Treaties in 1954, the story lines came closer again and German companies could think of getting back in the infrared business again.

The difficulties to continue the scientific and technical development in post-war Germany after 1945, besides the general lack of everything, were marked by the facts that all valuable equipment was confiscated and many German experts had left the country. Partly they were forcibly taken abroad or they were purposefully hired by special programs like the American "Project Paperclip" or they volunteered to go, succumbing to hardship. This fate befalls also some protagonists of infrared technology.

Edgar KUTZSCHER, for example, was taken to Great Britain in 1945 while his equipment found a way to Russia. At the beginning of the 1960s he joined LOCKHEED AIRCRAFT.

Paul Görlich and his group were delegated to Krasnogorsk near Moscow in 1946, to continue the lead sulfide (PbS) development there and came back to ZEISS in Jena only in 1952.

Heinz Schlicke, who had come voluntarily to the USA with U 243 in May 1945, got a legal status by "Paperclip" and was involved in several classified programs of official government organizations until 1950. Thereafter, he worked for the American industry.

To get an idea of the situation of the German industry a glance at the end of war in Jena and the new beginning of the company CARL ZEISS in Oberkochen might be instructive. Other German companies shared a similar fate. In his May Day speech in 1954, the then Federal President Theodor HEUSS saw the processes as likeness of the German fate: "This Zeiss plant symbolizes the fate of Germany like few other factories visible to the world ".

**On 13 April 1945,** the ZEISS factory in Jena was occupied by American troops. The importance attached to the company during this period, even in the United States, was evident not only in the headline of the New York Times[267]: "The Famous Zeiss Plant at Jena in Ruins!"

A detailed inventory of the factory, of which 60% had been destroyed, was immediately carried out by American special forces, who regarded technological expertise as valuable booty.

**On 8 May 1945,** World War II ended in Europe with the capitulation of Germany.

Because the capital Berlin was conquered solely by the Soviet Army, the Allied agreed an exchange. Thuringia, Vogtland and parts of Saxony should be given to the Soviet Army and conversely the western Allies should take part in the administration of Berlin. In view of the pending handover of Thuringia to the Soviet Army, a relocation of the entire ZEISS plant to the West was considered, but the estimated scale (roughly 600 railway cars) and cost of the proposed move made it impossible in the short time before the handover.

Thus, the new slogan became "We take the brain"[268] However, also the new plan needed some time to be implemented and therefore "the United States delayed its withdrawal from parts of the Soviet Zone until early summer 1945, and used its temporary authority in these areas to capture scientists and remove documents, especially from Thuringia and Saxony-Anhalt. The delayed withdrawal caused the Soviet authorities to lodge a protest, and the capture of scientists caused the Russians to request the return of both scientists and documents"[269] Still at the meeting of the Zone Commanders in Berlin-Karlshorst (Headquarters of Zhukov) in June 1945, no agreement on the withdrawal was reached, though Marshal "Zhukov apparently wanted an answer rather than a discussion"[270] on this topic.

**On 24 June 1945,** the preparations in Jena were completed and a convoy carrying 83 Zeiss employees with their families (1 truck per family) and a total of 1300 staff members from ZEISS, SCHOTT, SIEMENS and the UNIVERSITY OF JENA left Jena for the police academy in Heidenheim, a German town in the southern state of Wuerttemberg. The convoy also included 14 trucks with over 80,000 pieces of laboratory equipment and drawings, which however got lost somewhere between Jena and Heidenheim and vanished forever.

**On 1 July 1945,** American forces turned Thuringia over to the Russian occupation authorities. In return, the Western allies participated in the administration of Berlin.

The **winter of 1945/46** was a difficult period for the Zeissians and their families who had resettled in Heidenheim. In addition to being uprooted from their home, they had to endure extremely harsh living conditions while their prospects for the future were uncertain.

"The Soviet authorities sent a flood of letters to "Heidenheim scientists" … to attract them back to East Germany" in this situation of "general hopelessness". "After the escape of Zeiss scientist Herbert KORTUM, a member of both SS and NSDAP since 1931, from Heidenheim to Jena and his quick denazification by the Soviets, the American Military Government in Heidenheim / Wuerttemberg rejected all requests from KORTUM's wife to allow her and her family to return to Jena. These requests were turned down because "the return of his family might encourage other families to resort to the same method."[271]

**On 25 February 1946,** the military government approved the opening of an "optical repair shop". The OPTISCHE UND FEINMECHANISCHE WERKSTÄTTE GMBH was founded in the cramped premises of the former SCHÄFER cigar factory in Heidenheim, which also had to be shared with the SIEMENS staff. New facilities were thus urgently required and had to be found soon.

**On 29 April 1946,** the ALLIED CONTROL AUTHORITY issued the 'Control Council Law No. 25' "Control of Scientific Research", which states that "all technical military organizations are hereby dissolved". Among others, also infrared technology itself was restricted by this law. In Schedule "A" to the law "applied scientific research" related to "(VI) Electromagnetic, **infra-red** and acoustic radiation which has as its purpose (a) the detection of the position of vehicles, aircraft, ships,

49 – *Allied Control Authority, Control Council Law No. 25 (excerpt)*

**ALLIED CONTROL AUTHORITY**
**CONTROL COUNCIL**

## Law No. 25

### Control of Scientific Research

In order to prohibit for military purposes scientific research and its practical application, to control them in other fields in which they may create a war potential, and to direct them along peaceful lines, the Control Council enacts as follows:

**Article I**

All technical military organizations are hereby dissolved and prohibited. Equipment and buildings of a purely military character shall be destroyed or removed. Equipment and buildings having a possible peace time application may be utilized for that purpose with the permission of Military Government.

**Article II**

1. Applied scientific research shall be prohibited on:
   a. Any matter of a wholly or primarily military nature; or
   b. Any of the matters specified in Schedule "A" hereto.

2. Applied scientific research on any of the matters specified in Schedule "B" hereto shall be prohibited unless the written permission of the Commander of the Zone in which the research establishment is located is first obtained.

**CONTROL COUNCIL LAW NO. 25**
**SCHEDULE "A"**

**Prohibited Applied Scientific Research**

(VI) Electromagnetic, infra-red and accoustic radiation which has as its purpose:
(a) the detection of objects or obstacles; or
(b) the determination of the position of vehicles, aircraft, ships, submarines or missiles; or
(c) the remote and the automatic control of vehicles, aircraft, ships, submarines or missiles; or
(d) the destruction of living matter, except for specifically medicinal and public health purposes.

**CONTROL COUNCIL LAW NO. 25**
**SCHEDULE "B"**

**Applied Scientific Research Requiring Prior Permission**

(I) Electromagnetic, infra-red and accoustic radiation which has as its purpose:
(a) communication of intelligence by telephony or telegraphy; or
(b) provision of public broadcast or television services; or
(c) location of fixed transmitters by direction finding methods; or
(d) other applications not banned under Schedule "A".

submarines or missiles …" was prohibited. According to Schedule "B" also "applied scientific research" related to "(I) Electromagnetic, **infra-red** and acoustic radiation which has as its purpose (a) communication of intelligence by telephony or telegraphy …" was "requiring prior permission"[272].

Despite Germany being shattered, it should obviously be secured that it never could become a competitor in this field of technology again - this is also a way to acknowledge someone's extraordinary accomplishments.

At CARL ZEISS, however, people did not need to worry about this law because there were much more urgent problems to solve, for example the shortage of space.

**On 14 June 1946,** the Ministry of Commerce of Württemberg and Baden issued a manufacturing permit to CARL ZEISS, WERK OBERKOCHEN (for the Oberkochen plant), which began work in the former tool factory of Fritz LEITZ in Oberkochen.

**On 4 October 1946,** the OPTON OPTISCHE WERKE OBERKOCHEN GMBH was founded with working capital of RM 1,000,000. Remarkably, Zeiss in Jena contributed to this with RM 950,000 in the form of machines from Amberg and Coburg as well as a cash deposit of RM 450,000.

**On 15 November 1946,** ZEISS celebrated the 100th anniversary of its founding. In accordance with the difficult circumstances, only a small party was held in Oberkochen, while in Jena the dismantling of the factory began.

**On 1 June 1948,** ZEISS was formally expropriated in Jena.

**On 23 February 1949,** the new CARL ZEISS STIFTUNG (CARL ZEISS FOUNDATION) was founded, with its headquarters in Heidenheim.

**On 15 January 1951,** the commercial register entry was changed to CARL ZEISS. The company already had 2500 employees; many had come from Jena and had brought more than just goodwill with them. Despite this, the financial results revealed a loss of DM 1000/employee.

**On 1 May 1954,** the nationwide renown of the newly opened company was strikingly demonstrated when ZEISS succeeded in persuading the president of the Federal Republic of Germany, Dr. Theodor HEUSS, to give a May Day speech in Oberkochen.

ZEISS was making a profit, manufacturing 600 various products with 2850 employees. Since 1946, 11 million pairs of glasses, 750,000 camera lenses, 7,000 levels and 6,000 microscopes had been produced.

**On 26 February 1954,** the Bundestag passed a change in Basic Law (the German constitution) that permitted the development of the country's own armed forces in the Federal Republic of Germany.

**On 23 October 1954,** the Federal Republic of Germany became a member of the North Atlantic Treaty Organization (NATO), founded on 4 April 1949, under the terms of the 'Paris Peace Treaties'. In addition to the founding members, namely the United States, Canada, Iceland, Norway, Denmark, Netherlands, United Kingdom, Belgium, Luxembourg, France, Italy and Portugal, also Greece and Turkey have been members of the defense alliance since 1952.

**On 16 June 1955,** the Bundestag passed the 'Freiwilligengesetz', the law concerning the recruitment of military volunteers, which permitted the enlistment of 6000 soldiers.

**On 20 January 1956,** Chancellor Dr. Konrad ADENAUER welcomed 1500 soldiers from the Army, Air Force and Navy [273]: "The only goal of German rearmament is to assist in maintaining peace. We will have achieved this goal when the joint potential defensive forces of the Allies represent too great a risk for any would-be attackers at any time. No one can see a threat in military strength which is used only for our defense."

**In February 1956,** the SECURITY COUNCIL passed a resolution with 18 votes in favor and 8 against that the new armed forces should bear the name BUNDESWEHR. This was a proposal of the spokesman for defense of the FDP in the Bundestag and distinguished former Army general and tank commander Hasso von MANTEUFFEL.

Shortly after Germany had become a member of NATO upon conclusion of the Paris treaties, efforts were launched to create the technological prerequisites for the development of the new armed forces.

**On 1 December 1954,** CARL ZEISS put in its bid to provide the development and delivery of infra-red optical equipment of all kinds, particularly thermal pointers for ships and airborne targets. However, it took some more years until tangible orders did materialize.

50 – *Letter of application to supply thermal pointers for maritime surveillance*

*"Manufacturer: Carl Zeiss,Oberkochen*

*During the war, the company Carl Zeiss had developed a maritime thermal pointer with bolometers as sensors. With this device naval targets could be detected up to 20 km and tracked with an accuracy of 1/16°. The device consisted of a 60 cm-mirror in the focus of which the bolometer was mounted, and an electrical display unit that could be stationed in a distance up to 30 m."*

51 – *Letter of application to supply thermal pointers for air defence*

*"Manufacturer: Carl Zeiss, Oberkochen*

*The company Carl Zeiss had developed a pointer for aircraft which was equipped with lead-sulphide cells. With this device only objects with higher temperatures are comfortably detectable. The detection range against night fighter aircraft was up to 10 km."*

**Leistungsblatt Nr. 17**

Arbeitsgebiet
C 5
1. Dezember 1954

Gerät: Wärmepeilgerät für Seeziele

Hersteller: Carl Zeiss, Oberkochen

Die Firma Carl Zeiss hat während des Krieges ein Seezielgerät mit Bolometern als Empfängern entwickelt. Mit diesem Gerät konnten Seeziele bis zu 20 km festgestellt und auf 1/16° genau laufend verfolgt werden. Das Gerät bestand aus einem 60 cm-Spiegel, in dessen Brennpunkt das Bolometer eingebaut war, und einer elektrischen Anzeigevorrichtung, die in einer Entfernung bis zu 30 m aufgestellt werden konnte.

50

**Leistungsblatt Nr. 18**

Arbeitsgebiet
C 5
1. Dezember 1954

Gerät: Wärmepeilgerät für Flugzeuge

Hersteller: Carl Zeiss, Oberkochen

Die Firma Carl Zeiss hatte ein Peilgerät für Flugzeuge entwickelt, das mit Bleisulfidzellen ausgerüstet war. Mit diesem Gerät sind nur Objekte mit höheren Temperaturen bequem nachweisbar. Die Nachweisentfernung betrug gegen Nachtjäger bis zu 10 km.

51

# 5

# THERMAL IMAGING UNTIL 1975: HOW PICTURES LEARNED TO MOVE

„Cameralike Device Sees a Purple Cow"

*LIFE (1956)*[274]

„1966 could be considered the year that thermal imaging systems, as we know them today, were created."

*Gerald HOLST (2000)*[275]

„Conquest of Darkness"

*Motto of NIGHT VISION LABORATORY (1965)*

"Before 1939, very few" civil infrared devices like e.g. "IR spectrometers were available. Those that were, mostly hand built for laboratory research purposes, were time-consuming to operate and limited in their scope." The amount "of wartime advances in IR is probably the greatest single factor accounting for the numerous and rapidly increasing applications of IR techniques in modern industry." As civil evidence "It has been estimated that between the years 1943 and 1953 in the United States alone, over 1,300 accurate and rapid recording IR spectrometers were manufactured."[276]

"Despite their limited deployment during World War II, infrared devices showed sufficient merit to justify a strong postwar development supported by military funding. … As infrared matured to a recognized technology, the annual sales of infrared devices assumed significant proportions. … Most estimates assume that sales to the military constitute about three-fourth of the total infrared market."[277]

"The convincing World War II demonstration of the effectiveness of infrared techniques ensured continuing military support for developments in this field. Virtually all of the world's major military organizations now use infrared equipment in one form or another."[278]

The relevance attributed to the infrared also grew because of the "fundamental changes that are occurring in the way wars are fought." Rather than for a World War the military organizations planned for "an era of limited wars. It is not easy to define what is meant by limited war, but … DEITCHMAN[279] has carefully documented a total of 32 limited wars in the period from 1945 to 1962."[280]

The starting conditions in the post-war time for the beginning of the technological contest for the infrared were unequal also for the (erstwhile) allied victorious powers. Great Britain and the United States had a high technological level each. Great Britain, however, faced severe economic difficulties after the loss of the "British Empire". "Apparently the Soviets had no infrared capability until they acquired a large part of the German scientific know-how at the end of the European phase of World War II". This resulted in "the emergence of the Soviet Union as a major proponent of infrared equipment" "in the post-World War II period".[281]

## 5.1 Thermal Imagers with Lead-salt Detectors

In **1945**, "a much more interesting German equipment was captured at the end of the war" by the British Allies. "This was the Kielgerät airborne aircraft detection system. [282] The only complete installation in a JU88 was found by Air Commodore CHISOLM and flown to England, where it became the starting point for our development of IR-guided missiles. After preliminary examination, it was agreed to fly the JU88 to Defford for detailed flight trials of the equipment. Unfortunately, the aircraft was wrecked in a flying accident before this could take place. Luckily the apparatus was salvaged and fitted in a Tudor [AVRO Type 688 "Tudor"]. Flight tests then confirmed the German claims. Originally it had used a PbS cell cooled with solid carbon dioxide, but by this time more sensitive lead telluride (PbTe) detectors cooled with liquid nitrogen were being made at TRE. Even better performance was now obtained. In the original Kielgerät, a visual display indicated the bearing of a target. It was now decided to combine the German optical system with a lock-follow scanner taken from an aircraft interception system. This produced an infrared sensor that automatically followed its target. At this time, the development of guided missiles had just started. These experiments showed the way to design an infrared-based guidance system. The first successful IR homing missile was the first version of Firestreak, produced in 1953-54. It was also possible to devise an IR fuse, thus ensuring that a direct hit was not necessary for the missile to be effective."[283]

"In **1946** ASE Haslemere (the British ADMIRALTY SIGNALS ESTABLISHMENT) decided to build a similar apparatus "to establish whether thermal images of outdoor scenes could be produced, and to see if these would be of value for the detection and identification of objects at night. They immediately obtained encouraging results and arranged for a better version to be built by GEC, which was used for the first detailed study of thermal imaging by ARL from 1948. The scanning optics was a modified German coast-watching equipment and the detector was a thermistor bolometer designed by BTH based on German research. The imager had to use a very slow scan rate, taking about 20 minutes to form a picture. The purpose of it was to explore the nature of thermal pictures, for at that time it was not clear whether thermal imaging had a military use.[284] The conclusions reached from these trials depended upon who was viewing the pictures. To those with the vision to see the future advance of infrared technology they were very promising., but to those who wanted operational equipment now they at best were too far in the future."[285]

In **1946,** the first so-called line-scanner camera for military application was presented in the United States. As is indicated by the name already, this camera type could capture image points of a scene along a line, i.e. it could record only a one-dimensional image. "By putting many lines together, a two-dimensional image was created. A thermogram took one hour to produce. By adding a scanner (1954), the system could directly create a two-dimensional image. Even with this improvement, it still took 45 minutes to create an image."[286] According to other sources imaging with a thermopile needed a recording time of 20 minutes, while cameras with bolometer detectors were believed to be a little faster needing 4 minutes per image.[287]

In the post-war Unites States "the interest in research of lead salt detectors initially received less military emphasis." Only at the company PHOTOSWITCH some work was done to make the preparation of the thin PbS-layers more reliable.

As early as **1947** however, intensified development work was started at EASTMAN KODAK – now for the U.S. Navy – with the objective zu develop a PbS detector for the air-to-air Dove missile. "Upon successful demonstration that the infrared seeker could generate signals in response to detecting a 100°C target, the U.S. Navy awarded a contract to EASTMAN KODAK to develop PbS photoconductive cells."[288]

In **1947,** Marcel J. E. GOLAY[289] developed a pneumatic detector[290] for infrared radiation, in which a volume of gas is heated by the absorption of IR radiation that in turn is expanding in this process. The expansion is transferred to a flexible membrane and measured optically. The Golay-cell, as it is named after its inventor today, can be used at room temperature and is sensitive in a wavelength range from 1 µm up to several millimeters.

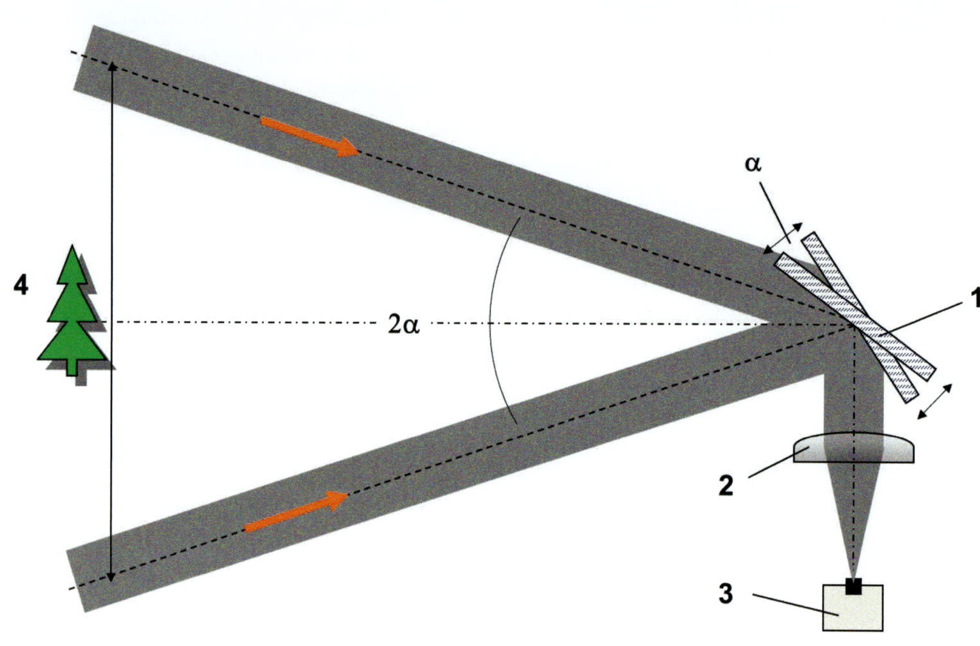


**52 – *Functional principle of a line-scanner camera***

*The imaging optics (2) and the infrared detector (3) make up the basic unit. By means of the oscillating plane scan mirror (1) the line of sight of the basic unit is deviated regularly. During one scan, the camera can capture a strip-type part of the scene (4). A mechanical rotation of the mirror (1) by an angle α causes a two times larger change of the line of sight by an angle 2α.*

On **3 February 1948,** a US patent[291] "Image Forming Heat Detector" was granted to William A. TOLSON working for the RADIO CORPORATION OF AMERICA in that time. The application had been filed in 1944 already. According to the invention, the detector consisted of a thin sheet metal blackened on one side on which thermal radiation was imaged. Due to the increased heating of the sheet at places where the radiation was more intense the back side of the sheet should exhibit changes due to thermal expansion. The changes should be made visible in the reflected light using the so-called schlieren-technique (see below).

In **1948,** "the NAVAL ORDNANCE LABORATORY (NOL) assembled a committee … to review the studies made by various scientists in the area of infrared technology, both locally and abroad. ...Within a decade, NOL scientists felt they had developed a thorough understanding of the photoconductive behavior of lead-salt detectors. As a consequence of the benefits received by the combined efforts of many scientists, symposia were sponsored by the INFRARED INFORMATION SYMPOSIA (IRIS) to exchange information on the research of lead salt detectors."[292]

On **4 April 1949,** the "North Atlantic Treaty Organization" NATO was founded as a defensive alliance and in reaction on the growing confrontation with the Soviet Union. Founding members were USA, Canada, Iceland, Norway, Denmark, Netherlands, Great Britain, Belgium, Luxembourg, France, Italy and Portugal. From 1952 on also Greece and Turkey belonged to the NATO.

## 5.2 Short Flourishing of Non-Electric Thermal Imagers

Despite all efforts it became clear that it would still take a long development time until thermal images with acceptable frame rate and quality could be recorded with the existing electrical detectors. Therefore, scientists in several countries increasingly turned to non-electronical methods hoping for success on this alternative way.

Early in the **1950s,** at SERVICES ELECTRONICS RESEARCH LABORATORIES in England a device called Edgegraph was investigated. "The edgegraph is an infrared-to-visible image converter which utilizes the thermal shift of the optical absorption edge in amorphous selenium to make an infrared image visible. In the simplest form, the transmission edgegraph uses a thin selenium film evaporated onto an edge-mounted supporting film of $Al_2O_3$ or nitrocellulose, and the assembly is mounted in vacuum for thermal isolation. The observer views a uniform source of monochromatic (sodium) yellow light through the membrane. The infrared image is also focused on the film assembly to create a corresponding temperature pattern. The sodium 'D' line wavelength falls in the middle of the selenium absorption edge at room temperature. Since the transmission at the D line falls as temperature increases, the visible image is the negative of the infrared image."[293] The inventors "reported detection of an 8 °C scene temperature difference using F/0.5 optics. A principal limiting mechanism was low contrast in the converted visible image."[294]

In **1952,** the first indium antimonide (InSb) crystals[295] were produced by Heinrich WELKER[296] who was head of the department for solid state physics at the SIEMENS-SCHUCKERT-WERKE in Erlangen and dealt with the development of new III-V mixed crystals.

In **1953,** the topic Evaporography was taken up again in Germany. Referring to the work of CZERNY, H. GOBRECHT and W. WEISS reported an updating of the well-known method by contemporary technical means.[297] Whether these activities led to practical devices is not known to the author.

Also in **1953,** a group of engineers at BAIRD Associates Inc. in Cambridge, Mass. (USA) filed an application "Method and Apparatus for long wavelength infra-red viewing" at the US Patent Office which was granted in 1958 as US Patent 2,855,522.[298] In their description, the inventors referred to CZERNY's Evaporography as "prior art" and claimed several improvements which mainly were related to details of the setup.

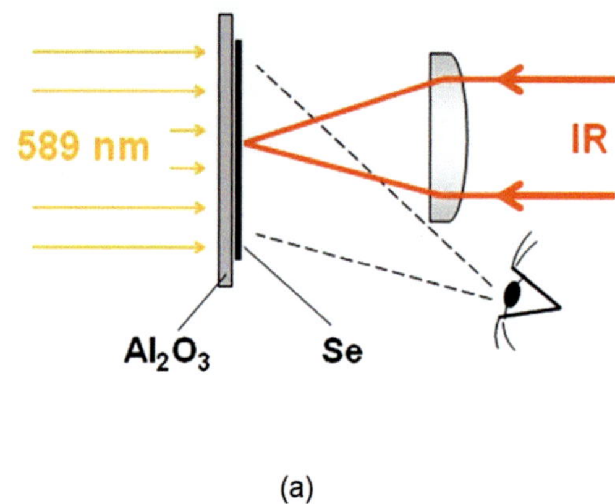


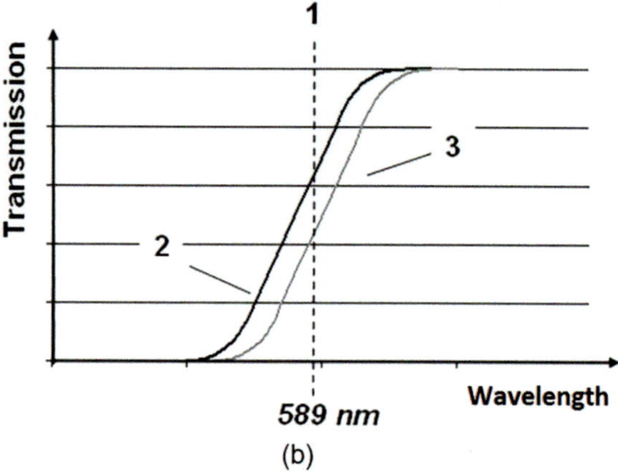


53 – ***Functional principle of a Transmission-Edgegraph camera***

*On one side of a thin carrier plate ($Al_2O_3$) made from aluminum oxide a film of amorphous selenium is applied which is heated up locally by thermal radiation (IR). When illuminated from the backside with yellow light (589 nm) of a sodium lamp the warmer parts appear darker to the observer due to the lower transmission of the selenium layer. As is depicted in Fig. (b), the yellow sodium D line (589 nm) lies nearly in the center of the absorption edge (2) for room temperature. The absorption edge of selenium (3) is shifted to longer wavelength with increasing temperature and therewith the transmittance for sodium light is decreased.*

The practical example given by the inventors was fully in line with CZERNY's functional principle. In addition, a method was described that was quasi an inversion of CZERNY's method. Therein a dry membrane at first should be illuminated by thermal radiation to be heated up locally. In a second step, it should be coated with a suitable oil resulting in differences of the thickness of the oil film between warmer and colder patches. This would create the anticipated thermal image. The devices later produced by BAIRD, used exactly this process.

In **1954**, researchers at the UNIVERSITY OF MICHIGAN's Engineering Research Institute developed an Interference Edgegraph "to provide high-contrast imagery while retaining the edgegraph principle. The four-layer image-converter membrane ... is formed on a thin edge-supported aluminum-oxide support film. A gold-black infrared-absorbing layer is formed on one face by evaporation through an inert gas at low pressure. A thin gold mirror layer is formed by high vacuum evaporation on the other face of the support film, followed by a ³/₄ wave thickness of amorphous seleni-

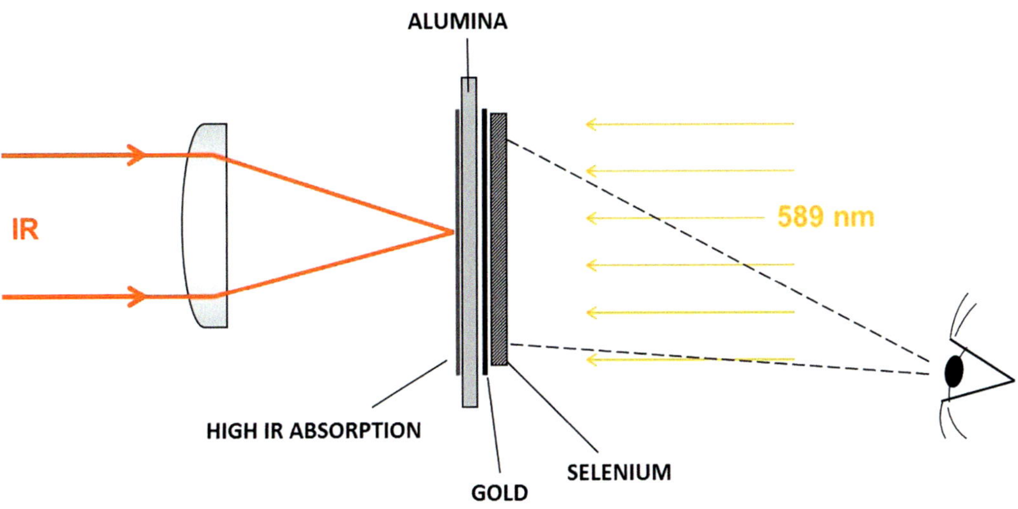


54 – *Functional principle of an Interference Edgegraph camera*

*The image converter comprises a carrier plate made from aluminum oxide ($Al_2O_3$), which carries an infrared-absorbing layer (IR Abs.) of black gold. On the side under observation a reflecting gold layer and a layer of amorphous selenium (optical thickness 3/4 of a wave-length) is applied. The surface of the selenium layer and the gold mirror beneath perform like a Fabry-Perot interferometer. Initially it is tuned to minimum reflection for the 589 nm of the sodium light. At those patches that are heated up due to absorption of the in-coming thermal radiation (IR) the interference is suppressed by the higher absorption of the selenium and the visible image appears brighter.*

um. The membrane is viewed at essentially normal incidence by specularly reflected sodium yellow light. The infrared image is focused on a gold-black absorber on the reverse side of the membrane. Where the membrane is cooler, transmission through the selenium is higher, and the first and second surface reflections cancel in part to give low brightness. Where the layer is warmer, corresponding to higher infrared irradiance, the second surface reflection is reduced by absorption in the selenium and the image brightness increases."[299] A comparison to the Transmission-Edgegraph showed "that the interference edgegraph permits the detection of approximately 2° to 3° temperature difference with more realistic F/0.7 to F/1.0 infrared optics. The advantages of the edgegraph appeared to be continuous operation, fast response (on the order of milliseconds) to follow moving images, and easy adaption to a sealed vacuum environment with no need for vacuum pumps. The problems were the need for reasonably precise film-temperature-control to keep the absorption edge at the D line wavelength. Work on this approach was apparently concluded in 1959."

55 – *Principle of the Evaporograph Model JZ 1 of Baird Atomic Inc.*

*The thermal radiation (IR) coming from the scene is imaged by the Newtonian telescope (1) onto the blackened side of a membrane (3) which is fixed inside a vacuum chamber (2). To get a visual presentation of the thermal image the backside of the membrane on which the evaporation and condensation processes take place is illuminated by a light source (4). The light reflected by the membrane is directed by a lens system into a photographic system (5) or can be viewed through an eye piece by an observer (6).*

At least since **1954,** the optical and photo-electrical properties of indium antimonide were investigated in detail. A group of researchers (D. G. AVERY, D. W. GOODWIN, W. D. LAWSEN und T. S. MOSS) at the RADAR RESEARCH ESTABLISHMENT RRE[300] in Malvern, Worcestershire in Great Britain conducted elaborate measurements of the refractive index and the absorption coefficients. They observed photo-conductive and photo-voltaic effects as well and so discovered the suitability of InSb as a detector material for wavelengths up to 7.5 µm. The photo-voltaic (pv) detectors produced around 1957 attained a measured quantum efficiency of 80% for the mid-wave range (MWIR 3 - 5µm) when cooled down to 80 K.

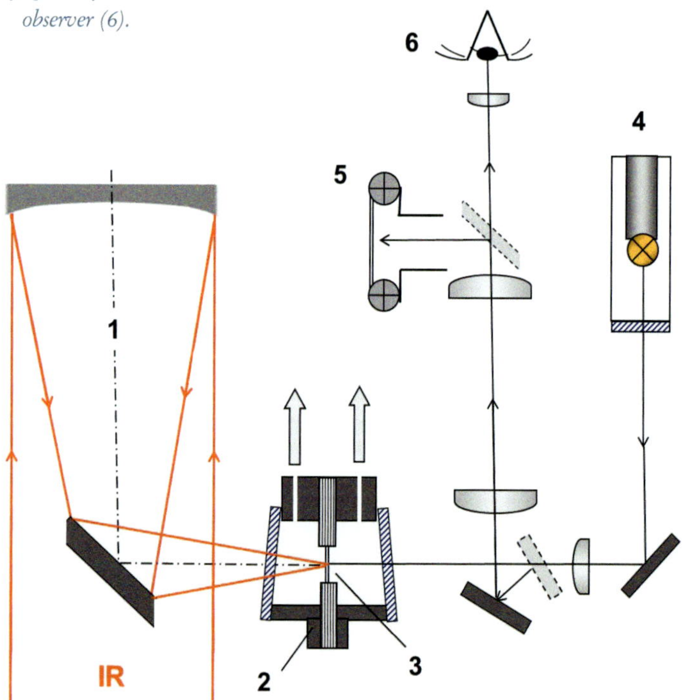


In **1955,** a group around C. Maxwell CADE at the KELVIN HUGHES Research Laboratories in Great Britain launched the Pyroscan program. "It started as a mathematical investigation to determine whether or not a short-range system for seeing through fog was feasible. We hoped to apply such a system to the docking of ships and to the landing of aircraft in foggy or misty weather."[301]

In **1956,** the US-American company TEXAS INSTRUMENTS began research work in infrared technology that led to several contracts for line-scanner camera development.[302]

Also in **1956,** BAIRD-ATOMIC at Cambridge, Mass. (USA) placed their first commercial thermal imager named Evaporograph on the market. The image formation happened in a dynamic process during which the thickness of an oil film on a membrane was steadily increasing. At the beginning a thin oil film was formed by condensation resulting in a uniform yellow color of the image. During further condensation, the thickness of the oil grew steadily and the color changed from yellow via orange, red, purple, blue and green back to yellow again. Where thermal radiation was impinging on the membrane, it was heated up. Thereby more oil was evaporated and the increase of the oil film was delayed and therewith the succession of the colors. Thus, at each point in time the membrane showed a multicolor image in which the cold objects were more advanced in the color cycle than warmer objects. In this fashion, the colors served as a measure for the temperature.[303]

BAIRD looked for application of their new evaporograph mainly in the industrial field, especially in electronics.[304] It was the first commercial camera available which could create a visible image from the thermal radiation emitted by objects at room temperature.

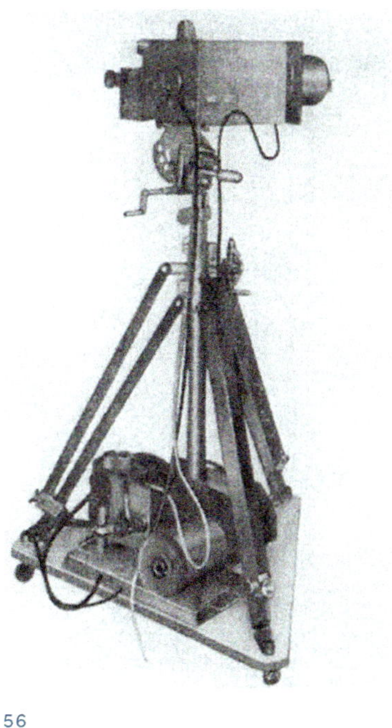

56

56 – *Evaporograph Model JZ 1 of Baird Atomic Inc.*

**Technical Data of Model JZ-1 Evaporograph**

| | |
|---|---|
| Detector | temperature dependent evaporation of a thin oil film |
| Wavelength range | IR |
| Cooling | none |
| Scanner | staring imager |
| Number of image points | approx. 160 000 |
| Integration time/Image | < 1 sec @ object temperature 315°C<br>≈ 15 sec @ temperature difference ± 17°C |
| Infrared optics | Newton mirror telescope |
| Focal length | 200 mm |
| Aperture | F/2.25 |
| Field of view | 5° |
| Image reproduction | visual |
| Resolution | |
| spatial | 10 to 14 lines /mm |
| thermal | ±0.3 K NETD @ 30 sec integration time |

In **1957** "a Canadian breast surgeon, Ray LAWSON, working at the ROYAL VICTORIA HOSPITAL in Montreal, was the first to use pictorial heat scanning medically, for which he coined the name Thermography. … He demonstrated that certain cancers of the breast promote a rise in the temperature of the overlying skin, and he argued that if one could scan the chest thermally one might be able to detect these cancers at an early stage, when treatment would be more effective. He first used the Evaporograph, an instrument developed by the BAIRD ATOMIC company (this was, in fact a sophisticated version of the lampblack and alcohol system devised by Sir William Herschel's son, John, in 1850)."[305]

A paper of the co-inventors Gene W. MCDANIEL und David Z. ROBINSON in the Applied Optics journal in 1962 gave insight in more technical details of the commercial Evaporographs made by BAIRD Ass. Inc.[306]

A specification of the performance characteristic by means of NETD und MRTD[307] is done in the same way today. The NETD (=Noise Equivalent Temperature Difference)

**Technical Data of Model KR-1 Evaporograph**

| | |
|---|---|
| Detector | temperature dependent evaporation of a thin oil film |
| carrier membrane | Nitro-cellulose, thickness 5 µm |
| front surface | Gold, black |
| back surface | siloxane polymere (Dow-Corning silicone 200) |
| Wavelength range | IR |
| Cooling | none |
| Scanner | staring imager |
| Number of image points | approx. 120 000 |
| Infrared optics | Germanium lenses |
| Focal length | 75 mm |
| Aperture | F/1.6 |
| Field of view | 14° |
| Image reproduction | visual |
| Resolution | |
| spatial | 10 Line pairs/mm |
| thermal | 1 K NETD at 20°C<br>10 K MRTD at 10 lp/mm |

**Technical Data of Far-IR Camera**

| | |
|---|---|
| Detector | flake thermistor |
| Wavelength range | IR to 25 µm |
| Cooling | none |
| Detector elements | 1 |
| Scanner | plane mirror tilted in 2 axes |
| Integration time | 2 to 15 min |
| Number of image points | 30 000 |
| Infrared-Optics | mirror optics, (8 inch) |
| Fields of view | 20° x 10°, 20° x 5°, 10° x 10°, 10° x 5° |
| Image reproduction | optical display via backside of scanner |
| Temperature range | -170 to 500°F (-112°C to 260°C) |
| Resolution | |
| thermal | 0.02 K |

indicates which temperature difference relative to the back ground an object must so that its generated signal is distinguishable from the noise of the thermal image. The MRTD (= Minimum Resolvable Temperature Difference) indicates which temperature difference is necessary to resolve an object with a given spatial structure (e.g. 10 line pairs per millimeter) on sensor surface. Several useful versions of the Evaporograph were marketed by BAIRD in the period from 1950 to 1968.[308]

In **1957** BARNES ENGINEERING COMPANY at Stamford, Conn. (USA) put their FAR-IR Camera on the market which very soon found broad application within the industry.[309]

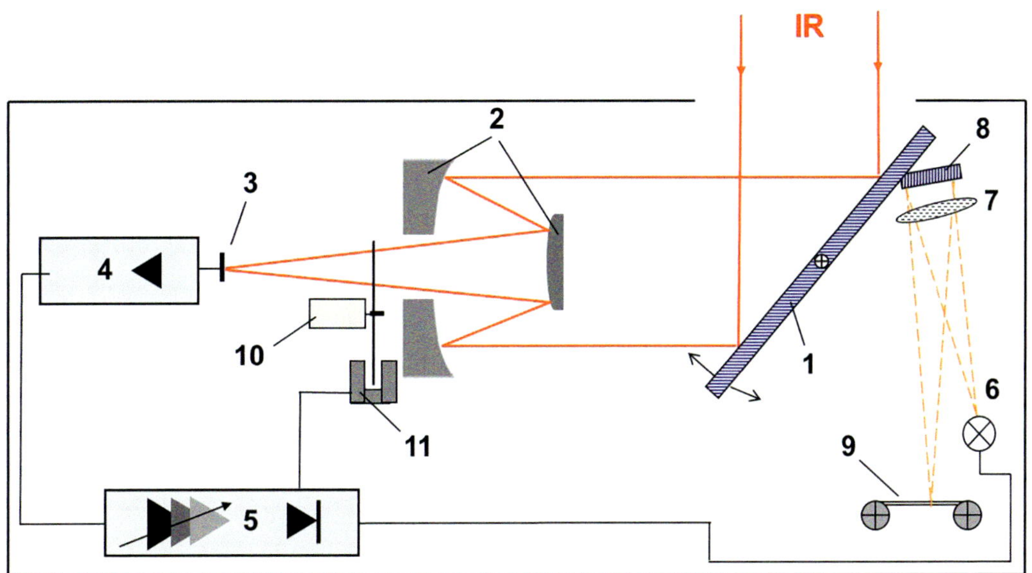


**57 – *Functional principle of the Far-IR Camera of Barnes Engineering***

*The thermal radiation (IR) coming from an object hits a scan mirror (1) which can be tilted in two axes. It is then imaged by a mirror objective (2) of the Cassegrain type on a thermistor detector (3). In between a chopper (10) is placed which also generates a reference signal (11) for synchronizing the signal processing. The detector signal passes the pre-amplifier (4), is further amplified in the electronic unit (5) (factor is scalable) and rectified with correct phase. With the DC voltage generated here a light bulb (6) is energized in the way that the brightness is varying proportionally to the incoming thermal radiation. The emitted light is collimated by a lens (7) and is directed by an auxiliary mirror (8), attached tightly to the scan mirror, over the photographic film (9) in a synchronous manner. By this process a visible grey level distribution is developed on the film exactly corresponding to the distribution of the thermal radiation of the scenery.*

The camera "consists of a standard BARNES 8-in. Optitherm radiometer with a flake thermistor detector, a scanning attachment, and a small camera using standard photographic film. A Polaroid Land camera is usually employed to give rapid results. The scan mirror consists of a large plane mirror actuated by a conventional gear cam system in order to scan over the object plane the small area subtended by the IR detector. The detector output is amplified and used to modulate the intensity of a glow tube. Light from the glow tube, focused by a collimating lens, is reflected by a small recorder mirror rigidly attached to the back of the scan mirror and moving synchronously with is, across the photographic film. The scanning mirror covers the object plane in a rapid horizontal scan, with a small vertical step deflection after each horizontal sweep. A blanking circuit cuts off the glow-tube output during each rapid horizontal return to avoid retrace lines on the film. A calibrated gray scale is automatically superimposed on the film by the camera mechanism. Calibration for a given target is achieved by adjusting the electronic gain so that the black area in the scale corresponds to a reference source at a known temperature, and the black-to-white range in the scale corresponds to a known temperature difference.... Thus, the temperature variations in the object under investigation can be determined directly from the photograph."[310]

58 – *Thermographic camera of Barnes Engineering*

From the technical point of view, it is remarkable that in this thermal imaging camera the infrared-optical image acquisition and the light-optical image reproduction were synchronized by using the front and back surface of the scan mirror. This principle was already demonstrated in the 1940s by the German company E. LEYBOLD'S NACHFOLGER in the thermal imager Potsdam (see also Chap. 3.1) and was patented[311] by Robert W. ASTHEIMER und Eric M. WORMSER from BARNES ENGINEERING in 1957. This ingenious and simple optical-mechanical synchronization became commonly known and a paragon for many new designs in the subsequent period. This method was replaced by digital-electronic control systems not until mid of the 1990s.

From **1959** on also the Canadian surgeon Ray LAWSON, who had established medical thermography with the BAIRD Evaporograph, now used the BARNES camera that was called Thermoscan at that time "and he was able to produce heat pictures displaying the temperatures he had noted with a thermocouple."[312]

In **1958**, mercury-cadmium-telluride (HgCdTe) was synthetized by a research group at the British ROYAL RADAR ESTABLISHMENT[313] in Malvern[314] for the first time. The inventor group of William D. LAWSEN, who had already participated in the development of the InSb detector some years before, comprised Alex S. YOUNG and Stan NIELSEN; Ernest PUTLEY joined a little later. "This work was the successful outcome of a deliberate effort to engineer a direct-bandgap[315], intrinsic[316] semiconductor for the long wavelength infrared (LWIR) spectral region (8-14 µm). Early recognition of the significance of this work led to intensive development in a number of countries including England, France, Germany, Poland, the former Soviet Union and the United States."[317]

"Soon thereafter, working under a U.S. Air Force contract with the objective of devising an 8-12-µm background-limited semiconductor IR detector that would operate at temperatures as high as 77 K, the group lead by KRUSE at the HONEYWELL CORPORATE RESEARCH CENTER in Hopkins, Minnesota, developed a modified Bridgman crystal growth technique for HgCdTe. They soon reported both photoconductive and photovoltaic detection in rudimentary HgCdTe devices. The parallel programs were carried out at TEXAS INSTRUMENTS and SBRC."[318]

"Several properties of HgCdTe qualify it as highly useful for infrared detection:

- Adjustable bandgap from 0.7 to 25 µm
- Direct bandgap with high absorption coefficient
- Moderate dielectric constant/index of refraction
- Moderate thermal coefficient of expansion
- Availability of wide bandgap lattice-matched substrates for epitaxial growth"[319]

Because of the first point even today HgCdTe is a valuable detector material being capable to detect infrared radiation in both atmospheric windows, the mid wavelength infrared (MWIR 3 - 5 µm) and the long wavelength infrared (LWIR 8 - 12 µm) as well. The HgCdTe detectors are mostly referred to as "CMT" (= Cadmium-Mercury-Telluride) or "MCT" (= Mercury-Cadmium-Telluride) detectors.

By adjusting the Hg / Cd-mixing ratio the sensitivity of the detector can be optimized to each of the wavelength bands:

| | |
|---|---|
| LWIR range 8 - 12µm | $Hg_{0.8}\,Cd_{0.2}\,Te$ |
| MWIR range 3 - 5 µm | $Hg_{0.7}\,Cd_{0.3}\,Te$ |

Whereas for LWIR detectors cooling to 77 K (e.g. by liquid nitrogen $LN_2$) is absolutely necessary, MWIR detectors also can be used uncooled in principle.

Until the early **1970s** HgCdTe was produced by crystallization of a liquid melt. Today HgCdTe is formed in very high quality by epitaxial[320] growth on substrates from cadmium telluride (CdTe) or cadmium-zinc-telluride (CdZnTe).

End of the **1950s**, the American company WESTINGHOUSE developed the Mesoscope, in part for the U.S. ARMY NIGHT VISION LABORATORY. This liquid crystal infrared viewer is a nonelectronic infrared image converter.

"Certain cholesteric liquid crystals reflect light in a narrow spectral region by scattering within the liquid crystal layer. Further, for certain of these mesomorphic materials, the wavelength of maximum reflection varies strongly with film temperature.[321]... A thin, edge mounted, plastic support-layer is coated with a gold-black absorbing layer on one surface, and with a mesomorphic crystal film on the other. For thermal isolation, the retina is enclosed in a cylindrical vacuum envelope, and the infrared image is focused on the gold-black layer through a sodium chloride window at one end of the cylinder, while the liquid-crystal layer is viewed through a glass window at the other. Two modes of viewing are used. If the liquid crystal is illuminated with white light, the infrared image is converted to a color-shift image, somewhat like the evaporograph. If illumination is with monochromatic light the shifting wavelength of the reflectance peak with temperature modulates the amount of reflected light. In either case, the wavelength of maximum reflectance must be tuned to the desired value by adjusting the operating temperature of the layer."[322]

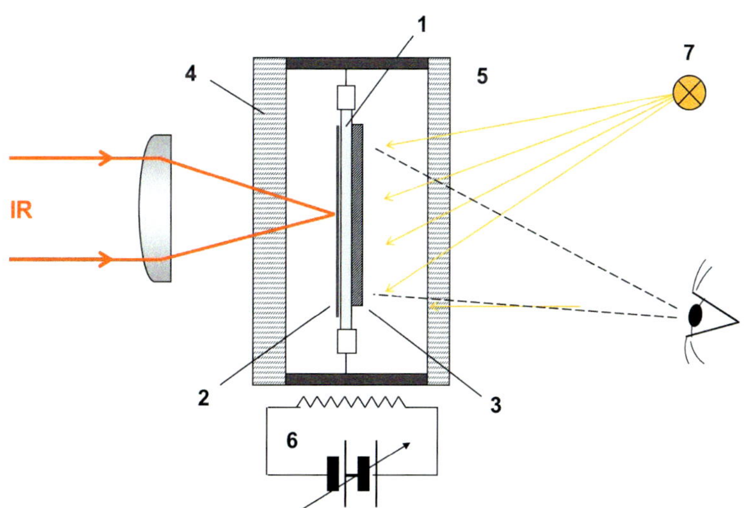


**59 – *Functional principle of a Mesoscope image converter***

*The image converter comprises a carrier foil (1) made from nitrocellulose, bearing a black gold layer on the front side (2) as an absorber. On the back surface, a layer (3) of cholesteric liquid crystals is applied. The coated carrier foil is fixed in a vacuum vessel that is closed by an infrared window (4) made from rock salt and a glass window (5). It is equipped with a heater (6) and illuminated by a source (7) of white light. The incoming thermal radiation (IR) heats up the image converter locally causing colour changes of the light reflected at these parts of the foil.*

"Mesomorphic layers can be made extremely sensitive, with spectral shifts as high as 100 nm $K^{-1}$. Such extreme sensitivity is required for observing fractional-degree differences in the scene, but proved a handicap in the mesoscope, since the working temperature range of the liquid-crystal layer was then only a few tenth of a degree. Further, the local retina temperature depended on radiation from the scene, and, if retina temperature and scene temperature differed significantly, was even affected by the normal off-axis vignetting of the optical system.[323]

In the early 1960s, WESTINGHOUSE made mesoscopic devices for studying far-field laser patterns and for thermal infrared viewing, with a typical diameter of 6 cm. The highest temperature-sensitivity liquid-crystals were physically unstable, however, and the general liquid crystal activity was sold."[324] Today flexible plastic foils with embedded liquid crystals are a very helpful auxiliary means in many optics labs for adjusting infrared laser beams which are not visible to the human eye, of course.

## 5.3 Medical Thermography as a Pacemaker

In the years **1957-58** John JOHNSON in the United States conducted comprehensive investigations about which conditions must be fulfilled for detection, recognition and identification of a target, and he framed the "Johnson Criteria" as they are named after their originator still today. His work was originally dedicated to the image intensifier systems, which were favored in the United States. The results, however, are of general importance for all imaging systems (direct view periscope, TV cameras, image intensifiers or thermal imagers). All commonly used range modelling is based on JOHNSON's results until the present day.

In **1958**, JOHNSON presented his results on the "1st Night Vision Image Intensifier Symposium".

In **November 1959**, the group around C. Maxwell CADE at the KELVIN HUGHES Research Laboratories in Great Britain built a thermal imager for the 5 µm wavelength range from "whatever odds and ends were available. Ex-War Department amplifiers and an old searchlight mirror were used – there was even a broomhandle which served as a crankarm"[325] The first recorded images of this device called Pyroscan immediately were of such a good quality that they were published in June 1960. Until January 1961 images through fog were recorded and also published, and "as soon as we made successful pictures more money was granted, and an improved machine was completed inside four months. Further pictures through fog were obtained, showing an increase over the visual range of between three and four to one. Rather to our surprise, little interest was shown in this invention. Potential customers said they could manage with existing navigational aids and a "vision-through-fog" machine was of no value unless it was cheap. Eventually we realized that the customer was right, at least as the shipping industry was concerned. Most large vessels lose at most only one or two days a year through fog delays."[326]

The first Pyroscan prototype was outright used to obtain thermal images (thermograms) also in the medical field and "was first used in Bath in 1959 and was used to image the increased heat over arthritic joints."[327] However, no further activities were reported afterwards.

CADE's group "then tried to interest government departments in the advantages of the Pyroscan camera for surveillance. We gave demonstrations of fog-viewing and of the detection of camouflaged men in fields on dark nights, and produced detailed schemes (published in February 1961) for airborne mapping and reconnaissance systems. These efforts raised enthusiasm, but no funds. Regretfully, we consigned the Pyroscan to the limbo of dead projects. However, by March 1961, our published work had attracted much attention. In particular, we were approached by surgeons from the Middlesex Hospital who wished to borrow the heat-scanner to investigate the diagnosis of breast cancer … A small sum was granted by the company to enable us to restore the scanner to working order and simplify its controls for use by unskilled staff.

Six month later, the Pyroscan camera had shown such outstanding promise as a clinical tool that the company decided to finance the design of a new machine specifically for this purpose."[328]

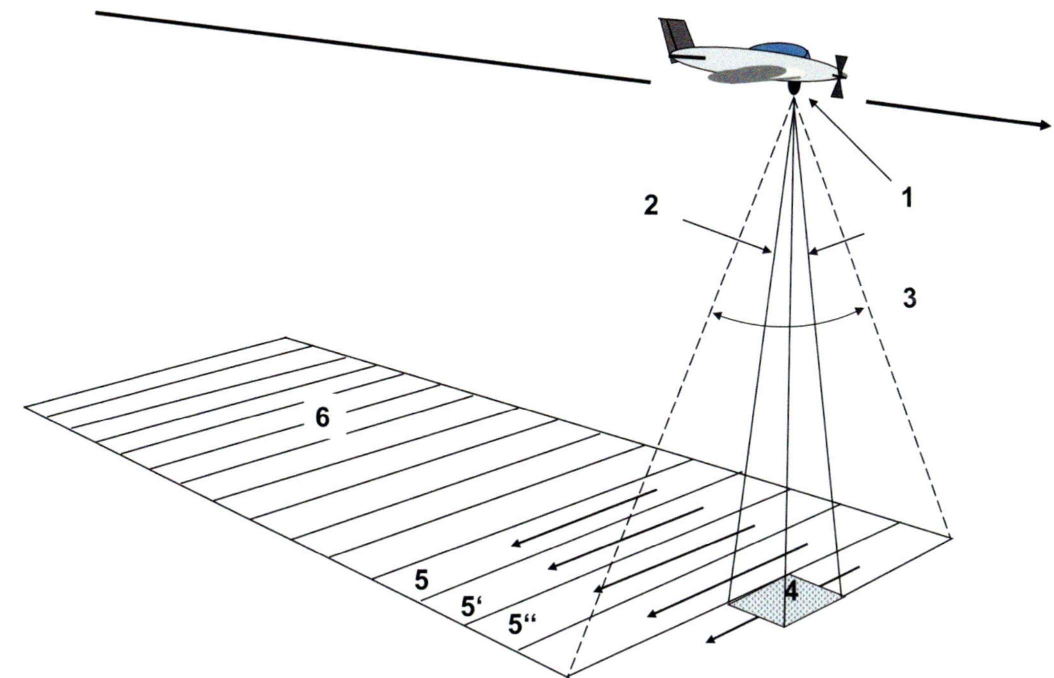


60 – *Functional principle of a push-broom sensor*

*A line scanner camera (1) is mounted at the belly of an aircraft and is directed forward and downward. The instantaneous field of view (2) of a detector element is very small and covers only a tiny area on the ground (4). The width of the field of regard (3) across the direction of flight is generated by the line-scan. Due to the aircraft movement one stripe-type image is strung to the other (5, 5', 5'', etc.), and as a whole a two-dimensional scene (6) along the flight trajectory is covered.*

"In the early 1960s … small and reliable cooling devices became available"[329] so that the long known improvement of the detector sensitivity by cooling could be utilized in practical applications eventually. Therewith, the recording time of a line-scan could be reduced dramatically, and an application of the line-scan camera in an aircraft as a so-called "push-broom" sensor was possible now.

A "push-broom" sensor can obtain a two-dimensional image by scanning one dimension with the line-scan camera while the second dimension is created by the aircraft movement over ground. The possible use of this sensor is obviously constrained by the fact that the line of sight of the line-scan camera must point downwards and that e.g. the pilot can see an object in the thermal image not before he has flown over it.

Around **1960**, the development of structuring semiconductors by photolithography began, and towards the end of the decade it should become of deciding influence on the detector technology for the infrared spectral range. In place of single detector elements or small groups of detectors at the end of the 1960s, linear detector arrays with 100 and more elements could be efficiently manufactured by photolithographic processes.

"In **1963** TI's [TEXAS INSTRUMENTS] Kirby TAYLOR began visualizing a real-time, two-dimensional infrared viewer, and was given $30,000 in company funds to investigate the possibility of adding another scan mirror to the single-axis line-scanner optics ["what engineer Kirby TAYLOR called a "little experiment" in infrared imaging"[330]]. The study determined the feasibility of the concept, and TI wasted little time in preparing and delivering an unsolicited proposal to the ADVANCED RESEARCH PROJECTS AGENCY, long recognized as the venture capital, visionary arm of the DEPARTMENT OF DEFENSE. The response came back quickly. The company's concept was interesting, the agency said, but no military requirement existed for such sensor. TI was disappointed but not willing to give up on the program. The company turned next to the AIR FORCE AVIONICS LABORATORY, which had been involved with TI on its infrared detector development, as well as the line-scanner source selection. The AIR FORCE decided that the infrared sensor was certainly worth studying further but insisted that the project be bid competitively. TI's original idea no longer was proprietary, and eight companies were battling for one of two contracts, each valued $250,000. Within months, all competitors but TI dropped out, and by the end of 1965, TI delivered a crude lab model. It was a simple prototype, nothing more, and certainly not ready to be tested in a critical airborne environment. The imagery was marginal at best."[331]

From the days of competition, TAYLOR told a story about the early demonstrations of the FLIR system to the AIR FORCE. During one demonstration, the competitor imaged rabbits hopping around. "Well, as you know …everything is bigger in Texas," Taylor said. "So we brought in some circus elephants to be subject of our demo… Nobody expected that!"[332]

This was the beginning of the development of the first, fast, airborne infrared system that could deliver two-dimensional thermal images. In contrast to the push-broom sensors available up to that time, the new infrared system could not only look downwards from an aircraft but also look in forward direction. This capability found expression in the name-giving "Forward Looking Infra Red" or FLIR as it is typically addressed. Detached from its original application, in the English-speaking world, this name is used wholesale for all thermal imagers providing a two-dimensional image.

In **1963**, the Pyroscan Mark II camera, which had been developed by S. SMITH & SONS specially for medical application, was utilized by several customers.[333] In the following years the device gained international distribution and was employed, for example, also at several American hospitals.[334] High hopes were particularly invested in applying the new tool to cancer diagnosis. Due to an enhanced metabolism, certain cancers promote a rise in temperature of the overlying skin; this was verified experimentally at the end of the 1930s.[335] "By a quirk of the nature (or by evolutionary design), it happens that the human skin, black or white, is in the thermal sense an almost perfect 'black-body' radiator. It cannot be made thermally blacker by addition of paints or pigments to its surface."[336] As there are color differences from the thermal point of view, and because a black body has the highest possible thermal emission, chances are great to make thermally anomalies of the body visible by means of a thermal imager. This was first demonstrated in 1957 by the Canadian breast surgeon Ray LAWSON who used the Evaporograph of BAIRD ATOMIC Company. He also coined the name "thermography" for this method and until today the employed thermal imagers are termed "thermographic cameras". The great hopes were harbored for some years and found also some public interest as is indicated by the fact that Pyroscan became an entry in a popular German technical dictionary, named DAS GROSSE HOBBY-LEXIKON (1967)[337]. The limitations of the method became obvious not until later.

The Pyroscan Mark II "uses a special cell, manufactured by MULLARD, the indium antimonide element of which is cooled with liquid nitrogen; it has a response time of less than one microsecond. Filters eliminate background radiation and enable the instrument to be used for spectroscopic studies. The radiation reaching the cell is chopped by a perforated disc rotating at 10 000 rev/min to provide a.c. signals. The optical system comprises a planar scanning mirror and a collector, which is a spheroid, 3 inches in diameter. The mirror is oscillated by a scanning mechanism to give a horizontal line scan; it also tilts the mirror vertically at the end of each line to create a 'raster'. This scanning action is synchronized with the display unit, which gives a permanent black-and-white record on 'electrochemical paper' and also provides an output to a 21-inch television-type monitor screen."[338]

**Technical Data of Pyroscan Mark II**

| | |
|---|---|
| Detector | Indium antimonide InSb ( Mullard ) |
| Wavelength range | 3 – 5 µm MWIR |
| Cooling | liquid Nitrogen $N_2$, 77 K = -196°C |
| Scanner | plane mirror tilted in 2 axes |
| Number of image points | 40 000 |
| Integration time | 30 sek |
| Infrared optics | spherical mirror 3 inch |
| Chopper | rotating pinhole disc, 10 000 revs/min |
| Image reproduction | electro-chemical paper or 21-inch monitor |
| Resolution | |
| thermal | 0.2°C (between 25°C and 40°C) |

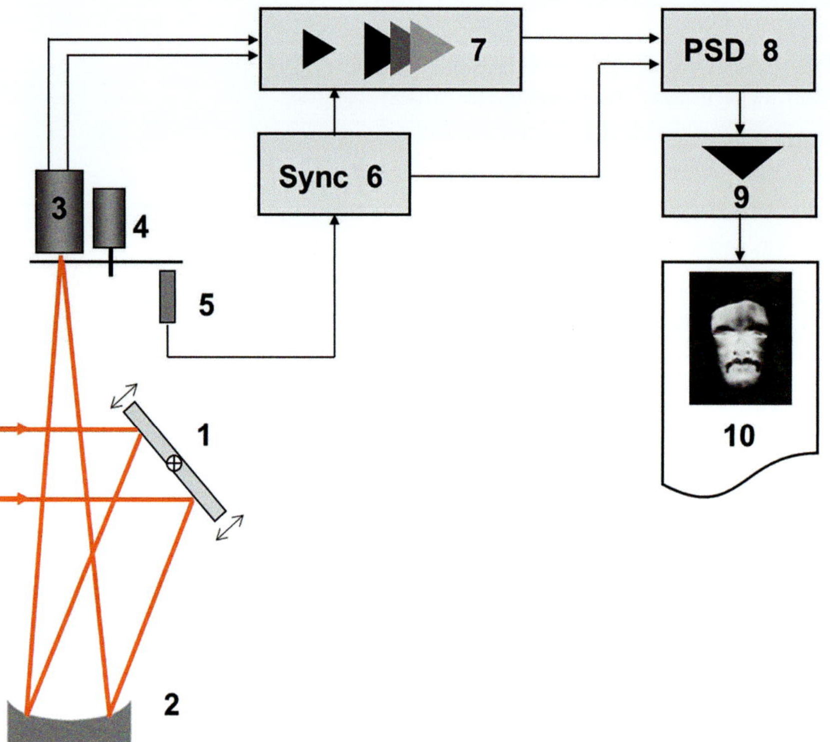


61 – ***Functional principle of the Pyroscan Mark II***
*The incoming thermal radiation (IR) is directed by a plane scan mirror (1) onto a concave mirror (2) which does the focussing onto an InSb detector (3). The radiation is modulated in front of the detector by a chopper (4) designed as a rotating punched disc. Thereby the detector is delivering an AC signal which is amplified by pre- and main amplifier (7). The phase sensitive detector PSD (8) is synchronized with the modulated radiation by the optical reference (5) and a synchronous pulse generator (6). Therewith the detector signal can be filtered very sensitively from the noise and transmitted via a DC voltage amplifier (9) to a display unit (10).*

As "a remarkable test of sensitivity … a 15-second picture" was obtained "showing the thermal imprint of a hand upon the cover of a book, taken 20 seconds after the hand had been removed."[339] The published picture showed the outline of a hand, somewhat blurred by noise but clearly recognizable.

On **2 November 1965,** the NIGHT VISION LABORATORY (NVL) was founded in the United States to support and coordinate all developments concerning night vision capability.

In **1965** the Thermovision 660 of the Swedish AB GASACCUMULATOR (AGA) company in Falköping was placed on the market. This device is commonly regarded as the thermographic camera in series production which had opened the civil market.

However, the technical base was a military development which was conducted early in the 1960s by the FÖRSVARETS FORSKNINGSANSTALT[340] FOA in Kista, a Swedish establishment for military research. Particularly the core of the thermal imager, a mechanical-optical scanner system with two rotating prisms was invented and patented in 1961 by Per Johan LINDBERG and Hans Gunnar MALMBERG[341], both working at the FOA.

The unique scanning system employed a first eight-sided polygon for horizontal and a second one for vertical deviation, both used in transmission. The polygons for the MWIR (3-5 µm) cameras were made from silicon (Si) and those for LWIR (8-12 µm) cameras from germanium (Ge). The frame rate was between 16 and 25 images per second. The following detectors were optionally installed:

| | |
|---|---|
| Indium antimonide (InSb) | spectral range 2-5.6 µm |
| CMT /HgCdTe | spectral range 8-12 µm |

The field of view of the basic unit was fixed by the scanner geometry, and interchangeable lenses were offered to adapt the camera field of view to intended applications. The total weight of the camera was 40 kg.[342]

"1966 could be considered the year that thermal imaging systems, as we know today, were created." This was retrospectively written by the American optoelectronic expert and author of several specialist books Gerald HOLST in his book[343] about thermal imaging technology in 2000. One may debate the exact year, but about facts and circumstances HOLST is right for sure. The devices presented here give evidence that mid of the 1960s thermal imagers with a fast, single detector and optical-mechanical, 2-axis scan apparatus[344] were ready for production in several counties[345].

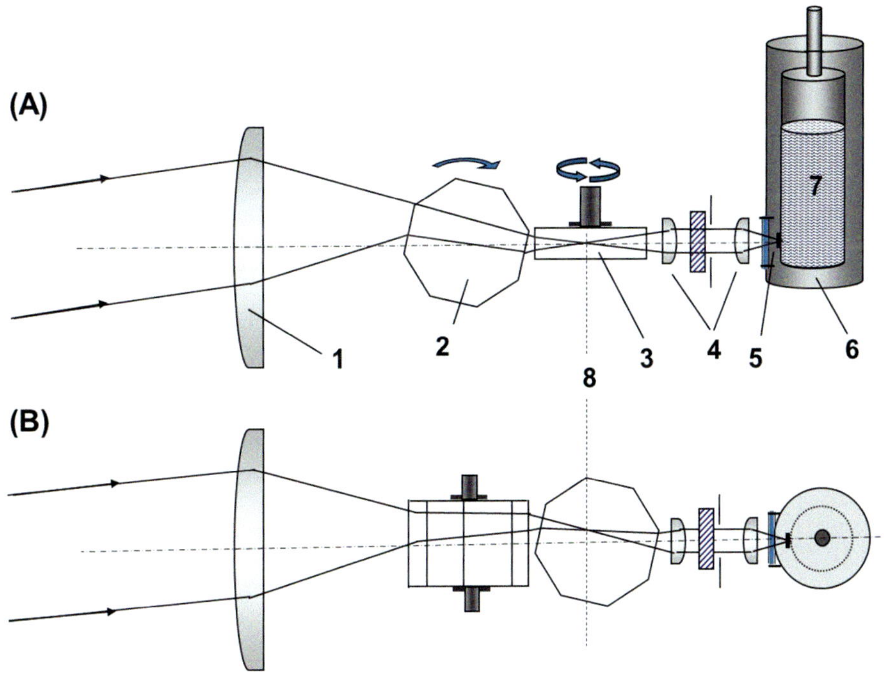

62 – ***Functional principle of the Thermovision 660 of AGA***

*Sketch (A) shows a lateral view of the system, Sketch (B) a top view. The incoming thermal radiation is imaged by an exchangeable lens (1) through the polygon (2) for vertical deviation into an intermediate image plane (8) which is located near the centre of the polygon (3) for horizontal deviation. After that the radiation is focussed onto the detector (5) by a relay system (4) which comprises also manually changeable filters and stops. The detector is mounted in a vacuum vessel and is cooled with liquid nitrogen (7) to 77 K. By means of the rotation of the polygons the optical image (8) is shifted vertically and horizontally with respect to the system axis, and thus the single detector measures all points of the chosen field of view.*

**State of the Art in Thermography around 1970**

| | |
|---|---|
| Detector | Indium antimonide InSb or Germanium-Gold Ge-Au |
| Wavelength range | 3 - 5 μm or 2 - 10 μm |
| Cooling | liquid Nitrogen $N_2$, 77 K (-196°C) |
| Detector elements | 1 |
| Scanner | plane mirror tilted in 2 axes |
| Image lines | 100 to 300 |
| Number of image points | 8 000 to 60 000 |
| Integration time | 1 to 2 sec (high image rate)<br>30 to 100 sec (high resolution) |
| Infrared optics | Cassegrain mirror |
| Field of view | 5° x 5° to 20° x 15° |
| Temperature range | 0° to 50°; optional sensitivity ranges 1°, 2°, 3°, 5°, 10° and 20°C |
| Image reproduction | cathode ray tube 5 to 7 inch<br>photographic: 35-mm or Polaroid film |
| Resolution | |
| spatial | 1 to 2 mrad |
| thermal | 0.1 K |

**Commercial Thermographic Cameras around 1970**

| Company | Country | Instrument |
|---|---|---|
| AEOL | Japan | Thermo-Ace |
| AGA | Sweden | Thermovision |
| Barnes Eng. | USA | Medical Thermograph M1-A |
| | | Medical Thermograph M1-B |
| | | Portable Thermograph T-60 |
| Barnes/Bofors | USA | Medical Thermograph M-101 |
| Bofors | Sweden | IR-Camera |
| Canon | Japan | Thermocamera CT-2 |
| Comp. Sans Fil | France | IR-815 |
| Dynarad | USA | Thermo-imager Model-201 |
| | | Thermo-imager Model-301 |
| Fujitsu | Japan | Infra-Eye |
| Infrared Industries | USA | Thermoscan |
| JEOL | Japan | Thermoviewer |
| NEC | Japan | Infravision IRV 2002 |
| San-ei | Japan | Thermoscope |
| Smith | Great Britain | Pyroscan |

Until the end of the 1960s several more commercial and military thermal imagers were put to market. One of the more prominent development stories happened at TEXAS INSTRUMENTS:

"By the end of **1965,** TI delivered a crude lab model. It was a simple prototype, nothing more, and certainly not ready to be tested in a critical airborne environment. The imagery was marginal at best. The AIR FORCE in

Washington, D.C. was excited about the technology and wanted to see what it could do in practice."[346]

From **28 July 1965** on, first tests in TI's own Douglas DC-3 were flown from the Airforce Base Eglin AFB. "Ignoring TI's protests, the model sensor was placed onboard a DC-3 named **Puff the Magic Dragon**" (the title of a contemporary pop-song of the trio Peter, Paul and Mary[347]). In the course of a test project named "Red Sea" operational test flights were scheduled "over the battle zones of Southeast Asia" - including North Vietnam - in a Douglas AC-47D Spooky[348] based at Bien Hoa AB in South Vietnam. [349]

The Douglas AC-47 was a cargo aircraft converted to a so-called "gunship". Besides various aiming sights, three machine guns firing sideways at a high rate were installed on the aircraft. During a mission, the aircraft was circling at a constant altitude above its target, e.g. a truck convoy or a tank, and fired a hail of bullets onto the target (up to 6000 rounds-per-minute) - a fire-spitting dragon indeed, when tracer ammunition was used. As the missions often had to be scheduled at twilight or under adverse visibility conditions, a FLIR promised a reasonable improvement in target recognition.

"For TI it was a bad engineering situation", because the employed device was only a prototype, and "Kirby TAYLOR flew the test missions to assure the system worked as well as possible." Despite all efforts it is not astonishing that the FLIR system had problems: with the climate, the humidity, with the terrain and with insufficient sensitivity. Hence "the official report was negative": "Red Sea was not successful". TI was maligned and the staff was appalled, and this "greatly eroded (the) support for TI's continued research in Forward Looking infrared (FLIR) development."[350] However, General BOLES, Director of JOINT RESEARCH & TEST ACTIVITY, advised to strengthen the FLIR development further. This happened against multiple oppositions.

In **1966,** despite of frustration about the test report, the first units of the new FLIR system could be produced at TI.[351]

The specific features of the new FLIR system were: [352]

- it had two moving mirrors to create a raster pattern
- the frame rate was 20 images per second
- the LWIR system used mercury-cadmium-telluride (HgCdTe) detectors
- the MWIR system used indium antimonide (InSb) or lead tin telluride (PbSnTe, LTT) detectors

A further issue, besides the high weight, was the high cooling power necessary to cool the focal plane array (FPA) down to a temperature as low as 26 K, which is a very low value compared to today's thermal imagers.

To the benefit of TI "some people in the DEPARTMENT OF DEFENSE also realized an important lesson from the test. During night missions, gunships could greatly benefit from having a sensor on board to locate and define targets in the dark terrain while the aircraft circled at a constant altitude and fired large-caliber guns"[353]

In **1967,** TI eventually won a development contract "Gunship FLIR Development Program" for equipping the AC-119K Stinger gunship, the follow-up model of the aged AC-47 Spooky. "The AIR FORCE ignored its own criticism of the lab model and awarded TI a million-dollar contract to further develop the idea of

63 – *Image intensifier Zeiss PST with Wärmeorter (thermal pointer)*
*For this purpose, a ZEISS PST, a very sensitive 3-stage image intensifier device, was equipped additionally with a single CMT (HgCdTe) detector and a simple linear scanner to measure the thermal radiation in 8–12 µm spectral range along a vertical line in the center of the image. During field tests the reconnaissance of this combination of image intensifier and target pointer against warm targets ("hot spots") was significantly improved.*

the FLIR." Temporarily many in TI were willing to eliminate or at least lessen the emphasis on the infrared program. It was that "Manager Ray McCORD believed strongly in the technology and was willing to battle for critical resources necessary to finance the research of infrared detectors." As also "key customers stood firm in their support of TI's research endeavors" McCORD's planning was supported "and the company began successfully producing a series of gunship FLIR systems that were capable of detecting radiated heat from an unseen, unknown target and providing operators a displayed image of heat sources."[354] The anticipated date of delivery was in June 1968. The following time line of the events, however, gives vivid evidence of how large the problems had been in fact:

| | |
|---|---|
| **June 1968** | Delivery plan of WRAMA (WARNER ROBINS AIR MATERIAL AREA) for the first AC-119K without FLIR |
| **August 1968** | Intended date of delivery of the first four AC-119 with FLIR |
| **October 1968** | Delivery plan cancelled by TI. All 18 AC-119K without FLIR. WRAMA accused TI, their plan had been "over committed" |
| **November 1968** | Discussion whether HUGHES Aircraft could be an alternative (HUGHES, however, was a year back in the development). |
| **December 1968** | New delivery plan of the WRAMA Tiger Team for 1969 (February: 1 piece, March: 2 pieces, April: 5 pieces, May: 4 pieces, June: 5 pieces, July: 4 pieces, August: 2 pieces, September: 3 pieces) |
| **16 January '69** | Announcement of a new TI delivery plan on 26 January |
| **24 January '69** | Total stop of the FLIR production at TI |
| **February 1969** | Communication of TI that 18 months and 5 Mio. $ would be necessary for fulfillment of the contract; as AC-119K and AC-130 were concerned, the USAF had no choice |
| **3 May 1969** | Delivery of the first AN/AAD-4 FLIR |
| **20 May 1969** | First flight tests |
| **April 1970** | Delivery of the last AN/AAD-4 FLIR |

In the end, the first stated delivery date was missed by approximately 20 months.

**In 1968,** also CARL ZEISS got an opportunity to enter thermal imaging technology again.

On **4 December 1968,** a contract was agreed with the Bundesverteidigungsministerium BMVg (federal ministry of defense) for a comprehensive study with the title "Zielerkennung bei Nacht" (target recognition at night). Admittedly, a major part of the study was dedicated to image intensifiers, but in a large chapter the employment of thermal imagers also was considered theoretically and practically. "For measurements in the course of this study three devices for image reproduction with InSb detectors and a Wärmeorter PST (thermal pointer) with a HgCdTe detector were available at CARL ZEISS-SONDEROPTIK."[355]

In the 1960s, the scientists and engineers at CARL ZEISS-SONDEROPTIK, as the department of military optics was named at that time, mainly worked with image intensifiers to improve the night vision capability. These devices used an electro-optical intensifier tube to intensify the residual brightness at night (moon, stars or artificial light sources) to create a visible image on a phosphorescent screen. As an interim stage to a pure thermal imager, it was tried to improve performance of the image intensifier devices by adding a thermal pointer. To circumvent the issues of a two-dimensional thermal image the plan was to use the already available image of an image intensifier and combine it with the spot measurement of the thermal radiation. The latter had been done already during World War II.

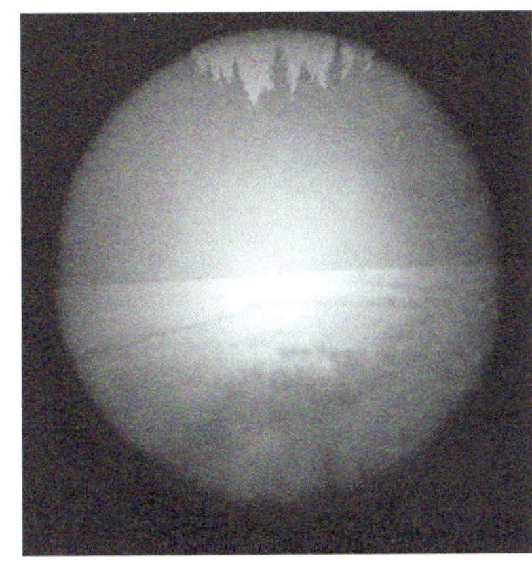
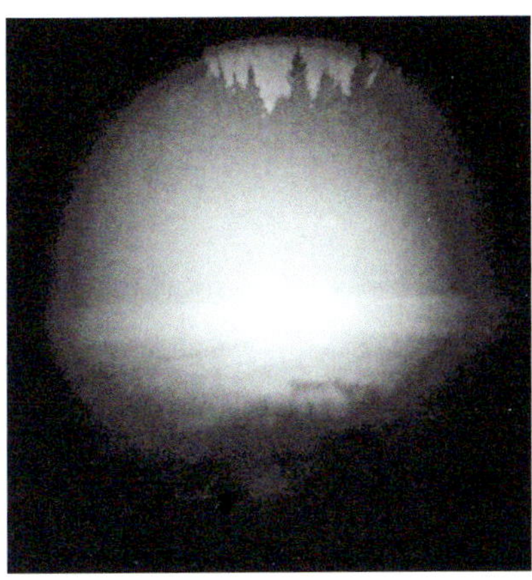

64 – *Observation of a passenger car with Wärmeorter (left) and without*

*In the center of the picture a passenger car in 600 m distance is spotted clearly by the thermal pointer (level of illumination 3 Millilux, mist).*

63

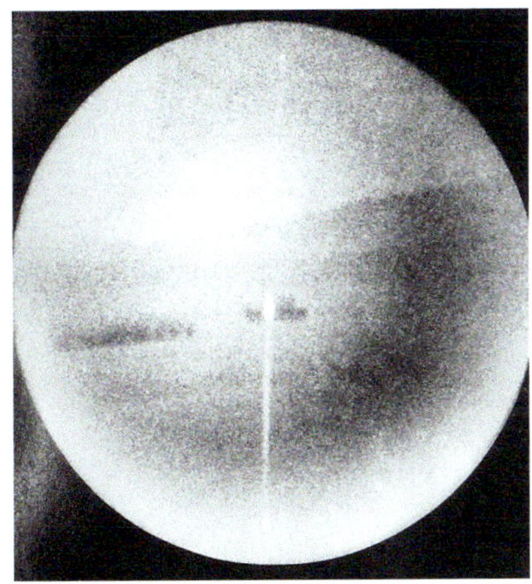
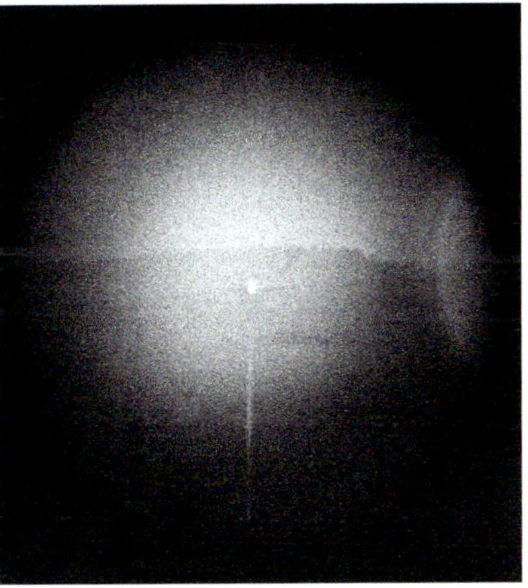

65 – *Reconnaissance against battle tank M 48 with Wärmeorter*

*The distance to the tank in the centre of the picture is 1200 m in the left and even 2000 m in the right picture. Over the large distance the tank would have been detected hardly without the thermal pointer (level of illumination 1 Millilux, drizzle).*

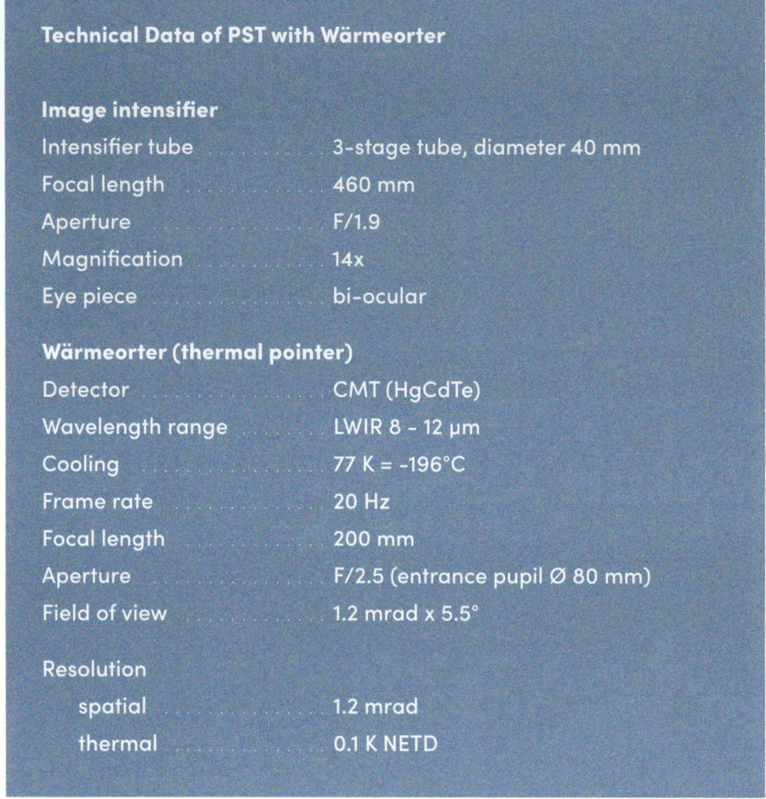

**Technical Data of PST with Wärmeorter**

| **Image intensifier** | |
|---|---|
| Intensifier tube | 3-stage tube, diameter 40 mm |
| Focal length | 460 mm |
| Aperture | F/1.9 |
| Magnification | 14x |
| Eye piece | bi-ocular |
| **Wärmeorter (thermal pointer)** | |
| Detector | CMT (HgCdTe) |
| Wavelength range | LWIR 8 - 12 µm |
| Cooling | 77 K = -196°C |
| Frame rate | 20 Hz |
| Focal length | 200 mm |
| Aperture | F/2.5 (entrance pupil Ø 80 mm) |
| Field of view | 1.2 mrad x 5.5° |
| Resolution | |
| spatial | 1.2 mrad |
| thermal | 0.1 K NETD |

**Technical Data of WBG 1/25**

| | |
|---|---|
| Detector | Indium antimonide InSb |
| Wavelength range | 3 – 5 µm MWIR |
| Cooling | 77 K = -196°C |
| Detector elements | 25 x 1 |
| Scanner | oscillating mirror |
| Image lines | 25 |
| Image rate | 25 images/sec |
| Number of image points | 1 660 |
| Infrared optics | lens optics |
| Aperture | Entrance pupil Ø 80 mm |
| Field of view | 8° x 3° |
| Image reproduction | optical display via backside of scanner and eye piece |
| Resolution | |
| spatial | 1 mrad |
| thermal | 0.5 K NETD |

## 5.4 First Thermal Imagers with Linear Detector-arrays

Around 1968 the new photolithographic technology was advanced so far that the production of linear arrays for the infrared wavelength range could be started. This was the beginning of a new era of thermal imaging, and the thermal imagers of the so-called "1st generation" based on this technology dominated the development programs in the 1970s.

From 1970 on the first photo-conductive (pc) CMT detector arrays were available. About 1970, CARL ZEISS-SONDEROPTIK also began to develop their first thermal imager with a one-dimensional (linear) detector array in the course of their "Zielerkennung bei Nacht" study.

At the start of the development, a prototype with a detector array with 25 elements was built and named WBG 1/25.

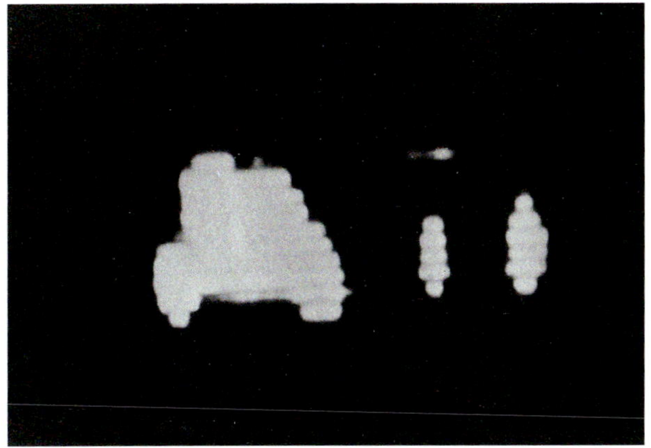

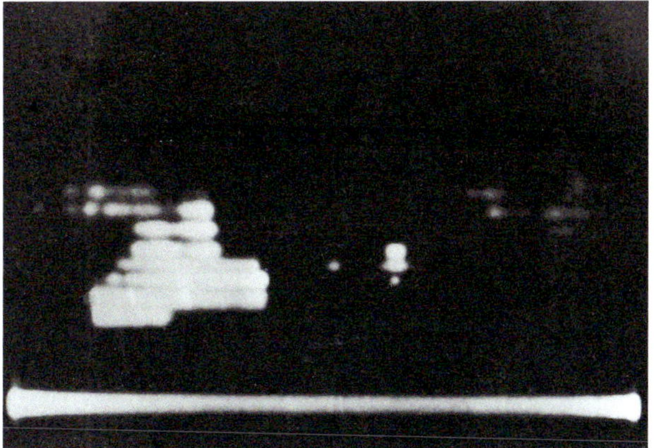

66

66 – *Thermal image of a tank obtained with a Zeiss WBG 1/25*

*The distance to the objects (battle tank M 48 and persons) is 300 m in the left and 550 m in the right picture.*

67 – *Hand held WBG 1/50 thermal imager with 50 image lines*

The WBG 1/25 used a cooled InSb detector array and could obtain "living thermal images" with 25 frames per second. Due to the small number of lines it was no wonder that the resolution was poor. In the thermal images, people always looked like the little rubber man ("Bibendum") of a well-known advertisement of a French tire company.

Consequently, the next development stage was to increase the number of lines. The 50-line WBG 1/50 "was a compact device comprising recording and reproduction units, featuring an extra low weight (2.4 kg)." The hand-held thermal imager was used for "pictorial display and detection, respectively, of temperature differences." The visible image could be viewed directly in an eye piece, or could be "transferred to a monitor with a TV adapter."[356]

67

68 – *Application of a WBG 1/50*

*The upper picture shows the application as a hand-held device; in the picture below, a combination with a contemporary TV camera (TV) is shown; the thermal image can be displayed on a monitor (M).*

In **May 1971,** first field measurements were made with the WBG 1/50 against maritime targets on the Jade water way to Wilhelmshaven. The WBG was placed onboard of the mine sweeper Westensee together with several image intensifier devices and the PST target pointer. In the night of 24 to 25 May observations against the boat Lindau as a target were performed under adverse wind conditions and a disturbing swell. Against Lindau, being soaking wet due to the spray, detection ranges were measured which were far below the expected values (rumors said the same about the seamanship of the operating personal). Some practical problems arose in conducting the tests. Due to the lack of a stabilization of the cameras against the swell it was virtually impossible to hold the target in the field of view. Furthermore, the observation of the thermal images through the eye piece was blurred by the tears in the observer's eyes, caused by the strong wind.

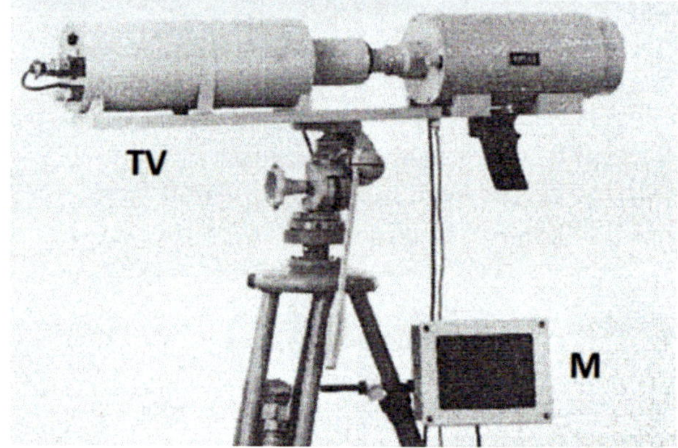


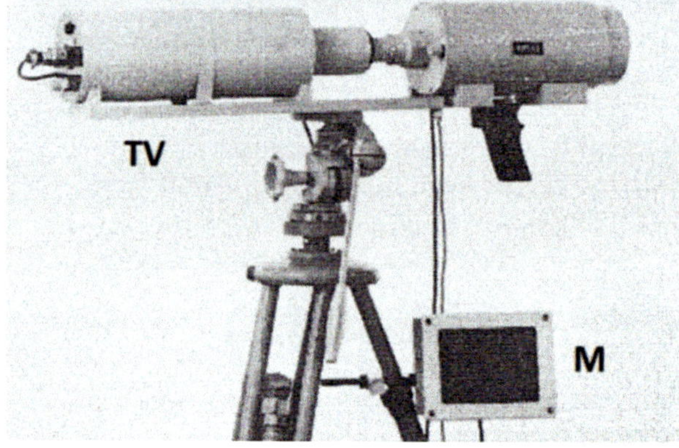

68

**Technical Data of Zeiss WBG 1/50**

| | |
|---|---|
| Detector | Indium antimonide InSb |
| Wavelength range | 3 – 5 µm MWIR |
| Cooling | 77 K = -196°C |
| Detector elements | 25 x 2 |
| Scanner | oscillating mirror with 2:1 interlace |
| Image lines | 50 |
| Frame rate | 35 images/sec |
| Number of image point | 5 000 |
| Infrared optics | IR lenses |
| Focal length | 230 mm |
| Aperture | F/2.3 (entrance pupil Ø 100 mm) |
| Field of view | 6° x 3° |
| Image reproduction | optical display via scanner backside and eye piece |
| Power consumption | 0.25 A (24 V=) |
| Dimensions | Ø 120 mm x 300 mm |
| Weight | 2.4 kg |
| Resolution | |
| spatial | 0.8 mrad |
| thermal | 0.3 K (NETD) |

On **17 May 1972**, further measurements with WBG 1/50 were performed in the Eckernförde Fjord, and it was obviously tried to avoid the problems of the Wilhelmshaven testing. The devices were arranged on solid ground at the WEHRTECHNISCHE DIENSTSTELLE WTD (military-technical establishment) to ensure the stability of the line of sight. Moreover, the WBG 1/50 was equipped with a downstream TV camera allowing for an easy viewing of the thermal images on a video monitor. Two boats with wooden hulls, Holnis and OT 2, served as observation targets and were recognized by the WBG 1/50 at ranges of more than 20 km under clear visibility conditions on the fjord.

In **1974**, the WBG 1/50 took part in benchmark tests at the military training area Münsingen in the Swabian Alb. As a target once again an American M 48 tank was used. Based on the data gathered in the various trials, simulations were conducted as part of the already mentioned target recognition study, to evaluate the possibly achievable ranges. With realistic assumptions regarding to the InSb detector and the infrared optical system the following values were obtained:

| | |
|---|---|
| Target properties | Temperature difference 3 - 5 K |
| | Target size 3m x 3m |
| Maximum range | 7.5 km |

Against larger targets of 6 m x 6 m longer ranges between 10 and 12 km were held in prospect. The positive numbers give the impression, at least to the author, that they were meant to overcome reservations regarding the performance of thermal imagers.

Approximately at the time of the first WBG 1/50 tests, the next development stage was pursued at CARL ZEISS-SONDEROPTIK with the 100-line WBG 1/100 thermal imager which used the optical system of the WBG 1/50. In 1974 the first prototype was ready for the field measurements in Münsingen, where again various devices with image intensifiers were also tested against ground targets.

**69 – *Thermal image of the ship Holnis captured with a Zeiss WBG 1/50***

*The thermal image shows the starboard side of the Holnis with her 36.5 m long wooden hull and was captured from 600 m distance. During the time of recording the hull was sunlit. In the test report is stated that the photographic picture was made from a video tape recording, and that the image quality of this recording was much worse than the visual observation of the thermal image.*

**70 – *Thermal image of a battle tank M 48 (Zeiss WBG 1/50)***

*The image of a driving M 48 was obtained from 150 m distance.*

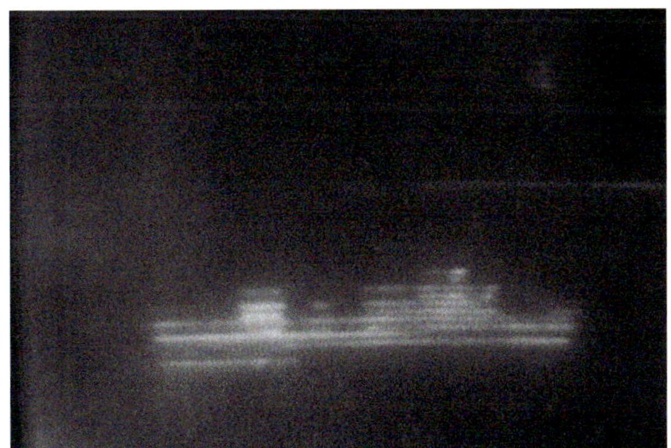

69

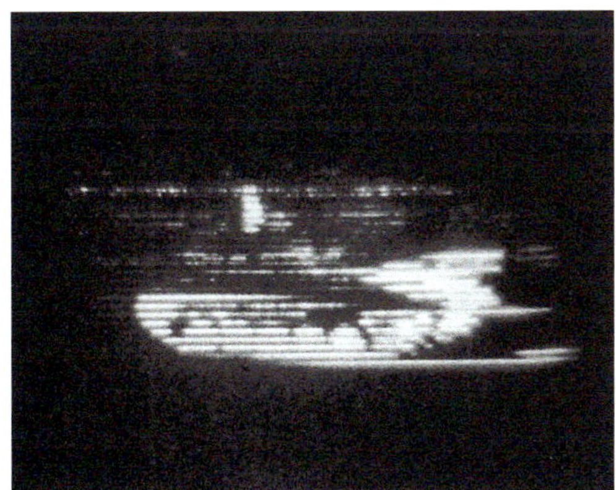

70

*71 – Zeiss WBG 1/100 thermal imager*

*72 – Back side of Zeiss WBG 1/100*

*At the back side of the thermal imager, an eye piece in form of a large-field magnifier (1) is arranged, through which the screen of an image intensifier can be viewed. With two turning knobs the gain (2) and the lower temperature limit (3) can be adjusted. Furthermore, with the switch (4) the noise bandwidth can be influence in two steps (narrow band / broad band). With an additional switch (5) the polarity can be inverted, i.e. it can be chosen whether a warm object shall be presented bright or dark in the visible image.*

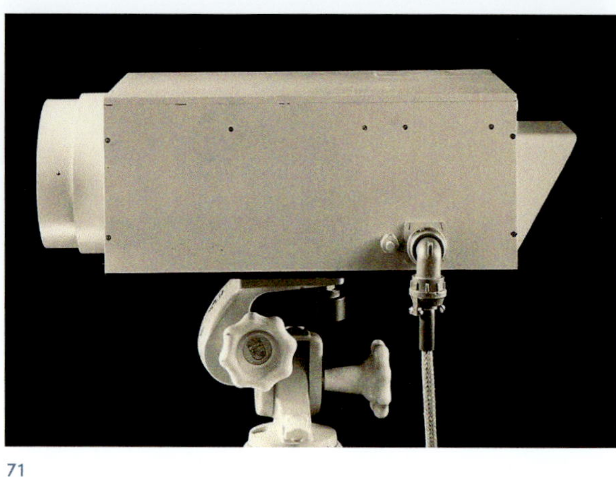

71

72

*73 – View into the Zeiss WBG 1/100*

*The thermal radiation (MWIR) is imaged by the IR imager (not shown in the picture) via the scan mirror (2) and a folding mirror onto a 100-element InSb detector array (1). The amplified signals of the detector elements are converted into red light signals by a LED array (3). The intensity of the light varies proportional to the intensity of the thermal radiation. The LED light is imaged by the lens (4) via the back side of the scan mirror (2) and an auxiliary mirror onto the entrance surface of an image intensifier (5). Using the front and back side of the scan mirror is a very simple means to synchronize the infrared and the visible image.*

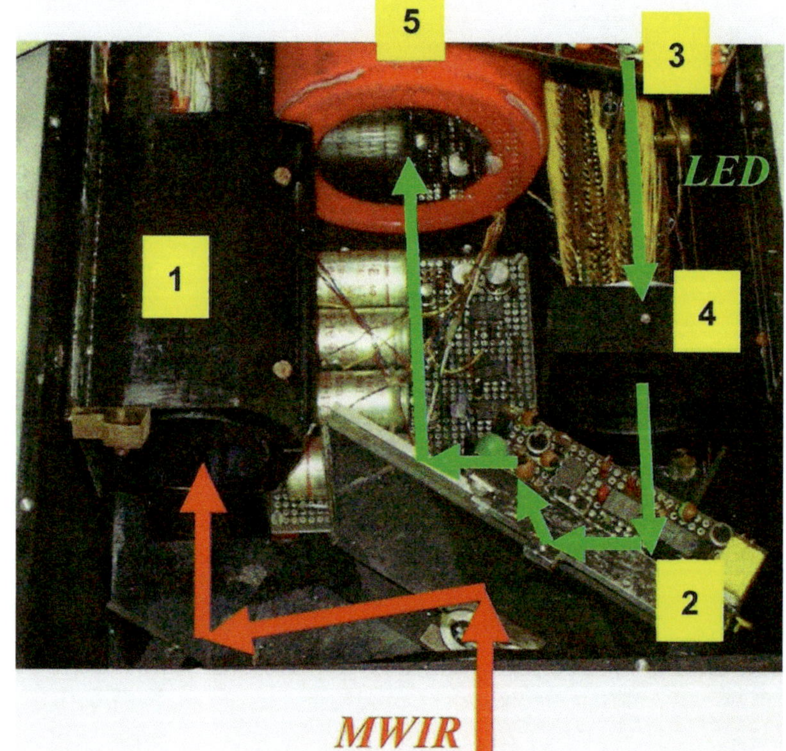


73

**Technical Data of WBG 1/100**

| | |
|---|---|
| Detector | Indium antimonide InSb |
| Wavelength range | 3 – 5 µm MWIR |
| Cooling | 77 K = -196°C |
| Detector elements | 100 x 1 |
| Scanner | oscillating mirror |
| Image lines | 100 |
| Image rate | 24 images/sec |
| Image format | 1.7 : 1 (horizontal : vertical) |
| Number of image points | 20 000 |
| Infrared optics | IR lenses |
| Focal length | 230 mm |
| Aperture | F/2.3 (entrance pupil Ø 100 mm) |
| Field of view | 6° x 3° |
| Image reproduction | optical display via scanner backside, image intensifier and eye piece |
| Power consumption | 0.25 A (24 V=) |
| Dimensions | 340 mm x 190 mm x 130 mm |
| Weight | 3.6 kg |
| Resolution | |
| spatial | 0.42 mrad |
| thermal | 0.5 K (NETD) |

74

*74 – Video electronics of Zeiss WBG 1/100*

*Opening the lid allows to look at the contemporary analog electronics. Each InSb detector element has its own amplifier and own potentiometers for individual adjustment of background signal and gain. The block of styrofoam (top right) can be taken as evidence that the imager was used in the lab because the Joule-Thomson cooler used in series was replaced by a styrofoam cup for cooling with liquid nitrogen.*

*75 – Thermal image of a battle tank M 48 with crew (Zeiss WBG 1/100)*

*The thermal image was captured from 200 m distance. The picture is a photograph taken through the large-field magnifier. The image quality (exposure, sharpness, distortion) is obviously limited due to this recording method.*

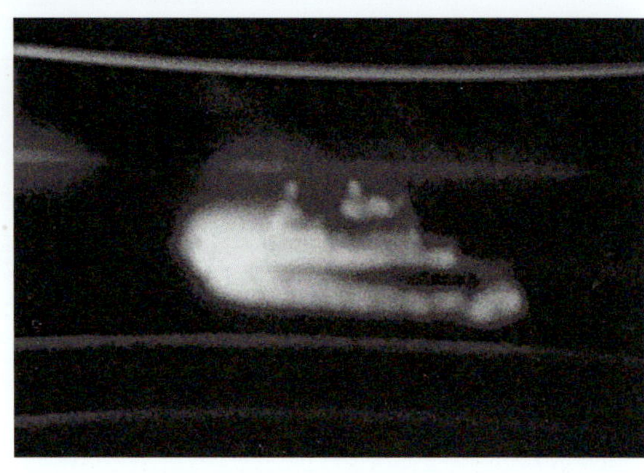

75

*76 – Thermal images of a Zeiss WBG 1/100 with different polarity*

*Both pictures show the same scene with M 48 battle tanks, a lorry and several persons in 400 m distance. In the left picture, warm objects are displayed bright as it is the common manner. In the right picture, the polarity has been switched to warm = dark.*

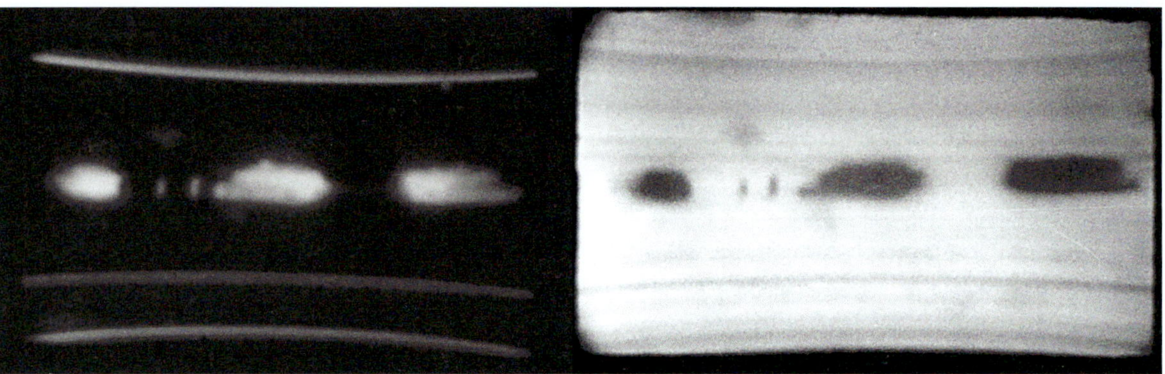

76

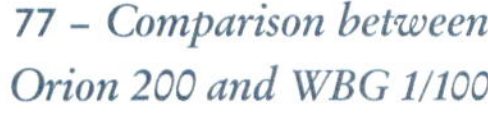

*77 – Comparison between Orion 200 and WBG 1/100*

*Both pictures show the same nocturnal scene with a battle tank M 48 behind a smokescreen. The left picture is taken with an image intensifier device Zeiss Orion 200, the right picture with a WBG 1/100 thermal imager.*

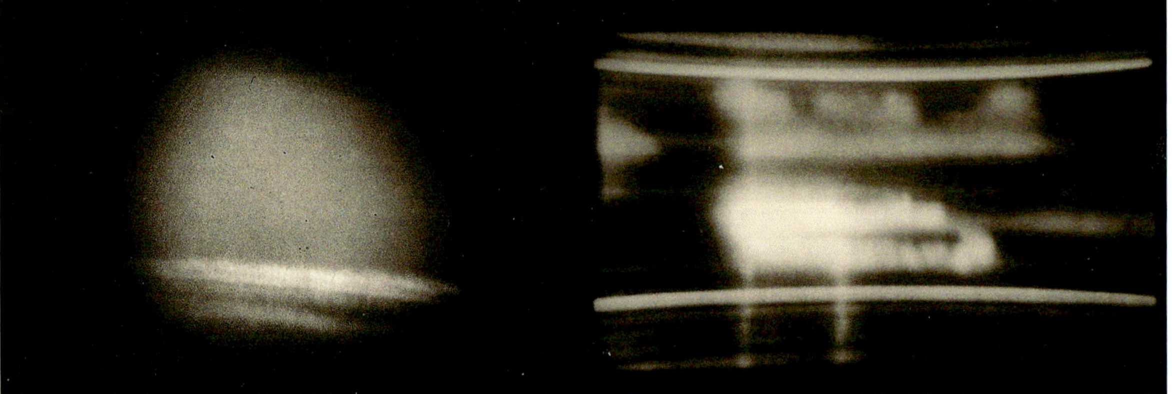

77

As a late representative of the non-electronical devices, the Swiss company GRETAG in Regensdorf developed the Panicon thermal imager early in the 1970s. The name was an acronym (**PA**ssive **N**onelectronic **I**nfrared Image **CON**verter) which indicated the functional principle, and was registered as a trademark in 1975.

The working principle was not based on differential evaporation but on the deformation of a thermo-plastic oil film. The film was locally heated by the impinging thermal radiation so that the initially planar surface was deformed due to the changing surface tension. Using a so-called schlieren-optics[357] the deformations could be made visible to an observer. The term "schlieren-optics" denotes a special optical assembly for imaging of very small density variations (German: Schlieren) of gases and fluids. With minor adaptions, this method can also be used to display deformations of a smooth surface in a grey scale image, e.g. the surface of the oil film in the Panicon.

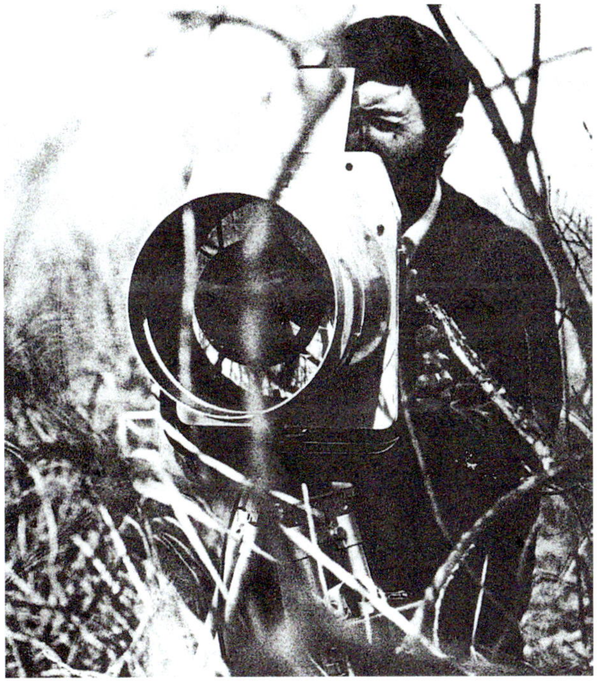

78 – *Swiss-made PANICON 7510 in action*

**Technical Data of PANICON 7510**

| | |
|---|---|
| Detector | thermoplastic oil film on a thin glass plate, blackened backside |
| Wavelength range | LWIR |
| Cooling | none |
| Scanner | staring camera |
| Number of image points | approx. 25 000 |
| Integration time | < 1 sec |
| Infrared optics | Cassegrain mirror |
| Focal length | 300 mm |
| Aperture | F/2.0 |
| Field of view | 11° x 7.5° (h x v) |
| Image reproduction | visual |
| Power consumption | 0.35 A (2 x 1.2 V=) |
| Dimensions | 620 mm x 340 mm x 470 mm |
| Weight | 12.3 kg |
| Resolution | |
| spatial | 1 mrad x 1 mrad |
| thermal | 0.3 °C temperature difference (time constant approx. 0.1 sec) |

78

In the data sheets the Panicon was promoted as "revolutionary device for passive night vision", and it was pointed out that it did not need "any of the typically used techniques which employ electronics, scanner or cooling systems". For these reasons it was likewise seen as "ideally suited infrared sight for military applications."

In the years **1973–75** detailed theoretical investigations on the method of image generation were conducted at the Laboratory for Solid State Physics of the EIDGENÖSSISCHE TECHNISCHE HOCHSCHULE to get a better understanding of the process and find means for optimization. This work was also funded by the Swiss government. Whether the results presented in two publications[358] were used practical is unknown to the author.

In any case the trademark Panicon was given up in 1982 what seems to be an indication that the work also was discontinued. This would be understandable as the electronical thermal imagers had made great advances at that time, and it became apparent that this was the technology for the future.

In **1973**, an attachable thermal pointer, the Wärmeorter AWO 10, was developed at CARL ZEISS-SONDEROPTIK and funded by the BUNDESAMT FÜR WEHRTECHNIK UND BESCHAFFUNG (BWB). It was designed to be attached to the proven ZEISS ORION 110, a sight with an image intensifier. Obviously, the objective was to create a cost-effective device that enabled the infantry man to detect heat targets and therewith improve his night vision capability.

The thermal radiation was imaged via a scanner on a short detector array. The scanning was performed in vertical direction. The detector signals were transferred to a LED array of equal length, arranged besides the detector. The LED light passed the same scanner and was optically coupled into the pupil of the ORION 110. In the ocular of the Orion the smaller thermal image could be viewed together with the image obtained by the intensifier tube.

**79 – *Wärmeorter AWO 10 mounted on an Orion 110***

*The picture shows how the thermal pointer AWO 10 is mounted on the Orion 110 image intensifier sight. The electronic control panel that is also shown here is an early version. Later devices were much more compact. The prize of an AWO 10 was estimated to be approx. 4000 DM.*

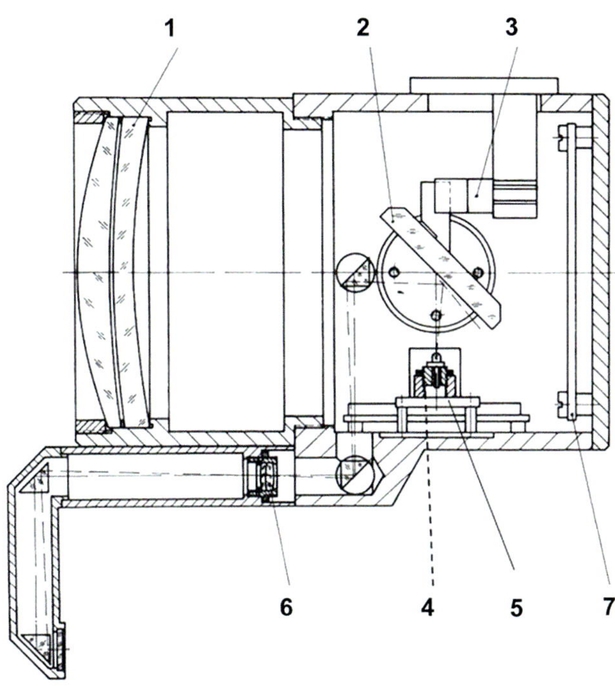


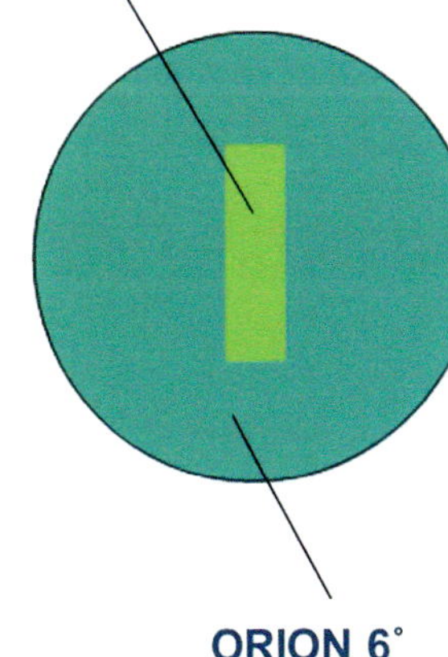


80 – *Sketch of the attachable Wärmeorter AWO 10*

*The thermal radiation passes the IR imager (1) and the scan mirror (2) to reach a 10-element MWIR detector (4). The electrical signals of the detector elements are transformed into light signals with corresponding brightness by LEDs (5). The LED light is imaged onto an image intensifier by the LED objective (6). The electronic board (7) for signal processing is attached to the back panel of the device. The sketch below shows how the field of view of the thermal pointer (AWO 1° x 3°) is arranged inside the Orion field of view (Orion 6°).*

**Technical Data of Wärmeorter AWO 10**

| | | | | |
|---|---|---|---|---|
| Detector | Lead selenide PbSe (pc) | | | |
| Wavelength range | 3.5 - 5.5 μm MWIR | | | |
| Detector elements | 10 | | | |
| Size of detector element | 75 μm x 75 μm | | | |
| Cooling | thermoelectric, switchable, 1.5 W | | | |
| Cool down time | approx. 10 to 20 sec | | | |
| Scanner | oscillating mirror, 100 Hz sinus shape, torque motor 0.6 W | | | |
| Image columns | 10 | | | |
| Image rate | 100 frames/sec | | | |
| Number of image points | 520 | | | |
| Infrared-Optics | Si & Ge achromatic lens | | | |
| Focal length | 75 mm | | | |
| Aperture | F/1.5 | | | |
| Field of view | 1° x 3° (h x v) | | | |
| Image reproduction | optical display via scanner frontside | | | |
| Power supply | 4 batteries 1.5 V, 2 Ah | | | |
| Operating time | > 10 h without detector cooling<br>> 1 h with detector cooling | | | |
| Weight | | | | |
| AWO | approx. 0.75 kg | | | |
| electronics | approx. 0.6 kg with batteries | | | |
| Resolution | | | | |
| spatial | 1 mrad | | | |
| thermal | with averaging over | | | |
| | 1 | 8 | 16 | scans |
| eff. Scan frequency | 100 | 12.5 | 6.25 | Hz |
| ΔT (uncooled) | 9.9 | 3.5 | 2.5 | K |
| ΔT (cooled) | 1.4 | 0.5 | 0.3 | K |

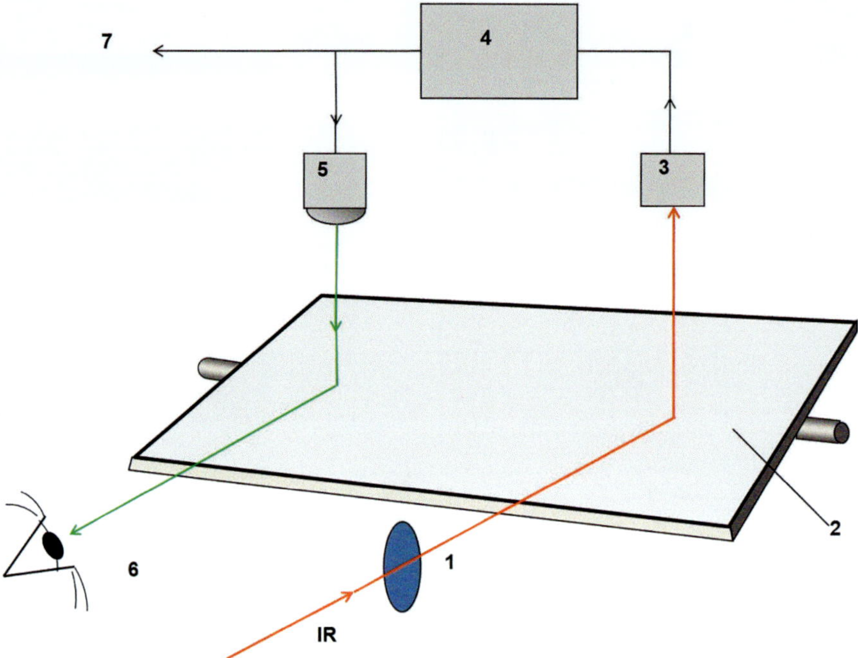


81 – ***Functional principle of the M.E.L. Thermal Pointer***
*The thermal radiation coming from the scene (IR) is imaged by a germanium lens (1) via the scan mirror (2) onto the IR detector (3). The electrical signals from the detector elements are led via an electronic unit (4) to a lamp (5) emitting modulated light that strikes the scan mirror again and reaches the observer (6). Furthermore, an acoustic alert (7) was possible too.*

In **1974** the M.E.L. Thermal Pointer was characterized by its manufacturer as follows:

"Role of the Thermal Pointer: The Thermal Pointer, which is under development for the British Army by M.E.L., indicates the presence of a potential target by detecting the thermal infra-red radiation naturally emitted by the target in the 8-13 micron wavelength range. It does not require a searchlight or starlight.

The Thermal pointer is thus able to distinguish targets which are slightly warmer than their surroundings and obvious examples are personnel and vehicles. The Thermal Pointer is used with a Passive Sight, a visible signal being injected into it. The signal indicates objects viewed in the Sight which are emitting an amount of thermal radiation above a threshold value chosen by the operator. Audio indication … can also be provided."[359]

In **1975,** CARL ZEISS-SONDEROPTIK offered the Oriotherm, an image intensifying device with integrated thermal pointer. This solution was much more compact and easier to use than the attachable AWO 10. Due to the limited space, the thermal scanning of the scene was performed vertically in one column only. To search a thermal target the device had to be panned horizontally.

As early as **1975,** schemes were afoot at CARL ZEISS-SONDEROPTIK to equip the contemporary PERI R 12 panoramic periscope with one of these thermal pointers. The possible applications of the concept were clearly demonstrated, but the project perhaps was not implemented.

82 – *Zeiss Oriotherm (left) and Orion 110*

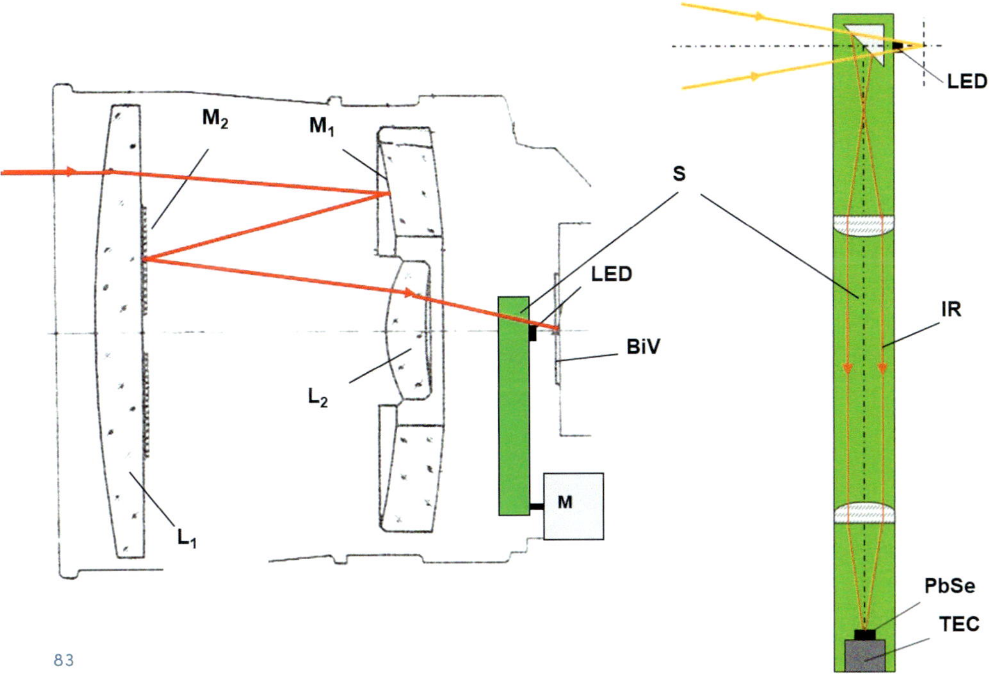


83

83 – *Optical principle of Oriotherm*

*The incoming thermal radiation is imaged by a so called catadioptric optic onto the entrance surface of an image intensifier (BiV). The optic comprises two mirrors (M1 & M2) and two lenses (L1 & L2) The lenses are made from calcium fluoride (CaF2) that has a high transmission in the near and mid wave infrared wavelength range. The image intensifier converts the near infrared radiation into a visible two-dimensional image. In front of the image intensifier, a slim swivel arm (S) is moved fast by an electro motor (M). The mid wave infrared portion of the radiation, which hits the aperture of the swivel arm, is focused by an IR imager (IR) onto the detector (PbSe) which can be cooled by a thermoelectric cooler (TEC). The detector signal is transformed into visible light by a small LED on the back side of the swivel arm. As the LED is close enough in front of the surface of the image intensifier a sufficiently sharp light spot is created without any additional optics as soon as the PbSe detector sees a warm object.*

*84 – Optical concept of the periscope Peri R 12 (Peri-Passiv)*

*85 – Night vision device NZG 200 of Eltro-Zeiss*

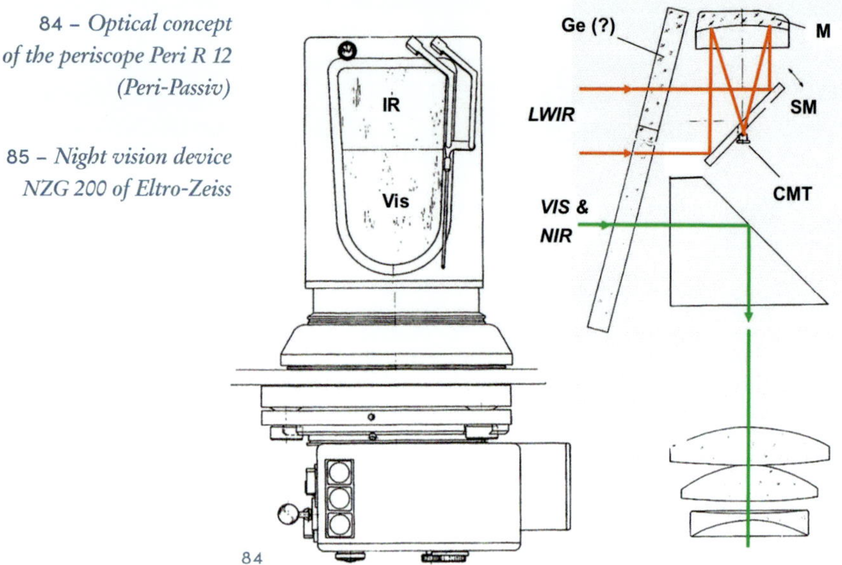


84

Mid of the 1970s, the German companies ELTRO in Heidelberg und CARL ZEISS-SONDEROPTIK in Oberkochen jointly developed a combined panoramic periscope that comprised visual direct view, image intensifier and thermal imager in one system.

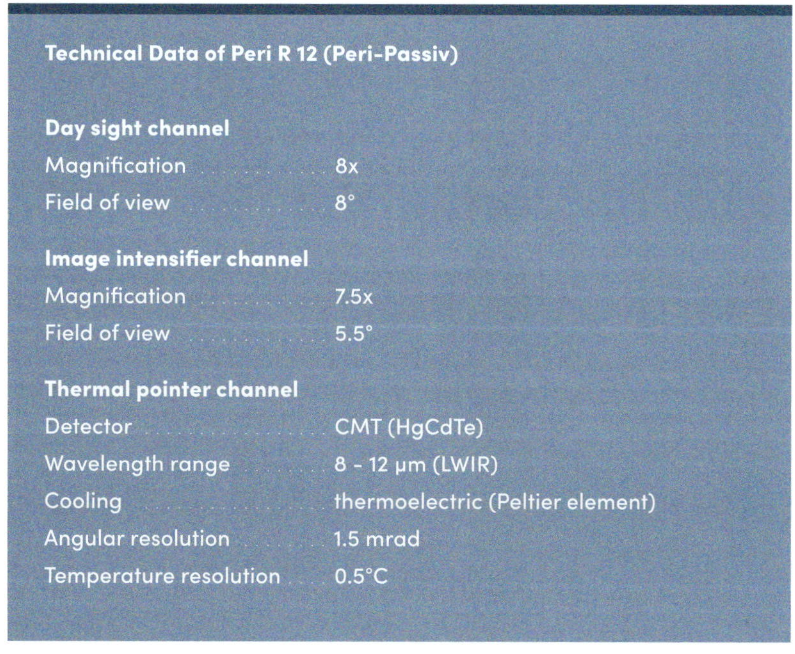

**Technical Data of Peri R 12 (Peri-Passiv)**

| | |
|---|---|
| **Day sight channel** | |
| Magnification | 8x |
| Field of view | 8° |
| **Image intensifier channel** | |
| Magnification | 7.5x |
| Field of view | 5.5° |
| **Thermal pointer channel** | |
| Detector | CMT (HgCdTe) |
| Wavelength range | 8 - 12 µm (LWIR) |
| Cooling | thermoelectric (Peltier element) |
| Angular resolution | 1.5 mrad |
| Temperature resolution | 0.5°C |

85

In **1976**, CARL ZEISS-SONDEROPTIK presented a special thermographic device named Ikotherm that was funded by the German Federal Ministry of Research and Technology BmFT. It was a radiometric thermal imager in the 8-12 µm wavelength region with high resolution, and with sophisticated signal processing and numerical evaluation compared to the existing devices. The intended fields of application were medicine, construction industry, car manufacturers, electronic industry, metallurgy, process technology, machine tools, environmental protection, biology, security technology etc.

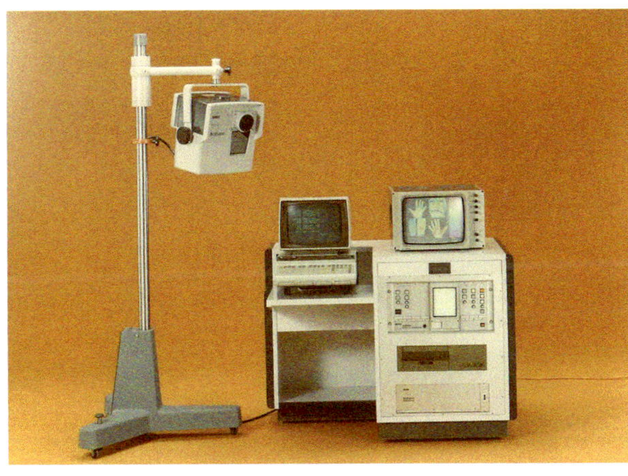

86 – *Thermographic camera Zeiss Ikotherm*

**Technical Data of NZG 200**

| **Periscope** | |
|---|---|
| Pointing range | n x 360° horizontal<br>-10° to +20° vertical |
| Equipment | 2-axis stabilisation<br>automatic BITE |
| Weight | 70 kg |
| **Night vision** | **Low Light Level TV** |
| Functional principle | Si-multidiode Vidicon with 1-stage pre-amplifier & automatic adaption to brightness up to 105 Lux |
| Mirror objective | 2.1/350 mm and 3.0/530 mm |
| Range | > 1000 m at 1 milli lux |
| Option | gated-viewing mode with pulsed light source |
| **Thermal pointer** | |
| Detector | Indium antimonide InSb |
| Wavelength range | 3 – 5 µm (MWIR) |
| Cooling | 77 K = -196°C<br>(Joule-Thomson cooler) |
| Number of lines | 80 |
| Image rate | 10 images/sec |
| Objective | Ge/Si achromatic lens |
| Aperture/focal length | 2.5/200 mm |
| Field of view | 3° x 5° |
| Angular resolution | 1 mrad |
| Temperature resolution | 0.6°C |

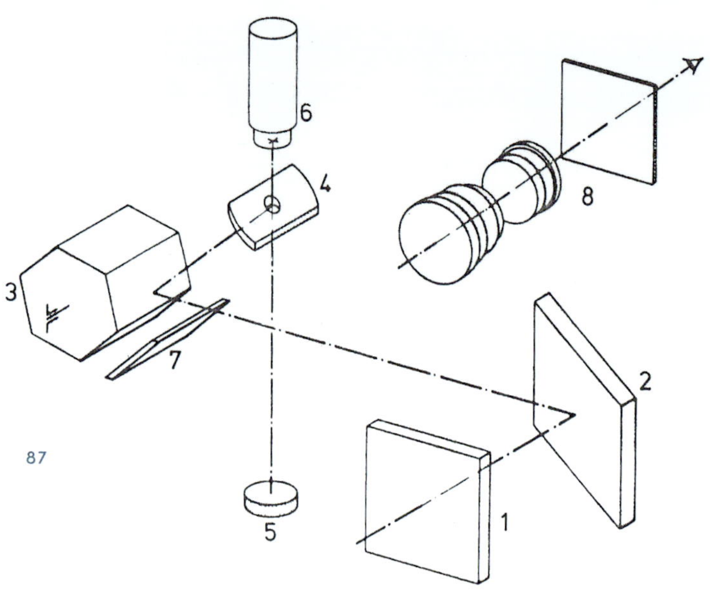


87

87 – *Optical principle of the Zeiss Ikotherm sensor head*

*The thermal radiation enters through a germanium window (1) into the sensor head. By means of an adjustable mirror (2) and a rotating polygon (3) the scenery is canned in horizontal and vertical direction, respectively. Via a folding mirror (4) and a parabolic mirror (5) with a focal length of 100 mm the radiation is imaged onto a single detector element (6). The reference surface (7) is used to calibrate the detector signals after each scan. Moreover, an optical view finder (8) is installed in the sensor head to make it easier to locate a desired detail in the scene.*

88 – *Data sheet Zeiss Ikotherm from 1983*

**Technical Data**

| | |
|---|---|
| Name | Zeiss Ikotherm |
| System description | universal high resolution thermal imager with radiometric measurement of absolute temperature |
| Scanning | in parallel ray path<br>horizontal: tilted mirror, can be fixed for scanning of single column<br>vertical: rotating polygon |
| Detector | HgCdTe or PbSnTe |
| Wavelength range | 7 ... 12 µm |
| Cooling | liquid nitrogen, consumption approx. 200 $cm^3$/ (2 to 4 h) |
| Temperature resolution* | 0.2 K at 30°C (NETD) |
| Temperature accuracy* | 0.3 K at 25 to 42°C<br>(at surrounding temperature 22±2°C) |
| Image construction time | approx. 1 sec |
| Columns per image | 300 (analog system)<br>256 (digital system |
| Image points / column | 300 analog / 256 digital |
| Field of view | 17.2° x 17.2° analog<br>14.8° x 14.8° digital |
| Weight | Sensor head: 19 kg<br>Operating & display unit: 26 kg |

* Temperature of black body ($\varepsilon$ = 1)

88

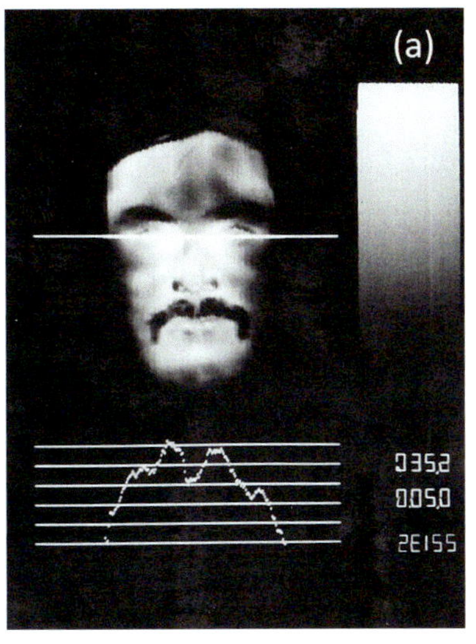


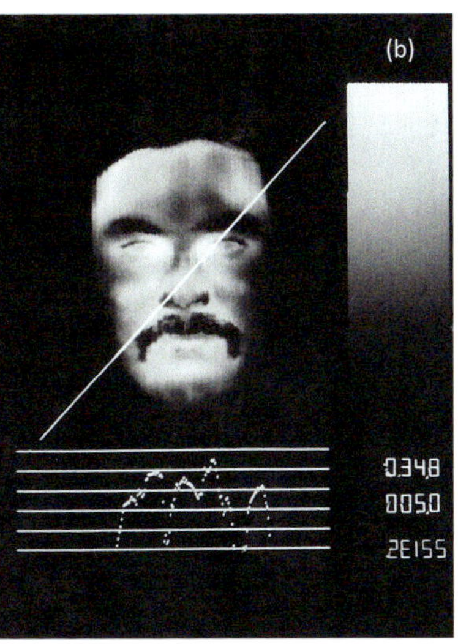


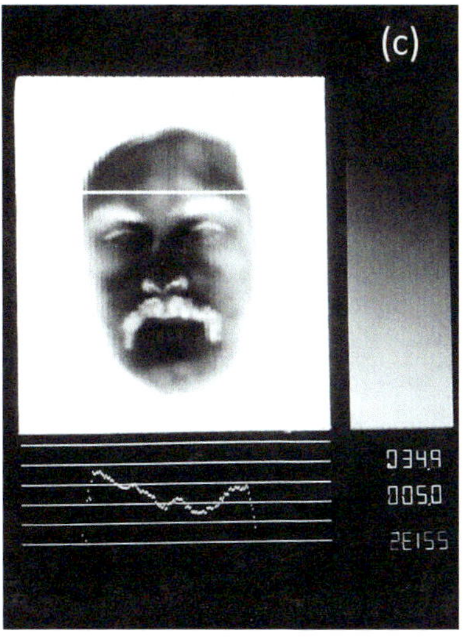


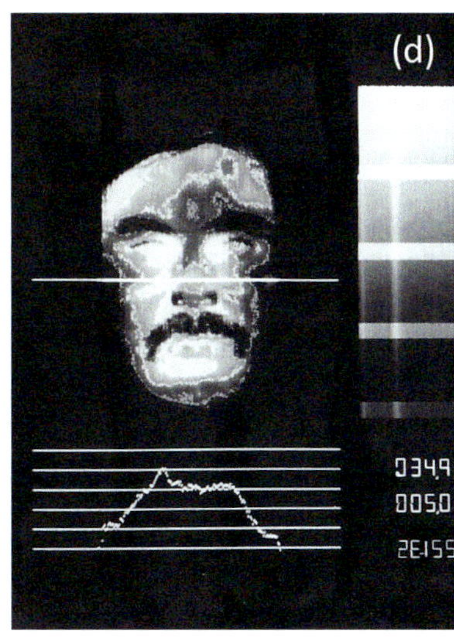


89

Besides the options of grey scale and false color display the Ikotherm provided the following functions to make the operator's life easier:

Maximum temperature
- automatic evaluation of the maximum scene temperature or
- manual adjustment of the maximum value

Temperature profile line
- displaceable in height; +-45° tiltable
- display of 6 lines / adjusted temperature range

Inverted grey scale (Polarity positive/negative)

Display of isotherms
- Marking of areas of identical temperature by beam modulation (automatic in 4 steps, or continuously adjustable)

In **1977**, the Ikotherm was used to support the military thermal imaging development. In a measuring campaign at the military barracks in Ellwangen the typical thermal signatures of several armored vehicles were recorded.

Starting in **1982**, various testing of the "Plataforma Solar de Almería" was conducted. Expressly the Small Solar Power System SSPS (92 heliostats á 40m², 43m-tower

89 – ***Functions of the Zeiss Ikotherm***

*Monochrome images with temperature profile (a) and (b), inverted polarity (c) and display of isothermal contour (d).*

90 – ***Thermal image of a tank obtained with Zeiss Ikotherm***

90

91

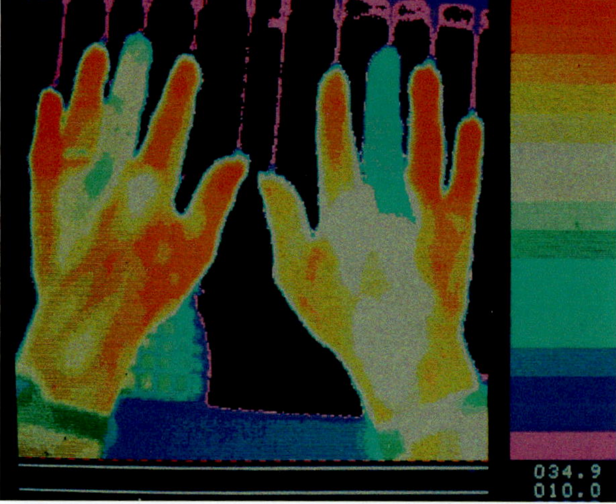


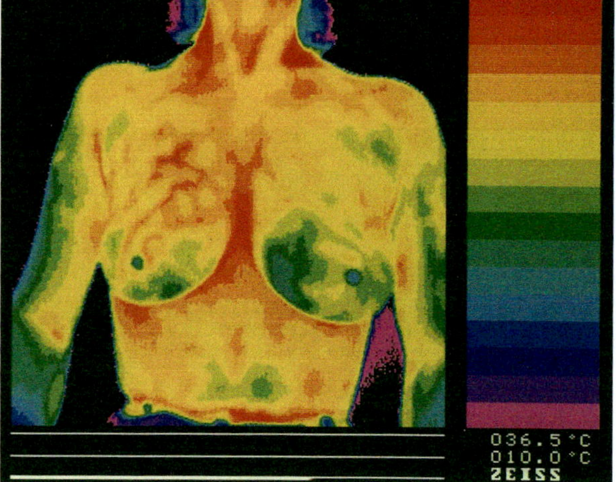


92

91 – *Application of Zeiss Ikotherm at the Plataforma Solar de Almeria*

92 – *Application of Zeiss Ikotherm in medical thermography*

with 2 platforms; power 2,7MWth), coordinated by the German aero and space organization DLR, was taken in operation. In this process, an Ikotherm was employed that was equipped with a InSb detector for the medium infrared region[360] to adapted to the high temperatures occurring at the solar towers.

Until the end of the 1980s, CARL ZEISS-SONDER-OPTIK kept on trying to establish the Ikotherm in medical thermography - alas all efforts were of no avail.

6

# COMMON MODULE DEVICES UNTIL 1995: SUCCESS BY NUMBERS

"FLIR allowed them to see through the darkness and own the night."

*Texas Instruments Incorporated (2007)*

"Carl Zeiss is sole licensee for the sale of FLIR Common Modules of TI on the territory of the BRD."

*Contract between Carl Zeiss and Texas Instruments (TI) (1976)*[362]

**Between 1964-72** more than 385 FLIR systems were produced by TEXAS INSTRUMENTS (USA) – an impressive success. Unfortunately, these devices had been produced in 55 (!) different configurations resulting in only 7 devices per model on average. This in turn meant a high development effort per device and production rates of each configuration being too low to achieve any economies of scale. The economic consequences followed on the problem's heels as we shall see in this chapter.

## 6.1 Emergence of the Common Modules

In **September 1970**, the ROYAL RADAR ESTABLISHMENT (RRE) advanced a proposal of a thermal imager as response to a specification of the British Army which had put a low-light level system out for tender. It was built by BARR & STROUD with a CMT detector which was cooled to 77 K (-196°C) by liquid nitrogen. Right away "the thermal imager easily outperformed the low-light entries" of the competition.[363] For the further application it was baptized to the code-name Shortie.

In **March 1972,** TEXAS INSTRUMENTS (TI) received a letter from the director of ADVANCED RESEARCH PROJECTS AGENCY (ARPA) explaining that the DEPARTMENT OF DEFENSE (DoD) could no longer afford the FLIR system. He also asked for the reasons of the high unit costs of "several hundred thousand dollars". TI answered by underscoring the enormous development costs but the DoD promptly rejected the explanation.

In **April,** a first specification of a "common-module FLIR concept" was subsequently worked out by TI. It was based on the premise that certain functions of an infrared sensor were not sensitive to the mission application and could be made universal from system to system. Such commonality should make custom design and development unnecessary and volume production possible. In consequence cost should greatly be reduced. Until end of the year much effort was spent by TI in convincing the US government of the economic viability of the concept. Some competitors thought – of course – that it was only a "marketing ploy" and also some customers were doubtful, but TI pressed on.

In **November,** TI already began to build the first Common Module (CM) prototypes.

**In 1972,** the thermal imager Shortie was deployed successfully by the British Army when the Troubles in Northern Ireland were at its height. The experience with this new technology compared to low-light systems was that "thermal imagers were much better at detecting targets, but the latter were better at recognition".

In the period following, Shortie was deployed successfully in further applications causing the same problems as in the United States: "the design effort required to produce all these models was becoming too much, and their cost was becoming excessive".[364] "The solution to the first problem was to copy the American idea of breaking the imager system into a series of subunits – 'common modules' – which could be combined in various ways." This led to first plans of a system of "Thermal Imaging Common Modules" (TICM).

In **1973,** all three US-military service chiefs declared support for the Common Module concept and TI began to win development programs using Common

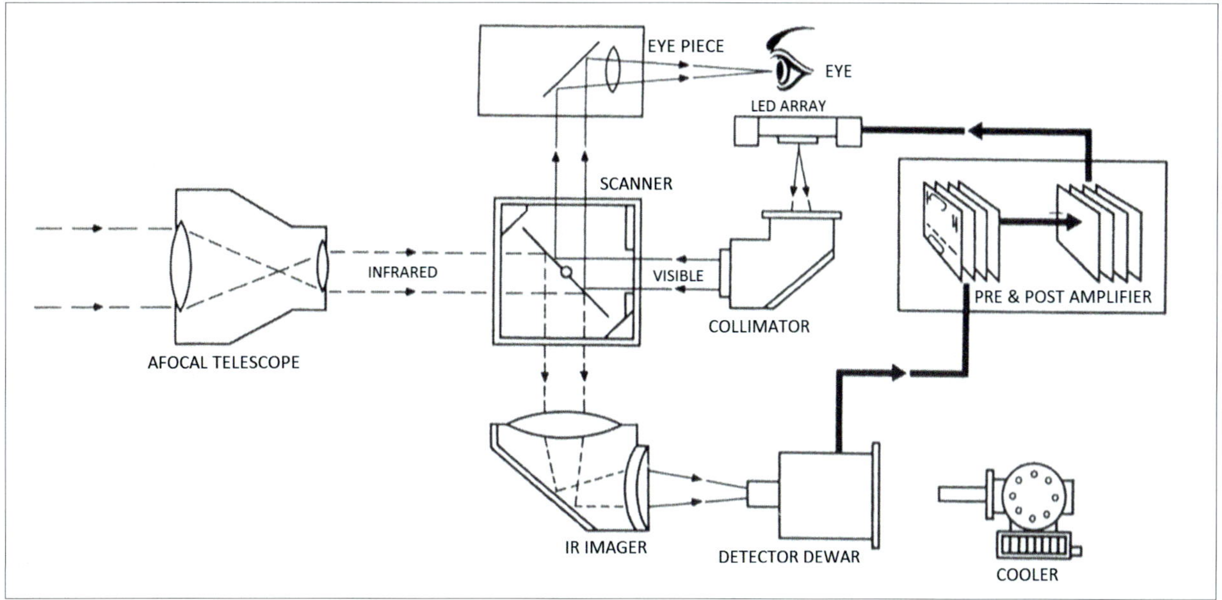


93 – *Functional principle of the US Common Modules*

*The thermal radiation passes through the afocal (a telescope to adapt the field of view) onto the front side of the scan mirror and is then deviated into the imager which produces a sharp image on the detector surface. The detector elements generate electrical signals which are proportional to the intensity of the incoming thermal radiation. The signal of each detector element is amplified and fed to the assigned light emitting diode of the LED array. The Light emitted by the LED again is proportional to the intensity of the thermal radiation. It is collimated by a collimator lens and reflected by the back side of the scan mirror into the ocular assembly which comprises an image intensifier and the eye piece. Eventually the observer can view the intensified image with the eye piece.*

Module designs. Concurrently TI was still involved in competitive developments.

The pressure also grew to develop second-sources for the modules even though TI had developed much of the Common Module technology on its own.

In **1973**, experts of the NIGHT VISION LABORATORY (NVL), Ft. Belvoir, USA presented a draft design of a "Universal Viewer for Far Infrared" which was widely equivalent to the Common Modules of TI.

In **1974**, the Department of Defense officially presented the Common Module concept. This was an important move because there were particularly promising developments in thermal imaging also outside the United States.

In **1974**, C. Thomas 'Tom' ELLIOTT developed the SPRITE technology for CMT detectors in Great Britain. SPRITE is an acronym for "Signal **PR**ocessing **I**n **T**he **E**lement". By means of a scanner the optical image of scene is moved across the large area detector with the same velocity as the charge carriers are drifting inside the detector material. Therewith the signal of an image point can be integrated over a longer time without losing spatial resolution. Compared to a conventional detector this results in a higher signal with the same resolution.

In this time, the British TICM thermal imagers were equipped with conventional CMT matrix detector 12 x 4 = 48 detector elements.

The new SPRITE technology, which – in honor to the inventor – was also called "Tom Elliott's Device" or in short TED, was to become the base of the new TICM II, the 2nd generation of the British Common Modules.

In **1974**, the WBG 1/100 thermal imager of CARL ZEISS-SONDEROPTIK went into trials. The technical concept was comparable to the US Common Modules. The imager used a linear indium antimonide (InSb) detector array with 100 detector elements.

In **1975** however, the German Ministry of Defense (BMVg) decided to use the US Common Modules in future development programs. In the time following, analog decisions were also made in Taiwan, South Korea and Denmark.

In **1976**, the Common Modules became official standard in the USA and TEXAS INSTRUMENTS began with the series production of the modules.

In the same year Carl Zeiss became sole licensee for the sale of FLIR Common Modules of TI on the territory of the BRD and obviously began to integrate the modules in the own development programs without delay.

The study "Thermal imagers & sights for Leopard 1, Marder and Luchs" presented in November already comprised a proposal for a "Thermal Imager with 120 Detectors and Common-Modules".

Also in **1976**, the DIRECTION GÉNÉRALE DE L'ARMEMENT (DGA)[365] in France perhaps animated by the American plans officially launched a national development of a modular system for thermal imaging that became known as "Système Modulaire Thermique" SMT. Already in the years before considerations had been conducted and experimental devices had been built by the companies TRT-PHILIPS (thermal imagers Luther and Thermidor) and SAT (thermal imager Chamois and line scanners Cyclope & super-Cyclope). With the guideline "20 kg / 25 l / 140 W" for weight, volume and electrical power consumption SAT und TRT-PHILIPS developed in cooperation the 1st generation of the SMT with the following properties:[366]

- Mercury-Cadmium-Telluride/CMT detector with Stirling cooler
- Spectral range: LWIR 8 – 12 µm, "bande spectrale 3"
- Detector elements: 11 x 4 → parallel scanning of 11 lines & summation of the signals of 4 elements
- Image lines: 500
- Image format: 4/3
- Frame rate: 25 Hz

Based on this module set (analog to the US Common Modules) many important French weapon systems were equipped with thermal imagers until the 1990s. Examples are

- SMT- Castor for main battle tank AMX30-B2
- Athos for main battle tank Leclerc
- Victor in a visor for helicopters
- Chlio for a stabilized platform for helicopters
- PDLCT in a stabilized reconnaissance pod for fighter aircraft
- Tango for a stabilized platform for Breguet-Atlantic 2

Many of these systems have been in service for a long time, some are working still today. A good example is the Breguet-Atlantic 2 with the stabilized reconnaissance platform which is on duty in marine research and ocean surveillance.

In **1977**, CARL ZEISS-SONDEROPTIK completed the WBG 1/200-10 "thermal imager of the FLIR type", which was developed under commission of the "Bundesamtes für Wehrtechnik und Beschaffung" (BWB), the German office for military technology. The technical concept was not very different from the US Common Modules. Particularly, both concepts relied on the reproduction of the optical image by reflection on the back side of the scan mirror, which was presented first by the German company E. LEYBOLD in 1944 (see also

Chap. 3.3). Unlike the US-Common Modules the WBG 1/200 did not use the particion of the beam path into afocal-scanner-imager but used a IR objective that also comprised the scanner. This layout was technically more complex but made possible a very compact thermal imager as is shown in the attached sketches and pictures. Because the German Ministry of Defense had decided earlier to use the US Common Modules, the WBG 1/200 was already superseded at the time of its presentation.

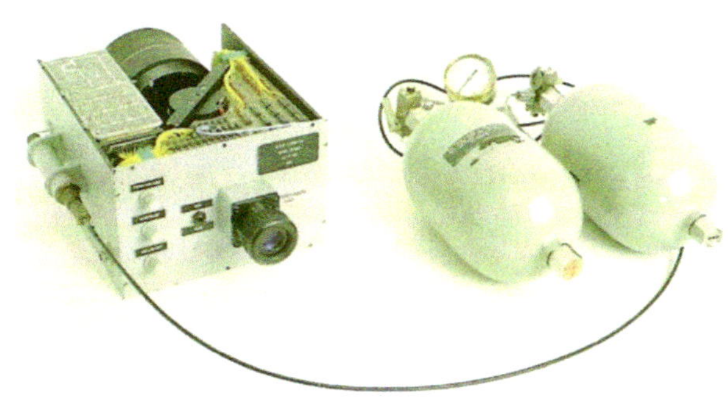

95

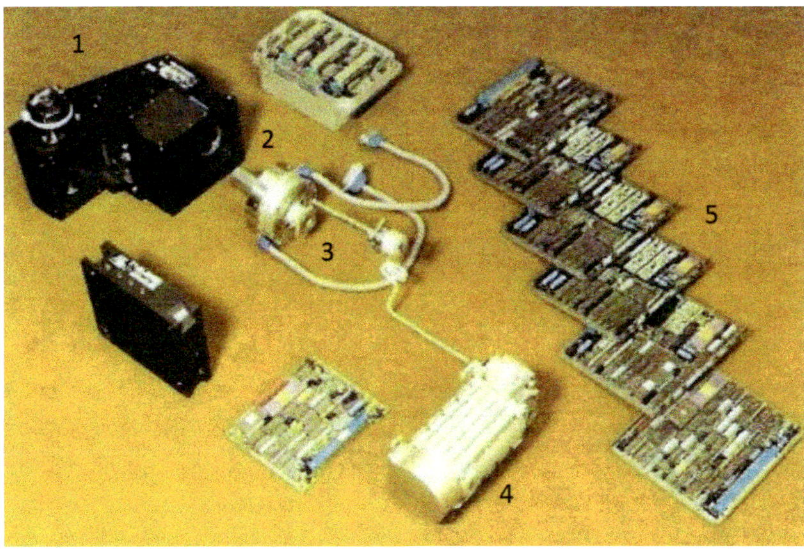


94

94 – *Modules of the French Système Modulaire Thermique (SMT)*
*Shown are scanner (1), infrared imager (2), detector (3), cooler (4) and the electronic amplifier boards (5).*

95 – *Zeiss WBG 1/200-10 with gas bottles for the Joule-Thomson cooler*
*The functional principle of a Joule-Thomson cooler is based on the effect that a gas under high pressure cools down when it is expanded. The achievable temperature decrease depends on the intramolecular forces. The cooler shown here uses nitrogen ($N_2$) compressed to 400 bar. The gas flows out of the bottles into a nozzle, cools down the surroundings of the nozzle and is then released into the environment. It is therefore important that the gas is inflammable and not hazardous to health nor to the environment. All requirements are perfectly fulfilled by nitrogen which makes up 80 % of the ambient air anyway.*

**Technical Data of WBG 1/200-10**

| | | | |
|---|---|---|---|
| Detector | PbSnTe (LTT=Lead-Tin-Telluride) | Field of view | 5.5° x 2.25° |
| Wavelength range | 7 – 12 µm LWIR | Image reproduction | optical display via scanner backside, image intensifier and eye piece |
| Cooling | 77 K = -196°C<br>Joule-Thomson cooler | Weight | 4.1 kg |
| Detector elements | 100 x 1 | | |
| Scanner | oscillating mirror with 2:1 interlace | Resolution | |
| Image lines | 200 | spatial | 0.2 mrad |
| Frame rate | 25 images / sec | thermal | 0.2 K |
| Number of image points | 80 000 | | |
| Infrared optics | germanium lens with 3 elements | Ranges | |
| Focal length | 250 mm | detection | max 1500 m |
| Aperture | F/1.9 | identification | max 600 m |

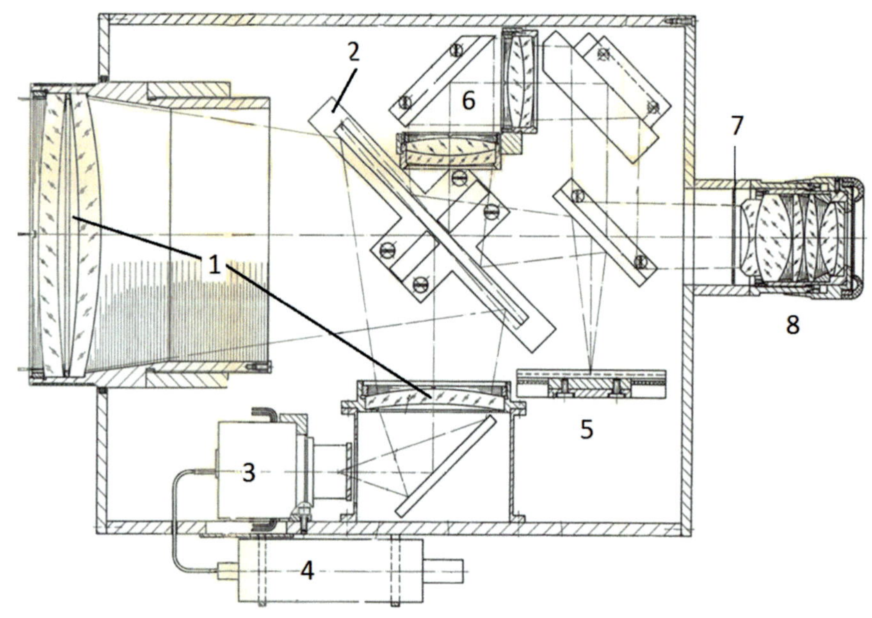


96 – *Drawing of the Zeiss WBG 1/200-10*

*The thermal radiation coming in from the left side finds its way through a 3-lens germanium objective (1) with 150 mm focal length and a scan mirror (2) onto an IR detector array (3) with 100 elements. The cooling of the detector is done by a Joule-Thomson cooler (4). The signals of the detector elements are electrically amplified and fed into a LED array (5) where they are converted into light signals of a brightness which is proportional to the intensity of the incoming thermal radiation. The light signals are imaged by a glass optic (6) into an intermediate image plane (7) that can be viewed by an observer through the eye piece (8).*

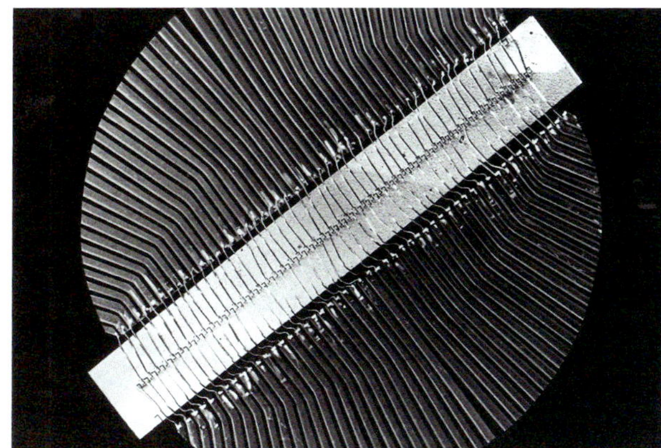

97 – *LTT detector with 100 elements of the WBG 1/200-10*

*The abbreviation LTT describes the mixed crystal lead-tin-telluride (PbSnTe) used as detector material. The picture shows clearly that each detector element is separately bonded. This was the typical design of that time. (Manufacturer: Plessey)*

In 1977, a detailed record of TEXAS INSTRUMENTS showed that a reasonable number of programs with almost 100 FLIR systems were already initiated. The main customer was the US Army - with respect to the number of programs and to the quantities as well.

By this time, a digital video recording system was developed for the LANTIRN program which was capable to convert the signals of the IR detector elements directly into TV compatible signals - the first "Digital Scan Converter" DSC.

| COMMON MODULE PROGRAMS AT TEXAS INSTRUMENTS | QUANTITY | CUSTOMER |
|---|---|---|
| Chaparral Target Acquisition Sensor | 1 | US Army |
| Airborne Laser Locator Designator | 6 | US Army |
| RPV | 3 | US Army |
| Multipurpose Observation Device | 2 | US Army |
| Night Observation Device, Long Range | 2 | US Army |
| Dragon | 4 | US Army |
| TOW Night Sight – Phase I & II ED | 12 + 28 | US Army |
| Tank Thermal Sight – AD & ED | 2 + 16 | US Army |
| MK-68 Fire Control Sensor | 4 | US Army |
| A-7 TRAM | 6 | US Army |
| Air Force FLIR Sensor | 4 | US Air Force |
| P-3 FLIR Sensor | 10 | US Army |
| ISTAR (LOHTADS) | 1 | US Army |
| Hydrofoil FLIR | 1 | US Army |
| V-10 FLIR/LRD | 3 | US Army |

On **23 March 1977**, CARL ZEISS-SONDEROPTIK presented a new proposal for a "Thermal Imager with 60 Detectors and Common-Modules" based on the study from 1976 which had been amended several times already. A development contract was signed which also included the delivery of two devices for field testing. In competition, two more German vendors also were assigned.

In **1979**, TEXAS INSTRUMENTS and CARL ZEISS-SONDEROPTIK agreed upon a cooperation for the German programs, especially for the thermal imagers needed to fulfill the "German Tank Thermal Sight" (GTTS) programs. Besides of some of the Common Modules, CARL ZEISS-SONDEROPTIK would manufacture the afocal telescope with 2 fields of view and be in charge for assembly, initial startup and qualification of the completed thermal imager.

Therewith TEXAS INSTRUMENTS had also fulfilled the requirement of a "second source" and in similar fashion installed a factory for Common Module detectors and coolers at AEG TELEFUNKEN in Heilbronn.

On **5 March**, the German Ministry of Defense (Bundesministerium für Verteidigung BMVg) decided to use the technology of optical image reproduction, where the light emitted by a LED array is reflected by the back side of the scanner into the eye piece unit. A corresponding order was placed at CARL ZEISS-SONDERTECHNIK.

In **August**, a second order followed of a series planned for the new main battle tank Leopard 2.

In **1980**, the first experimental thermal imager with a single SPRITE detector was put into operation. At a next step, an 8-element SPRITE detector followed having a performance almost comparable to 48-element matrix detector which was in use so far. Later, the final stage of development was the 16-element SPRITE detector which became the base of the British TICM II.

In **1980**, TEXAS INSTRUMENTS settled further "second source" arrangements in the United States and Germany. In the following years a remarkable rank of prestigious manufacturers qualified for producing the Common Modules. As the list shows clearly, the Common Modules had virtually become a pure American-German affair and sometimes even were addressed as "US-German Common Modules".

Between **1980 and 1995** at CARL ZEISS-SONDEROPTIK development work was done on a very exotic version of a Common Module thermal imager, at least compared with the rest of the pack. The optronic fire control sight OFRIS ("Optronische Feuerleitrichtsäule") was designed for upgrading the German anti-aircraft tank Gepard and should provide night-vision capability by means of a thermal imager and range finding capability using an eye-safe laser at 1.5 µm wavelength. The available space was narrow as usual and furthermore very high elevation angles up to 70° were specified. Therefore, in the conceptual study of 1981 some specific solutions had to be presented to meet the requirements. Visual and infrared beam path were nested and in part also the beam path of the laser rangefinder was integrated into the same optic. OFRIS was a real multispectral instrument. However, the demanding concept led to a very complex optical layout. The manufacturing of the components was very complicated, and the adjustment of the beam path was next to impossible. Shortly after 1995, all work in the program was stopped after approximately 10 prototypes had been built.

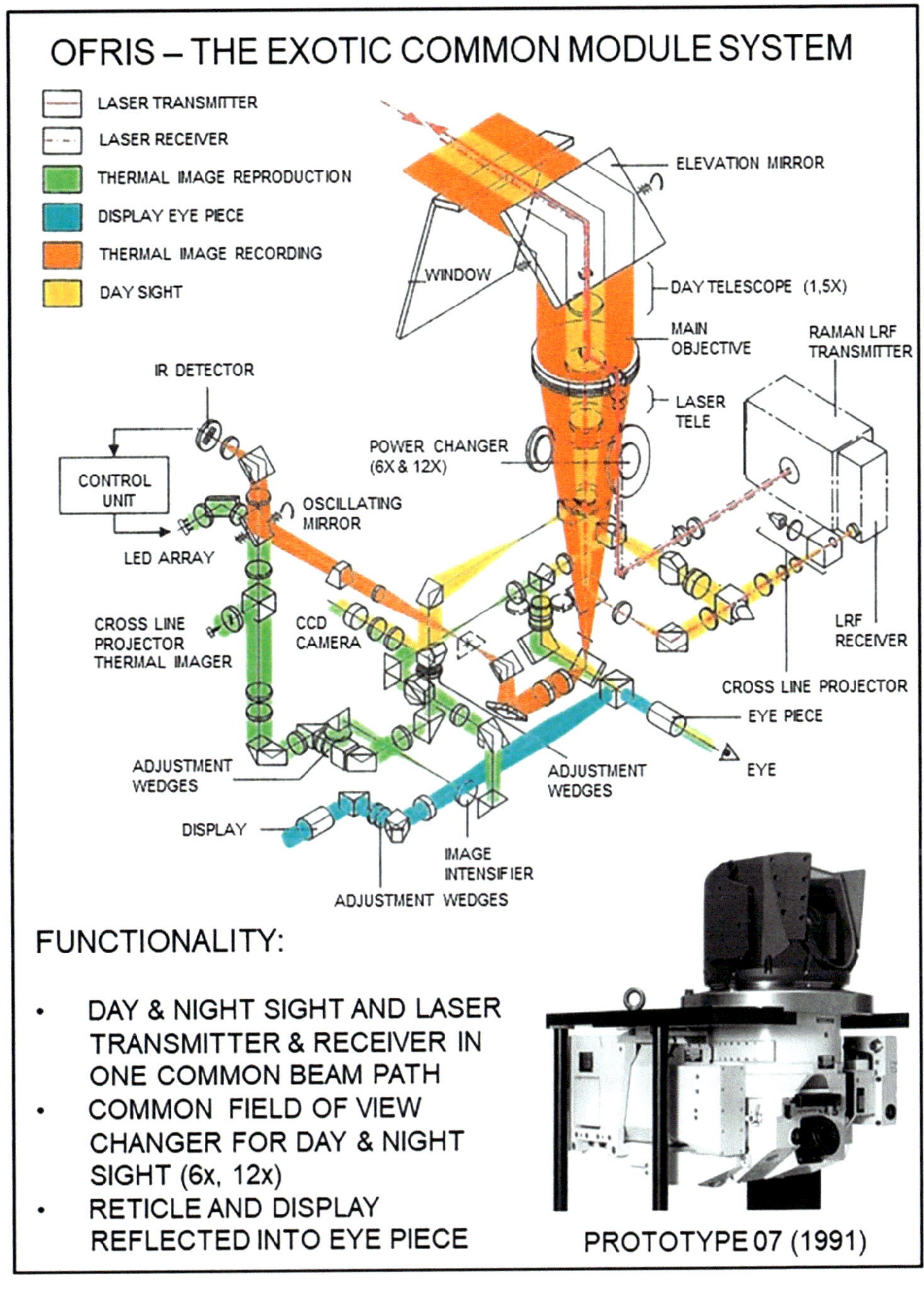
OFRIS – THE EXOTIC COMMON MODULE SYSTEM
LASER TRANSMITTER
LASER RECEIVER
THERMAL IMAGE REPRODUCTION
DISPLAY EYE PIECE
THERMAL IMAGE RECORDING
DAY SIGHT
ELEVATION MIRROR
WINDOW
DAY TELESCOPE (1,5X)
MAIN OBJECTIVE
RAMAN LRF TRANSMITTER
IR DETECTOR
LASER TELE
POWER CHANGER (6X & 12X)
CONTROL UNIT
OSCILLATING MIRROR
LED ARRAY
CROSS LINE PROJECTOR THERMAL IMAGER
CCD CAMERA
LRF RECEIVER
CROSS LINE PROJECTOR
EYE PIECE
EYE
ADJUSTMENT WEDGES
ADJUSTMENT WEDGES
DISPLAY
IMAGE INTENSIFIER
ADJUSTMENT WEDGES
FUNCTIONALITY:
• DAY & NIGHT SIGHT AND LASER TRANSMITTER & RECEIVER IN ONE COMMON BEAM PATH
• COMMON FIELD OF VIEW CHANGER FOR DAY & NIGHT SIGHT (6x, 12x)
• RETICLE AND DISPLAY REFLECTED INTO EYE PIECE
PROTOTYPE 07 (1991)

| COMMON MODULE | QUALIFIED MANUFACTURER (1990) | COUNTRY |
|---|---|---|
| Detector | Texas Instruments<br>AEG-Telefunken, Heilbronn<br>Honeywell<br>Santa Barbara Corp. | USA<br>D<br>USA<br>USA |
| Cooler | Texas Instruments<br>AEG-Telefunken, Heilbronn<br>Hughes-Torrence<br>CTI | USA<br>D<br>USA<br>USA |
| Visual collimator & IR-imager | Texas Instruments<br>Carl Zeiss, Oberkochen<br>OEC<br>Magna Vox<br>Martin Marietta | USA<br>D<br>USA<br>USA<br>USA |
| Scanner | Texas Instruments<br>Eltro, Heidelberg<br>Aero Flex | USA<br>D<br>USA |
| Printed circuit boards | Texas Instruments<br>Magna Vox<br>Hughes Aircraft<br>Martin Marietta | USA<br>USA<br>USA<br>USA |
| Preamplifier (only) | AEG-Telefunken, Heilbronn | D |
| LED | Texas Instruments<br>AEG-Telefunken, Heilbronn<br>Spectronics | USA<br>D<br>USA |

| MODULE | MIL - STANDARD |
|---|---|
| Scanner | MIL-S-49166 A |
| IR-Imager | MIL-I-49170 A |
| Detector | MIL-D-49172 |
| Cooler | MIL-C-49175 A |
| LED-Array | MIL-L-49169 A |
| Visual Collimator | MIL-C-49174 A |
| Preamplifier | MIL-P-49163 A |
| Biasregulator | MIL-R-49176 A |
| Postamplifier | MIL-P-49164 A |
| Auxiliary electronic | MIL-A-49167 A |
| Scan-interlace drive | MIL-S-49177 A |

In **April 1981,** CARL ZEISS-SONDEROPTIK delivered the first series thermal imagers for the battle tank Leopard 2 A1. This was the begin of installing thermal imagers into the German tanks.

In **1982,** the Common Module specifications of the NIGHT VISION AND ELECTRO-OPTICS LABORATORY (NVEOL) became MIL standards as shown in the table.

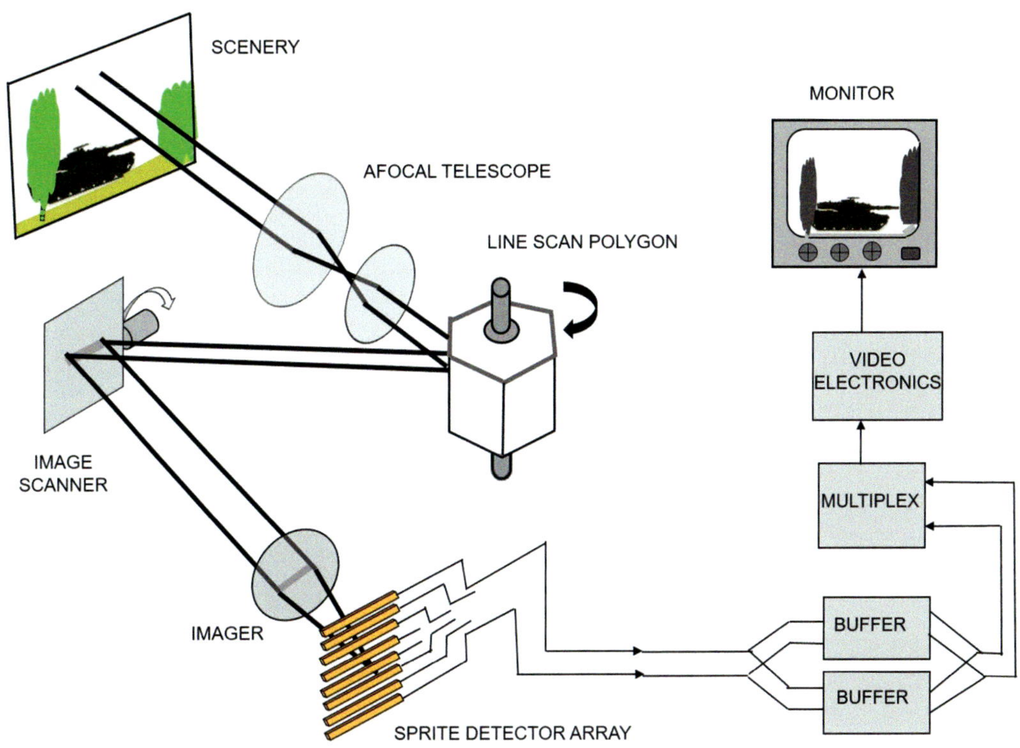


98 – *Functional principle of the Thermal Imaging Common Modules (TICM)*

*The British Thermal Imaging Common Modules TICM use a combination of sequential and parallel scanning. The thermal radiation coming from the scenery passes an afocal telescope and reaches a polygon scanner which moves the optical image of the scene horizontally over an 8-element detector. Thereby 8 image lines are scanned in parallel. The subsequent tilting mirror shifts the optical image stepwise in vertical direction and enables the scanning of a sufficient number of lines for a two-dimensional image. The signals of the detector elements are combined by means of a buffer memory and a multiplexer to a standard video image suitable to be displayed on a customary monitor.*

By **1984**, in cooperation of TEXAS INSTRUMENTS and CARL ZEISS-SONDERTECHNIK already more than 2300 systems had been delivered for the German programs. TI alone had delivered even more than 13,000 FLIR systems for its domestic market.

Around **1984**, the series production of the British TICM thermal imagers was started at PILKINGTON OPTRONICS. The necessary SPRITE detectors came from BAE SYSTEMS. At first, the imagers were divided into 3 classes, what was reduced to two later:

- Class I: hand-held models with low weight for direct vision
- Class II: More sensitive thermal imagers of medium performance with electronic image reproduction on a monitor
- Class III: High-performance thermal imagers for long range reconnaissance and night flight

The Class I devices, having a weight of approx. 3 kg, were significantly lighter than the thermal imagers applied in the Falklands War weighing 12 kg. Until mid of the 1990s, approximately 3000 thermal imagers were built mostly by THORN-EMI, and approximately 1000 devices thereof were exported.

The Class II devices were installed in battle tanks, ships and aircraft. Examples are the ground-air missile Rapier, the Tornado fighter aircraft and the AV-8B Harrier II of the US Marine Corps. In total, more than 3500 systems were sold world-wide of which GEC MARCONI had the greatest part. During the First Gulf War in 1991, the impact of these thermal imagers was “in the words of the troops, to turn night into day”.[367] Because the Class II devices exceeded all expectations they could overtake also the tasks of the Class III devices and the Cass III development thus could be given up.

In **1987**, the number of delivered US-German Common Modules had amounted to more than 25,000 towards the end of the year, the greatest part of more than 20,000 came from TEXAS INSTRUMENTS alone. In a contemporary diagram, the series ramp-up is impressively depicted.

Until **1988**, the US based company KOLLSMAN could deliver another 9,500 Common Module thermal imagers.

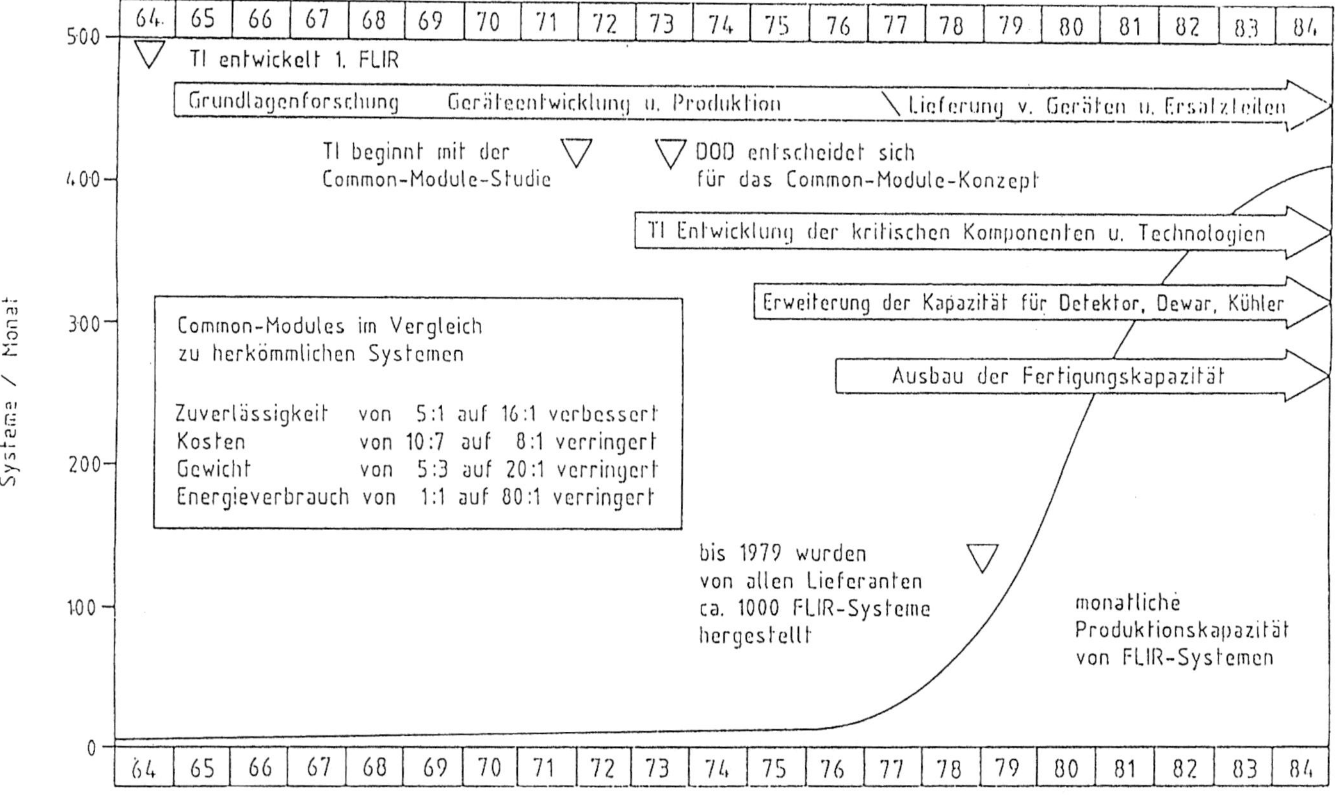


**99 – *Ramp-up of Common Module series production***

*The increase of production numbers, some milestones of the development and the achieved improvement over the previous systems are given in the fields of reliability, cost, weight and power consumption.*

Until **1998,** TEXAS INSTRUMENTS had produced more than 30,000 FLIR systems according to own records. The production rates of the historic FLIR systems had been at some ten units per year and they were around a few thousand per year in the age of the Common Module FLIR systems.

On **22 October 1998,** PILKINGTON OPTRONICS exposed in a memorandum for the British Parliament that until that moment in time 9,300 thermal imagers of the British TICM of the 1st generation were sold, and that the greater part thereof had been exported.[368]

## 6.2 Technology of US-German Common Modules

The functional design followed the concept of optical image reproduction presented in 1944 by the German company E. LEYBOLD: The electrical signals of the IR detector elements were used to control the brightness of a visible light source (e.g. a light emitting diode) the light of which was collimated and reflected into the direction of the observer by the back side of the scan mirror. By means of the reflection on the scanner backside the visible image was automatically synchronized with the infrared image.

The subassemblies of the basic thermal imager (scan mirror, imager, detector array, amplifier boards, LED array and visual collimator) were standardized and designed for a fixed 20° horizontal field of view. The adaption to the needs of a military application was done by an afocal telescope. The trade-of between a good overview (→ wide field of view) and high spatial resolution (→ narrow field of view) is mostly solved by using two or more magnifications to generate the anticipated fields of view.

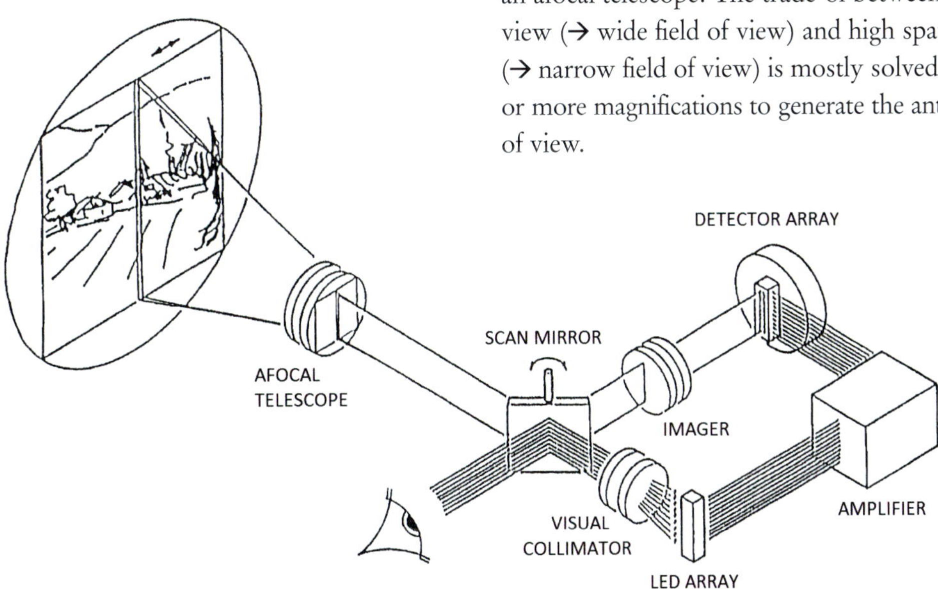


100 – *Functional principle of a Common Module thermal imager*

*The thermal radiation passes through the telescope into the basic unit and is directed by the scan mirror to the IR imager which images the radiation onto the surface of a detector array. The detector elements convert the impinging radiation into an electrical signal that is amplified and fed into the assigned diodes of a LED array. The red light emitted by the LED array is collimated by a collimator and reflected by the back side of scan mirror in the direction towards the observer. Detector array and LED array can only capture and reproduce a single column of the image at a time. By means of the rotation of the scan mirror around the vertical axis the image columns are captured sequentially one after the other.*

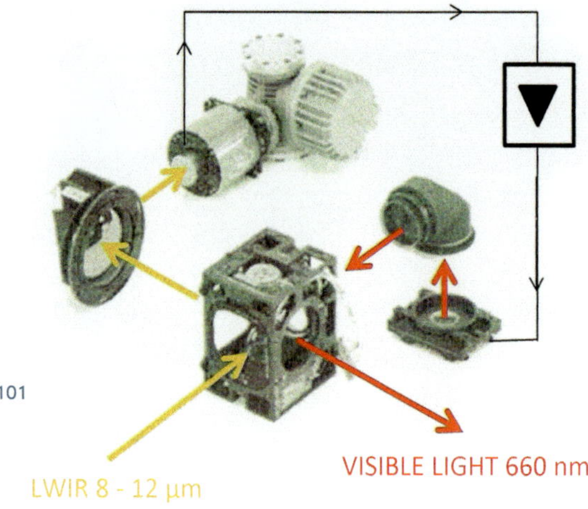


101

101 – *Interaction of the Common Modules in the basic thermal imager*

*The thermal radiation (yellow arrows) reaches the cooled detector via scan mirror and imager. The detector signals are amplified and given to a LED array the red light of which (red arrows) reaches the observer via collimator and the back side of the scan mirror.*

102 – *Common Module detector geometry and picture of a detector array*

*The picture shows the detector array mounted on a vacuum vessel made from glass. The bonding wires can be seen as faint lines on the long sides of the array. Each detector element is bonded to two connecting wires.*

In the following, each of the Common Modules is discussed in detail together with a list of the significant technical data.

## Common Module Detector Array

The detector used Mercury-Cadmium-Telluride (CMT) as a detector material sensitive to thermal radiation. Even today this material is the first choice for the longwave infrared wavelength range (LWIR) from 8 to 12 µm.

CMT is a semiconductor alloy with a band gap that can be adjusted by the mixing ratio of the constituents. For the LWIR range the band gap is adjusted to the photon energy of the 8-12 µm thermal radiation which is roughly around 0.1 eV. When photons with this energy hit the detector, electrons are lifted from the valence band up to the conduction band so that an electrical current can flow. The amplified current is the measuring signal that is proportional to the intensity of the thermal radiation. As the incoming radiation changes the electrical conductivity of the detector, this type of detector is referred to as "photo-conductive" (pc).

To prevent that unwanted thermal radiation from the surroundings could reach the detector surface,

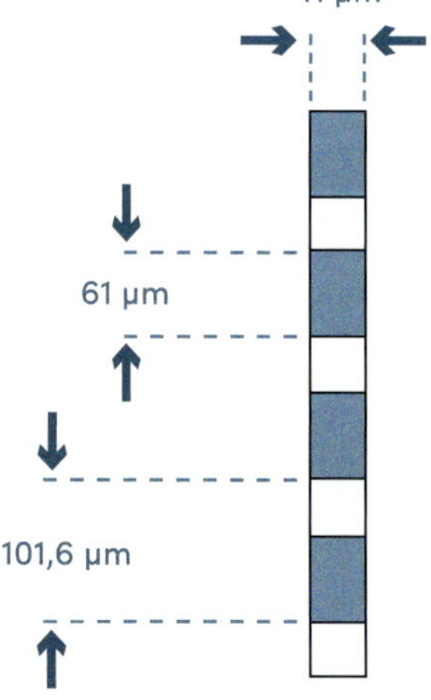

102

small shields were fixed sideward and between the detector elements (like blinders for a horse). This was also the reason for relative large gaps of approximately 40 µm between the detector elements.

The detector array was mounted on a vacuum vessel made from glass, known as Dewar[369], to achieve the best thermal isolation possible. In case of a bad isolation, a bigger cooler with higher power consumption would have been necessary, which also would produce a lot of disturbing waste heat. Two configurations of the Common Module Dewar were available: DT-591 for 60 detector elements and DT-594, which was originally designed for 120 detector elements and was used for 180-element detectors as well. To keep the vacuum for a long time a so-called getter was installed in the Dewar which could ionize the residual gas in an electrical discharge and tie the ions to the metal of the wall.

**Common Module Detector MIL-D-49172**

| | |
|---|---|
| Detector material | Mercury-Cadmium-Telluride CMT |
| Detector type | photo-conductive (pc) |
| Wavelength range | 7.7 - 11.8 µm (LWIR) |
| Detector elements | 180 (or parts thereof: 60, 120 etc.) |
| Height of array | 18.3 mm |
| Element size h x v | 41 µm x 61 µm |
| Pitch | 101.6 µm |
| Angular field h x v | ±23° x ±18° (cooled stops) |
| Detectivity D* | $1.1 \cdot 10^{10}$ cm $\sqrt{Hz}$ $W^{-1}$ |
| Time constant | < 2 µs |

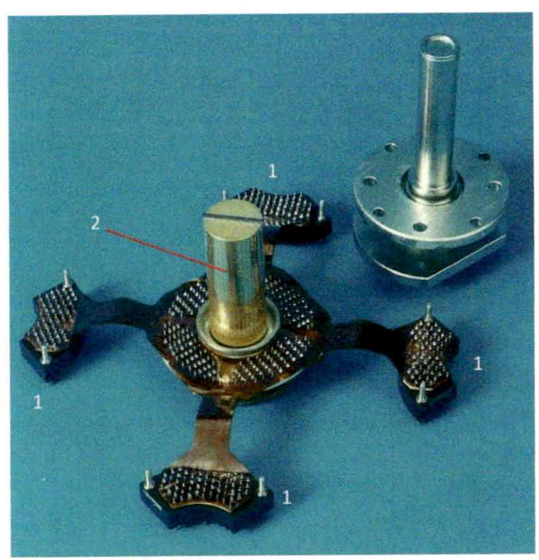

103

103 – *Internal wiring and picture of the detector*

*The left picture shows the Dewar (2) with the detector array on its flat top, surrounded by 4 groups of plug contacts. In the background of the picture the cold finger is to be seen, which is mounted into the glass Dewar from the backside during the final assembly.*

*The right picture shows a fully assembled detector unit. A thin, anti-reflection coated germanium plate (IR) with high IR transmittance is used to seal the front side of the vacuum vessel.*

*104 – Cooler - detector assembly and principle of the Stirling cooler*
*The cylindrical body with the marked cooling fins in the left picture is the electro motor (Em) that drives the compressor (V). To take the photograph of cooler and detector, the detector (D) is mounted on an auxiliary plate. The sketch shows the compressor piston (1) which presses the working gas helium through a transmission pipe into the cold finger (4). Therein it flows through a metal mesh (2) that acts as a heat exchanger and absorbs a great amount of the heat generated in the compression process. The gas expands into the space given free by the displacer (3) and therewith is cooling down the detector (6). In a last step of the Stirling cycle, the displacer pushes the gas back into the compressor (1). The detector is mounted inside a vacuum vessel (5) also known as Dewar. It prevents effectively that heat can flow from the warm surroundings to the cold detector.*

## Common Module Cooler

The Common Module cooler used a thermodynamic cycle patented in 1816 by the Scottish Referend and inventor Robert STIRLING[370] for a heat engine, and which is well-known as STIRLING cycle. The cooler worked with the reverse process: While the engine generated mechanical energy from a temperature difference, the cooler generated a temperature difference from mechanical energy. The cooler was a heat pump driven by an electric motor, which pumped the heat from the cold detector to the warm heat exchanger.

The cooler comprised an electrically driven compressor and a cold finger which was mounted in the cavern of the Dewar directly underneath the detector. In the mounting process of the cold finger it was important that a good thermal connection to the detector was established by using the right amount of heat paste, otherwise the detector could not reach the necessary deep temperature.

**Common Module Cooler MIL-C-49175 A**

| | |
|---|---|
| Functional principle | Stirling process |
| Working fluid | Helium |
| Cooling | 77 K (-196 °C) |
| Thermal load | cooling required by detector: 0.4 W |
| Power consumption | 50 W |
| Cool down time | 5 to 10 min |
| Leak rate | $2.5 \cdot 10^{-6}$ cm$^3$/s |

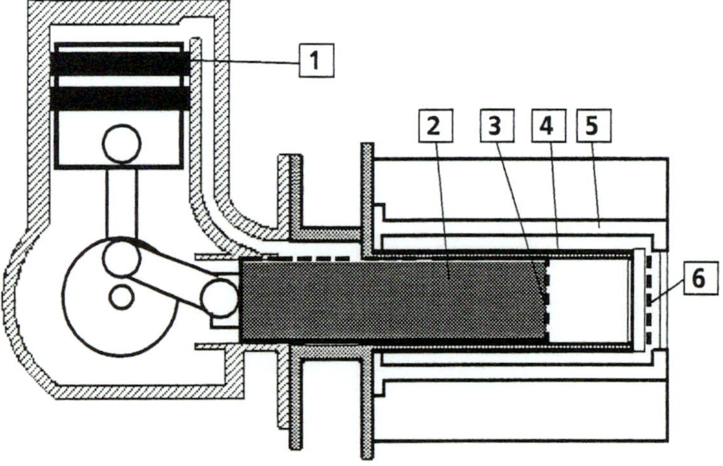


104

## Common Module Scanner

The task of the scanner was to shift the optical image in horizontal direction over the vertically oriented linear detector array so that a two-dimensional image could be captured. This so called parallel scanning could reach a high efficiency of 80% in case the data were collected during forward and return motion of the scanner.

However, in the interaction with the Common Module detector the problem arose that due to the gaps between the detector elements horizontal strips were missing in the image. To prevent this the return was shifted vertically by half of the detector pitch as shown in the scanning scheme. This method is commonly known as "interlace".

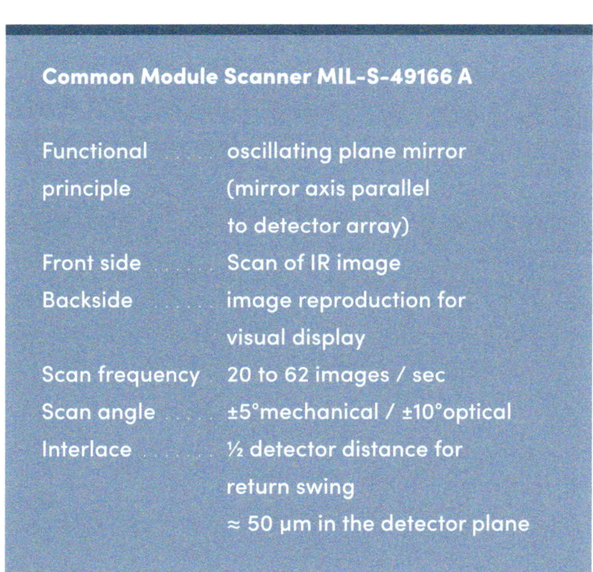

**Common Module Scanner MIL-S-49166 A**

| | |
|---|---|
| Functional principle | oscillating plane mirror (mirror axis parallel to detector array) |
| Front side | Scan of IR image |
| Backside | image reproduction for visual display |
| Scan frequency | 20 to 62 images / sec |
| Scan angle | ±5°mechanical / ±10°optical |
| Interlace | ½ detector distance for return swing ≈ 50 µm in the detector plane |

105

105 – *Scan pattern of the Common Module scanner*

*Top: The incomplete covering of the object (a tree) is a result of the gaps between the detector elements.*

*Center: Using a pure parallel scan would result in an image with empty strips lacking any image content.*

*Bottom: Applying the interlace concept with a shift of half of the detector distance fills up the empty strips during the return travel.*

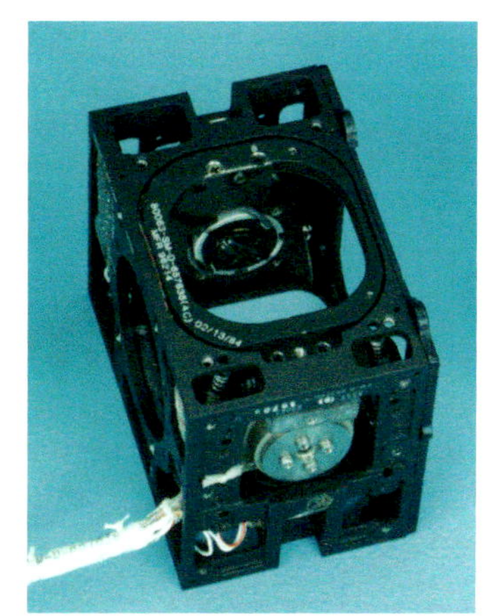

106

106 – *Common Module Scanner*

*107 – Common Module Imager with optical beam path*

*As shown in the sketch, the imager comprises two lenses made from germanium (Ge) and a zinc selenide (ZnSe) lens which performs the color correction. The ray path is folded by a plane mirror (M). The optical image on the detector surface (Det) is flat and the light bundles fall on the detector surface almost perpendicular. The imager can be focused by axial shift of the Ge-ZnSe ront lens. This is practically performed by turning a large grooved ring that can be seen on the right hand side in the picture of the imager.*

## Common Module Imager

The imager was laid out as a 3-lens objective for imaging the thermal radiation coming from the scanner onto the detector. Additionally, a folding mirror for 90° deviation was integrated to allow a compact beam path of the overall imaging system. The imager optic was color corrected for the wavelength range 8 - 12 µm and designed to generate a plane optical image. Furthermore, it was designed in a way that the chief rays of all bundles hit the detector surface almost perpendicular to prevent clipping of the bundles (loss of light) by the towering cold shields (the blinders).

**Common Module Imager MIL-I-49170 A**

| | |
|---|---|
| Objective type | Petzval lens (telecentric lens, plane image) |
| Wavelength range | 7.6 – 11.8 µm |
| Focal length | 67.8 mm |
| Aperture | F/1.2 |
| Correction | diffraction limited |
| MTF (on-axis) | 71% at 10 Lp/mm |
| Transmission | > 85% |
| Focussing | manual |

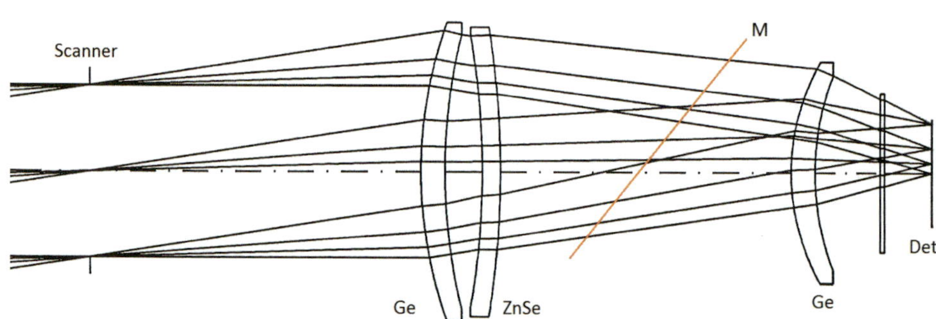


107

## Common Module LED Array and Visual Collimator

The LED array was a linear array of red light emitting diodes (LED) that were 1:1 assigned to the infrared detector elements. Depending on the signal strength of the detector element the assigned LED did shine with variable brightness, and so converted the intensity of the incoming thermal radiation into the brightness of the visible red light: an IR image point getting much thermal radiation was shining bright in the visible image.

**Common Module LED-Array MIL-L-49169 A**

| | |
|---|---|
| Principle | visual counter part to IR detector |
| Wavelength | 660 nm (red) |
| LED material | Gallium arsenide phosphide |
| Element size h x v | 21 µm x 91 µm |
| Number of elements | 180 |
| Radiation power | $2 \cdot 10^{-6}$ to $21 \cdot 10^{-6}$ W |

**Common Module Visual Collimator MIL-C-49174 A**

| | |
|---|---|
| Principle | visual counter part to IR imager |
| Wavelength range | 660 nm (red) |
| MTF (on-axis) | 93 % at 10 Lp/mm |
| Distortion | < 2 % |
| Transmission | > 85% |

108

108 – *Common Module LED-Array (top) and Visual Collimator*

The visual collimator was a collimator lens that collected the red light emitted by the LED array and directed it to the backside of the scan mirror. Its function in the image reproduction path was corresponding to that of the IR imager in the thermal beam path but reversed, of course.

109 – *Electronic boards of the Common Module system*

*Besides the sensor and the electronic unit in the center the picture displays the boards with pre-amplifiers (1), post amplifiers (2), bias generator (3), scan electronics (4) and auxiliary electronics (5).*

## Common Module Electronic Boards

The boards with the necessary electronic circuits also were part of the standard modules whereby we today should make us aware that the development of these boards took place entirely in the age of analog electronics. The high number of analog amplifiers is amazing to today's standards: Each of the 180 detector elements had its own pre-amplifier and its own post amplifier, both had to be balanced in relation to offset and gain to achieve a homogeneous and strike-free image.

All boards had exactly specified interfaces so that a customer could mix the boards from all qualified vendors in his application

In detail, the following boards were standardized:

**Preamplifier** MIL-P-49163 A: Board for 20 detector channels with a gain of 71,5 V/V at 1 kHz
**Postamplifier** MIL-P-49164 A: Board for 20 detector channels with a minimum gain of 18 · $10^3$ V/V
**Biasregulator** MIL-R-49176 A: Generation of bias-voltage 5 ±0,5 V
**Auxiliary electronic** MIL-A-49167 A: Electronic board for adjusting brightness, contrast and polarity
**Scan-Interlace drive** MIL-S-49177 A: Scan controller for 20 to 62 Hz and interlace controller (optionally synchronized)

109

## 6.3 Common Module Zeiss WBG-X Thermal Imager

As stated already above, the participating companies could make use of the set of standard modules to design their own thermal imagers for various applications. This allowed for efficient solutions because not only the own modules could be applied but the modules of all other partners too.

As outlined in Chap. 6.1, CARL ZEISS-SONDER-OPTIK used this opportunity to develop a thermal imager, named Zeiss WBG-X, for their panoramic tank periscopes and gunner sights.

Besides the already discussed US-German Common Modules (IR imager and visual collimator were in-house made, the rest came from other vendors), additionally an afocal telescope with two fields of view and an image intensifier for the reproduction of the visible image were used for the WBG-X.

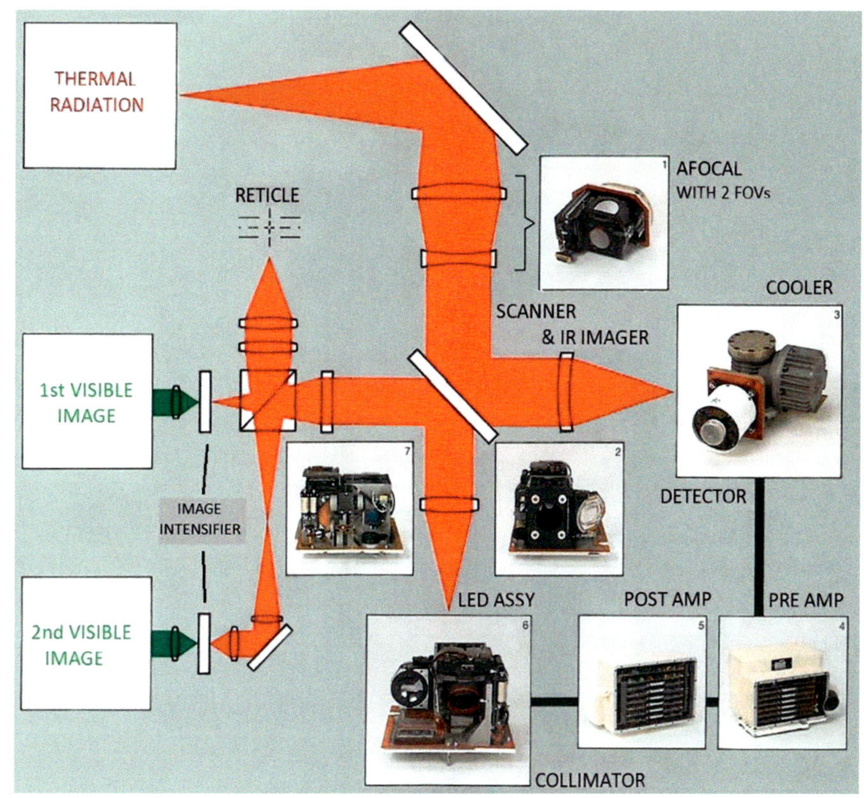


110 – *Set up of Zeiss WBG-X based on Common Modules*

*Additional to the original US Common Modules an afocal telescope in front of the basic unit is used and an image intensifier in the image reproduction channel.*

### Afocal system

The afocal[371] was used to realize the specified fields of view, a wide field of view (WFOV) for a good overview and a narrow field of view (NFOV) for high resolution of structural details. The optical principle of the afocal was that of a Galilean telescope comprising a collecting and a diverging lens which are arranged in the way that the focal points were coincidental. To change from wide to narrow field of view the magnification was switched from 1.33x to 3.9x. This was done by swinging out the ocular lens for magnification 1.33x and replacing it by a 2-lens assembly with a smaller focal length. The simultaneous movement of the lenses was driven by an electric motor, both ocular groups are moving at the same

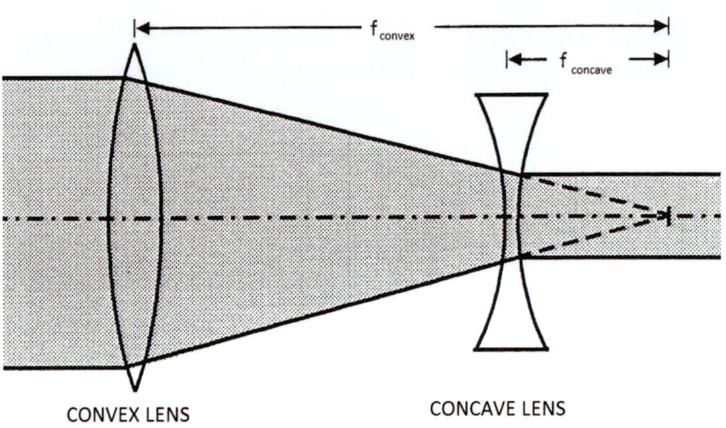


111 – *Afocal front optic of the Zeiss WBG-X and the optical principle (right)*

*In the picture of the afocal the lens for the wide field of view (WFOV) is swung out and the 2-lens system for the narrow field of view (NFOV) is put into the beam path. The optical principle is that of a Galilean telescope that comprises a convex collecting lens and a concave diverging lens which are arranged in the way that both focal points coincide. The system has no image plane instead an incoming parallel bundle is emitted as a parallel bundle as well. Beam direction and diameter are changing according to the magnification which is defined as the ratio of the focal length values.*

time. An additional electro motor was used for focusing between objects in 50 m distance and infinity by axially shifting the whole ocular assembly.

## Image Intensifier

For enhancement of the image in the reproduction channel, an image intensifier tube was used to transform the red image generated by the light emitting diodes into a green image with a higher brightness.

The use of the image intensifier tube brought several advantages for the visual observation:

- The absolute intensity of the light became higher
- The green image of the intensifier tube is generally perceived brighter and sharper by the human eye
- The image of the intensifier tube was phosphorescent and did flicker less than the LED image

## 6.4 Zeiss WBG-X in Active Service

As already outlined in Chap. 6.1, in the time of the Cold War the US-German Common Modules conceptualized by TEXAS INSTRUMENTS were applied in many military and para-military programs also in the Federal Republic of Germany: main battle tank Leopard, missile hunting tank Jaguar, die infantry fighting vehicles Marder and Luchs and submarine search periscopes BS und SERO 14.

The table shows that the tank programs were dominating in numbers and in the produces quantities as well. In the 1980s and 1990s, CARL ZEISS-SONDEROPTIK managed to produce more than 10,000 thermal imagers. Without the cost and time saving standardization this is hardly conceivable. Moreover, the governments of the countries as customers would not have been able to raise the financial resources for so many large programs.

| DEVICE | PERIOD | QUANTITY | APPLICATION |
|---|---|---|---|
| WBG-X | since 1981 | ≈ 250 | Leopard 1 A4 |
| EMES 15 | 1982-92 | 2125 | Leopard 2 A0–A4 |
| CM WBG | 1982-88 | 7 | SERO 14 |
| AN/TAS | 1983-85 | 165 | Jaguar 2 |
| WBG-X | 1984-89 | 1462 | Marder 1 A2 |
| WBG-X | since 1985 | 408 | Luchs A2 |
| HEOS | about 1985 | 36 | HAWK PIP-II |
| EMES 18 | 1986-92 | 1339 | Leopard 1 A5 |
| Pamir | 1990-93 | ca. 100 | ECR Tornado |
| EMES 18 | 1992-94 | 110 | Leopard 1 A5 DK |
| EMES 18 | 1994-95 | 132 | Leopard 1 A5 BE |
| EMES 18 | 1994 | 78 | Leopard 1 NO |
| MEOS | since 1994 | ? | Frigates & speed boats |
| Border surveillance TI | 1994-96 | 120 | BMI mobile IR-surveillance system |
| EMES 18 | about 1995 | 120 | Leopard 1 A5 DK |
| EMES 18 | since 1999 | 123 | Leopard 1 A5 C2 |

115 – *Contemporary logos for WBG-X and FLIR*

*"Zeiss thermal imagers for armored vehicles" & "Forward-Looking Infrared"*

116 – *Typical components of a modern panoramic periscope*

*The light enters the periscope through a window (AF) and is directed downwards by the rotatable elevation prism (EP). The telescope, comprising objective (Ob) and ocular (Ok), is imaging a magnified image of the scene into the observer's eye. The panoramic head (RK) can be rotated around the vertical axis without limits, so that the observer can look to all directions without any restrictions. The image erecting unit (BA) prevents that the optical image is rotating in this process. When the panoramic head is rotated by an angle ω a compact prism (SPP) known as Schmidt-Pechan prism is rotated by half of the angle ω/2 to compensate the image rotation.*

115

To discuss every military program of that time is beyond the scope of this book, of course. Therefore, in the following the technically most interesting systems are presented – applied by army, navy and air force.

## EMES 15 for main battle tank Leopard 2

After a challenging begin in the Great War, the armored force made great technological and tactical advances in World War II accompanied by a rapid increase of the number of employed vehicles. The optical equipment, however, remained widely untouched. The gunner had a sight adapted to the available space and combat range of the tank and the commander had to be content with the eye slits when he did not want to raise his head above the turret. Only in the post-war era the first panoramic periscopes came into use, and allowed the commander to look around safely from a covert position.

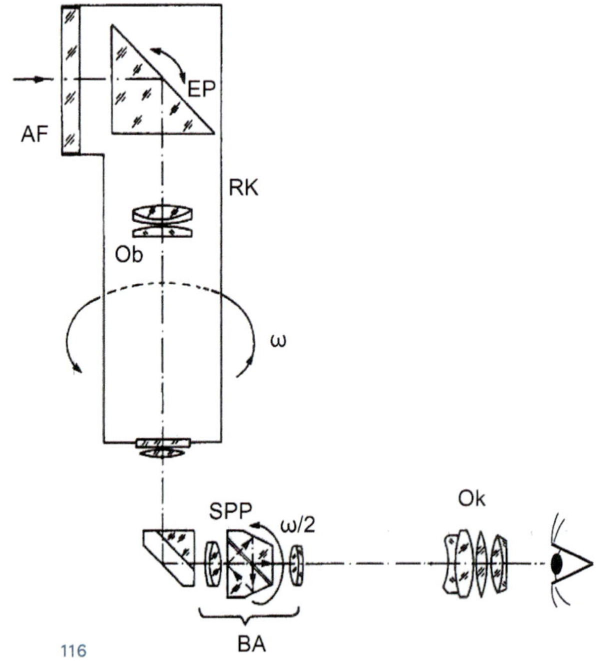


116

Initially, the new devices were designed for day-time vision. In upcoming nightfall commander and gunner could see less and less, and virtually were blind at night. Already towards the end of the war, engineers had tried to overcome this nuisance. The first available image converter tubes, which could generate an image visible to the human eye from the invisible near-infrared radiation (NIR, approx. 0.8 to 1.3 μm), were used together with also invisible NIR-searchlights to develop night-vision systems, e.g. the German aiming device Zielgerät 1229 Vampir. This development path was followed further in the post-war period. But the size of the searchlights, the glare sensitivity and the limited range were still an issue. The principle problem was, however, that the searchlight could be spotted by anyone who had an image converter. The user reveals his own position because his device was actively emitting radiation. Therefore, a purely passive device would be highly desirable - something like a thermal imager! With the upcoming of the Common Module thermal imagers this wanting could become reality.

In the years **1982-92**, the German main battle tank Leopard 2 A4 was equipped with the new gunner sight EMES 15 after thermal imaging technology had already found its way into the older model Leopard 1 A4 in the years before. The equipment of the Leopard 2 A5 consisted of the best contemporary sensor technology had to offer:

- Main gunner sight (Hauptzielfernrohr HZF) EMES 15 with
- 2-axis stabilization (in elevation and azimuth)
- Day time-vision channel with magnification 12x and 5° field of view
- Nd:YAG laser rangefinder (Manufacturer: Krupp Atlas Elektronik)
- Thermal imager WBG-X (Manufacturer: CARL ZEISS)
- Commander periscope Peri R17 A1 (Manufacturer: CARL ZEISS)
- 2-axix stabilization (in elevation and azimuth)
- Day time -vision channel with magnification 8x and 2x
- Gunner sight (Turmzielfernrohr TZF) FERO-Z 18 (Manufacturer: Leitz)

The commander periscope Peri R17 A1 had an own thermal imager not jet, but was optically connected to the WBG-X in the EMES 15. The technical data of the WBG-X were the same as given in the table in Chap. 6.3.

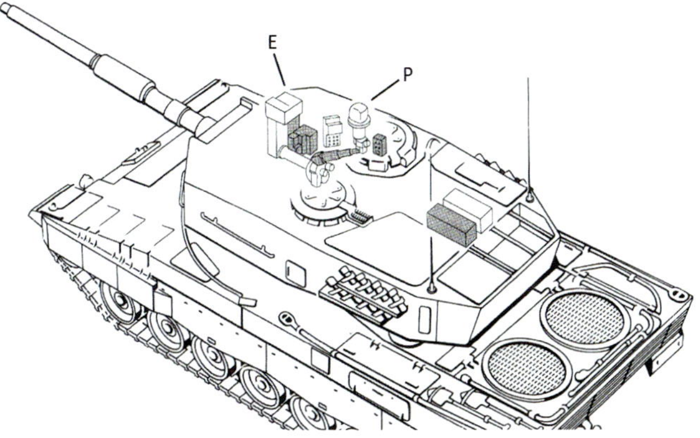


**117 – *Sensor units of the Leopard 2 A4 battle tank***

*All sensor units are installed in the turret of the tank. EMES 15 (E) and Peri R17 (P) are optically coupled. Therewith the commander can view the thermal images captured by EMES 15 in the eye piece of his Peri R17. This is advantageous because at that time the commander's Peri R17 A1 did not have an own thermal imager.*

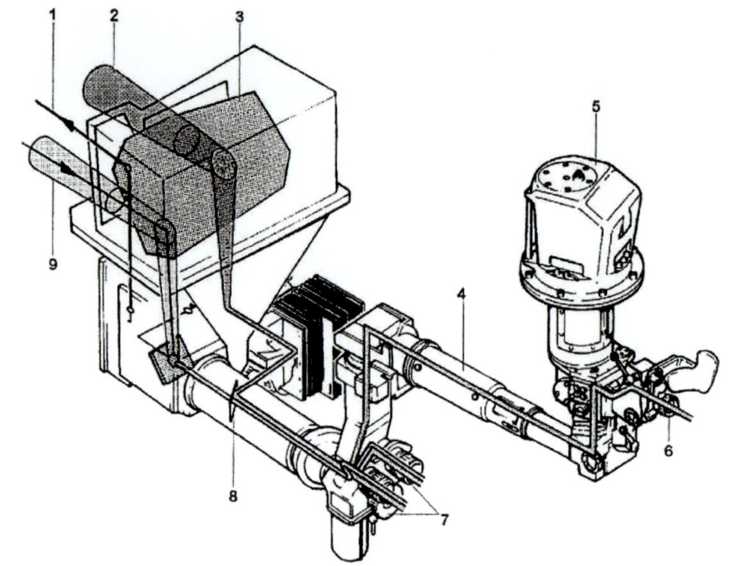


118 – *Optical coupling of EMES 15 and Peri R17 A1*

*The thermal radiation (2) entering the EMES 15 is directed into the thermal imager of EMES 15 by a stabilized elevation mirror (3). The thermal images obtained by the detector are made visible in the direct sight adapter and are coupled into the ocular assembly of the EMES 15 by a switchable mirror (8) to make them visible to the gunner in his two oculars (7). By means of a beam splitter in front of the gunner's ocular the thermal images are coupled into the transmission optics (4) of the Leopard 2 and reach the Peri R17 A1 (5) where the commander can view them in his eye piece (6). The sketch also shows the beam paths of the gunner's day sight (9) and of the laser range finder (1).*

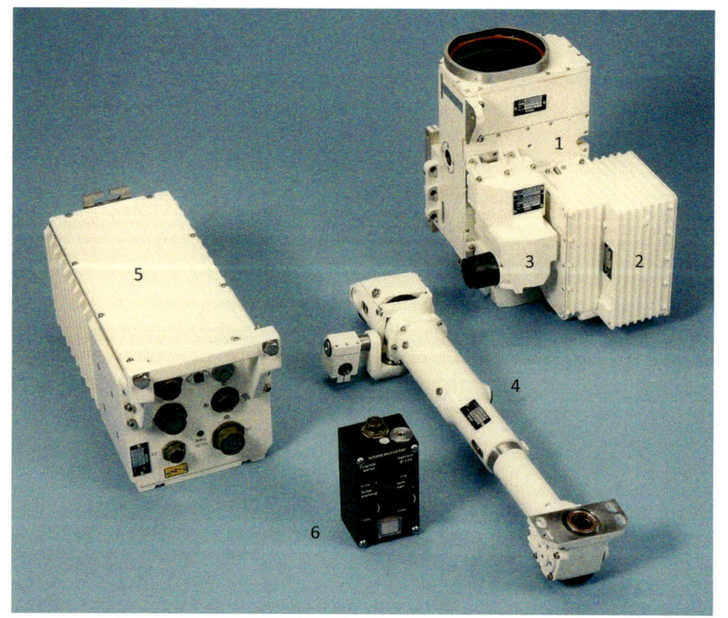


119 – *Components of WBG-X for Leopard 2 A4*

*The WBG sensor unit (1) is blocked to a unit (W) with the sensor electronic (2) and the WBG direct sight adapter (3). The small picture (top left) shows how this assembly is mounted on the EMES 15. The transmission optic of Leopard 2 (4) connects EMES 15 with the commander's panoramic periscope Peri R17 A1. Using an additional control panel (6), the commander can also control the thermal imager of EMES 15.*

On **29 December 1993,** the Federal Republic of Germany and KRAUSS-MAFFEI signed a contract about the upgrade of the first 225 battle tanks Leopard 2 A4. At the same date, the Netherlands also signed a contract to upgrade 330 Dutch tanks. Part of this upgrade was a modified commander periscope Peri R17 A2 with an own thermal imager. Surprisingly, the WBG-X was not chosen as would have seemed natural for logistic reasons because it was already installed in several weapon systems (gunner sights EMES 15 and EMES 18, fighting vehicles Marder and Luchs). The decision was made - as likely as not for political reasons - for the thermal imager TIM (Thermal Imaging Module) of the Israeli company EL-OP, a subsidiary of ELBIT SYSTEMS. In later years other countries like Sweden and Denmark followed this decision.

**Technical Data of TIM**

| | |
|---|---|
| Detector | CMT (Hg-Cd-Te) |
| Wavelength range | 8 – 12 µm LWIR |
| Cooling | Stirling cooler |
| Scanner | oscillating mirror with Interlace |
| Infrared optics | IR lenses |
| Image format | 2:1 |
| Fields of view | WFOV: 18° x 9°<br>NFOV: 4.2° x 2.1° |
| Image reproduction | optical display |

### Submarine search periscope Zeiss SERO 14

The technology of the submarines was defined in great parts in the era of the Imperial Navy and lasted until the early 1980s. This also holds for the submarine periscopes which were also referred to as the “eyes of the submarine”, without which the submarine would be blind during cruising at periscope depth. The periscopes must be adapted in size and design to the requirements of the submarine. The periscopic length, for example, must be large enough that the submarine can cruise under the water surface in a save depth. The so-called mast of a typical submarine periscope is built from a stable stainless-steel tube that is between 8 and 12 m long depending on the height of the boat (the retracted periscope should disappear completely inside the submarine). Furthermore, it should be as slim as possible to make it harder to be spotted by the warring observer.

Much like the tank periscopes discussed above, the submarine periscopes for a long time were devices for daytime vision only, and were depending on brightness sufficient for visual observation. Therefore, a tank and a submarine commander had in common that they were virtually blind at night. It was therefore coherent to set about installing the now available Common Module thermal imagers also in the submarine periscopes.

The effort necessary to fit the Common Modules in the narrow periscope head, to transmit the reproduced visible image through the long mast down into the boat and to make the entire system resistant against the maritime environment was significantly higher than anticipated.

As early as in the mid of the 1970s, first considerations started at CARL ZEISS-SONDEROPTIK for an optronic attack and search periscope which got the working title AS 80 / BS 80.[372] A whole lot of studies

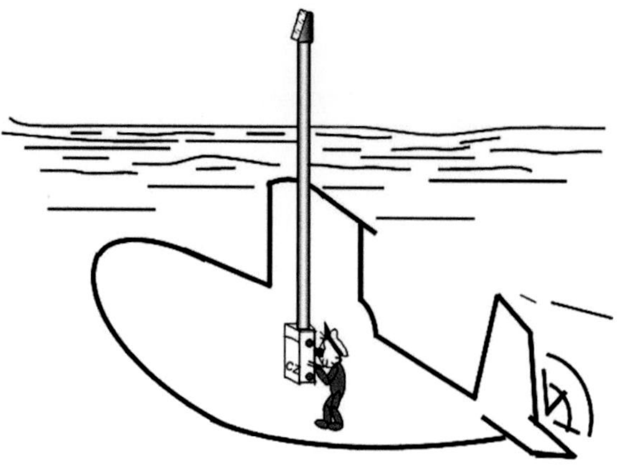

120 – *Functional principle of a submarine periscope*

and sub-projects were necessary to find satisfactory solutions for all issues. The prototypes were tested on the German submarine U1 one after the other. The following survey illustrates the activities for the search periscope BS 80 (later re-named to SERO 14) related to thermal imaging alone:

**09/75** 1st quotation for development and construction of an experimental periscope system AS 80 / BS 80

**12/76** 2nd quotation for development and construction of an experimental periscope system AS 80 / BS 80

**06/77** Contract "Entwicklung eines Uboot-Sehrohr-Systems für die Uboot-Klasse 208 und Bau je eines Entwicklungsmusters AS 80/BS 80" (Development of a submarine periscope system for Class 208 submarines and construction of one prototype each for the AS 80 / BS 80)

**03/79** Agreement between Germany and Norway on armament collaboration (standardization, workshare N: fire control instrument, D: periscope & torpedoes)

**10/80** 2nd agreement amending the contract on development of the AS 80 / BS 80 (device carrier)

**06/83** Delivery of the BS 80 (installed on U 1)

**09/83** Contract "Entwicklung und Bau eines neuen B-Sehrohrkopfes Entwicklungsmuster SERO 14-Mod 1" (Development and construction of a new search periscope head prototype SERO 14-Mod 1)

**04/84** Removal of BS 80, reinstallation of AS 80 on U1

**10/84 - 12/86** Study entitled "Festigkeits-untersuchung an Germanium-Abschlußfenstern SERO 14" (Strength analysis on germanium windows of SERO 14)

**12/84** Delivery of SERO 14-Mod 1 (test model)

**08/85** Contract for "Weiterentwicklung und Serienreifmachung SERO 14/15 (A-Kopf, B-Kopf, SW-Doku etc.)" (Further development and preparation for series production of the SERO 14/15 (A head, B head, SW documents, etc.))

**03/86** Study entitled "Infrarot Zieldarstellungs- und Auswertegerät (IRAD)" (Infrared Analysis and Description Device)

**12 March 1986** Norwegian decision to purchase SERO 14/15

**27 March 1986** Order from TNSW placed with CARL ZEISS for 6 SERO 14/15 optronic periscope systems

**03/89** Delivery of the first SERO 14/15 system

**1989-92** delivery of 6 SERO 14 periscopes for the Norwegian Class 210 (ULA class) with

- 2-axis stabilization
- Magnification 1.5x, 6x und 12x
- Elevation range -15° to +75°
- Periscopic length 9420 mm
- Weight approx. 1000 kg

In the end, it took more than 10 years until the ambitious undertaking to give true night-vision capability to a submarine periscope could be made reality. Without the decision of the Norwegian Navy - at approximately half time - to equip the new ULA class with the new optronic periscopes, the program might not have seen a successful end.

Particularly as there had been doubts about the principle suitability. When the first prototype of the search periscope BS 80 was tested on U 1, it was the first periscope with a thermal imager at all. It was equipped with a Common Module thermal imager with 60 image lines, the electronic was analog according to the time with the detectors being AC coupled. Because with this type of coupling the offset of each detector element (= one image line) could vary depending on the signal strength, there were concerns that the horizon could possibly not being distinguished with certainty. Therefore, in the first system (at least) the thermal imager was rotated by 90° so that the lines ran vertically and thus perpendicular to the horizon.

Besides the big issues there were, of course, also a bunch of (supposedly) little problems. A good example is the window of the thermal imaging channel. Because of the required high mechanical strength, the window had to have a relatively large thickness of several centimeters whilst its transmission for thermal radiation in the LWIR wavelength range at 8 to 12 µm also had to be sufficiently high. The only solution for this was a window made from the infrared material germanium. Unfortunately, the semiconductor germanium is attacked severely be seawater [373], and therefore a protective coating is a necessity. Fortunately, such a coating became available in that time: the DLC-coating[374] (Diamond-Like Carbon) which had been developed under participation of the Fraunhofer Institute in Freiburg, Germany.

One of the more challenging problems naturally was the arrangement of the two sight channels side-by-side in the periscope head.

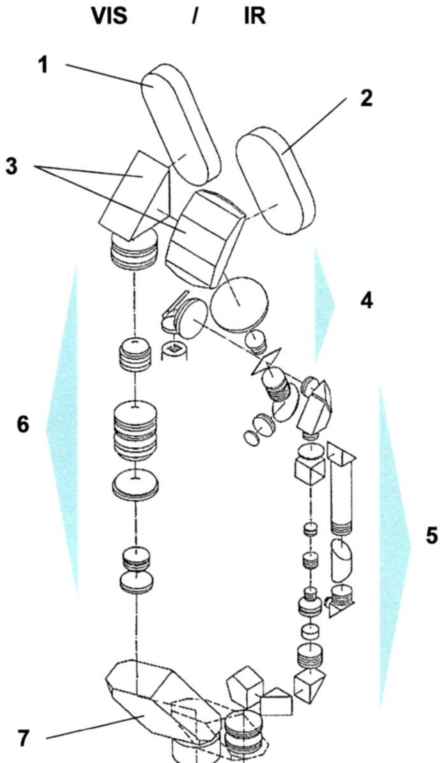


121 – *Optical principle of the Norwegian SERO 14 periscope head*

*The sketch shows the principal components of the complex SERO 14 head with the visual observation channel (VIS) besides the infrared channel (IR): Visual window (1), IR window (2), elevation prism and IR mirror (3), thermal imager (4), visual reproduction of the thermal image (5). In the visual beam path, a so called pancratic system (6) is shown. This is a telescope with three fields of view: an internal lens group can be shifted axially between defined positions to change the magnification between 1.5x, 6x and 12x. With a swiveling prism (7) the observer can switch between visual and infrared channel.*

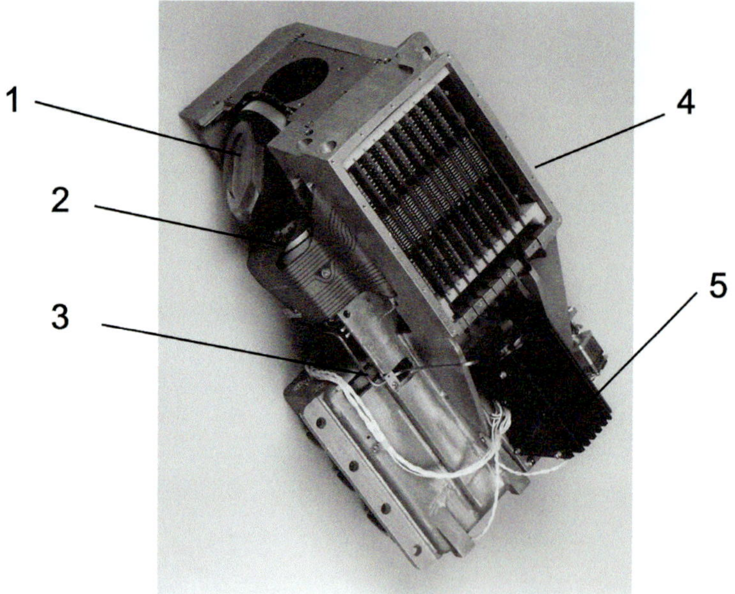


122 – *Common Module thermal imager of SERO 14*

*The specially designed thermal imager of SERO 14 N is based on the US-German Common Modules: IR imager (1), detector (2), transmission pipe between cooler compressor and cold finger (3) and amplifier boards (4). In preference to the standard Common Module cooler, a space-saving linear cooler with remotely arranged, electric linear motor (5) is installed which has the benefit of a lower running noise.*

To make the Common Module thermal imager fit into the periscope head the modules had to be arrange differently compared to the WBG-X configuration that already had been installed in tank periscopes many times. The detector was cooled by a so-called split-cycle STIRLING cooler driven by a linear motor, and the cold and warm sections could be mounted separately. Both sub-units only were connected by a transmission pipe that was flexible to a limited extend. This gave more freedom for the mechanical design. The silent running linear motor also was advantageous with respect to a low noise generation, which is a crucial requirement of submarine periscopes.

By using a 180-element detector for the SERO 14 in series, the image format was changed to 4:3 and thus was different from that of the WBG-X. The fields of view were chosen to match the narrow field of view NFOV with that of magnification 12x.

**Technical Data of CM Thermal Imager of SERO 14**

| | |
|---|---|
| Detector | CMT (Hg-Cd-Te) |
| Wavelength range | 8 – 12 µm LWIR |
| Cooling | Stirling cooler with linear motor 77 K = -196°C |
| Detector elements | 180 x 1 |
| Scanner | oscillating mirror with 2:1 interlace |
| Image lines | 360 |
| Frame rate | 42 Hz (60 Hz) |
| Number of image points | 172 800 |
| Infrared optics | IR lenses |
| Window | germanium substrate with sea water resistant DLC-coating |
| Field of view | WFOV: 14° x 11° NFOV: 4° x 3° |
| Image reproduction | optical display of the image via image intensifier & direct observation through mast and eye piece |

124

124 – *Thermal image of a maritime scenery obtained by SERO 14*

*The sails of the submarines with erected periscopes and people are clearly to be seen in the thermal image. Rumors say that during one of the first tests at night and under bad weather conditions the Norwegian commander was much surprised by the quality of the thermal images. He climbed on top of the sail of his submarine to convince himself that the presented scene was real – not that someone tried to fool him with a video clip from the day before.*[375]

*(Photo: Thomas Dietrich)*

123

123 – *Prototype SERO 14 Mod 1*

*Left picture: Periscope head with two windows side by side. The IR window (left) is made from germanium and is protected against sea water corrosion by a very durable DLC coating. The coating changes the shining metallic surface of germanium to a dark hue and reduces the reflection in the IR wavelength region, resulting in a higher intensity of thermal radiation on the detector. DLC stands for Diamond-Like-Carbon and relates to the diamond-like hardness of the coating. The window on the right is for the visual channel and made from tension-free fused silica. It is also chemically durable and will not be corroded by salt water.*

*Right picture: Display model in the periscope tower of Hensoldt Optronics in Oberkochen.*

## FLIR for Panavia Tornado ECR

In **1981**, the delivery of in total 357 fighter bombers MRCA PA-200 Tornado to the BUNDESWEHR (German armed forces) began with the basic configuration Tornado IDS (Interdiction Strike). In parallel the LUFTWAFFE (German air force) pushed a planning ahead for a special configuration called Tornado ECR (Electronic Combat Reconnaissance). The tasks were defined to be "localization, identification and combat if necessary of radar systems and radar-guided air defense systems "[376]. Besides the Emitter Location System (ELS) for radar systems the Luftwaffe also demanded a FLIR system.

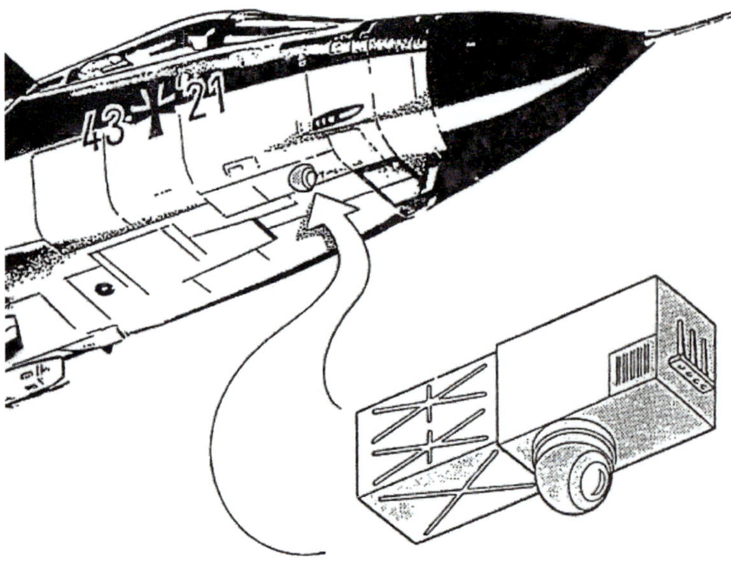


125 – *Installation of the FLIR in the Tornado ECR*

**In the early 1980s**, some manufacturers like TEXAS INSTRUMENTS, HUGHES and CARL ZEISS-SONDERTECHNIK as well worked on an upgrade of the Common Modules. One topic was the shortcoming that the number of image lines was much smaller than the number of image points recorded along a horizontal line. According to contemporary insight, this led to a decline in the achievable observation range.

In **1985**, CARL ZEISS-SONDEROPTIK followed a call from the German Tornado manufacturer MESSERSCHMITT-BÖLKOW-BLOHM (MBB) and submitted a detailed "Proposal for a Forward-Looking Infrared (FLIR) System for ECR Tornado"[377]. For this proposal, CARL ZEISS-SONDEROPTIK could use the results of an internal study "4 fach Zeilensprung für CM-WBG"[378] (4:1 Interlace for CM Thermal Imagers) finalized only a short time before, in which was investigated how the performance of the Common Module thermal imagers could be improved with a higher number of image lines by means of a new 4-fold scanner and some auxiliary measures. TEXAS INSTRUMENTS, following the same principles, submitted a very similar proposal with identical title[379] and technical solutions differed in detail only.

In **1986**, an order of 35 ECR FLIR thermal imagers was placed at CARL ZEISS-SONDEROPTIK, and the development and preparations for production could be started. The development was costlier with respect to time and money compared to just adapting the well-known WBG-X to another vehicle. Not only that a new thermal imager with better performance was to be developed, it had to be compliant to all legal regulations for aeronautical equipment too. The effort was indeed comparable to the SERO 14-thermal imager, but could fortunately managed before the series production of the Tornado ECR started.

At the aircraft itself, also modifications were necessary that took some time. The cockpit had to be changed to arrange the displays and control panels the FLIR system and the Emitter Location System (ELS) of RAYTHEON TI SYSTEMS as well. In the fuselage, the housings of the FLIR sensor unit and the FLIR power supply, both being not that slim, had to be installed and connected to the air condition. Moreover, the more powerful Version MK 105 of the TURBO-UNION RB199-34R jet engine also was installed in the Tornado ECR. Rumors spread that during practicing of aerial refueling a Tornado ECR pilot had to show the weakness to radio to the tanker aircraft, and asking to fly somewhat slower because otherwise he would not be able to follow the high-flying tanker.[380] The reason for this curiosity perhaps had not been the high weight of the FLIR sensor, as some people mocked, but that the RB199 jet engines were specially designed for a high thrust at low-altitude flight.

On **13 October 1989,** PANAVIA[381] issued a "Specification for Forward Looking Infrared (FLIR)" for the series production of the ECR system beginning shortly afterwards. The specified FLIR system should look forward and downward under an exactly defined angle. The thermal image of the scenery flown over should be displayed on the head-up display of the pilot. The magnification was defined in the way that the displayed infrared scene was of the same scale as the scene in front of the aircraft perceived visibly by the pilot. The objection was to support the pilot during night and under adverse weather conditions without major adaptions necessary. As turned out later, this worked out so well that even take-off and landing on unlit airfields not only were possible but also were successfully demonstrated.[382]

In **1990,** the production of the Tornado IDS came to an end and manufacturing of the Tornado ECR equipped with the ECR-FLIR of CARL ZEISS-SONDEROPTIK could start.

Until **1992,** all 35 Tornado ECR were delivered to the Luftwaffe (German air force), at first to the Jagdbombergeschwader (fighter bomber wings) JaBoG 38 in Jever and 32 in Lechfeld. From 1993 on all Tornado ECR were concentrated in the JaBoG 32.

Around **1992,** also the AERONAUTICA MILITARE (Italian air force) decided to convert 16 aircraft out of the purchased 100 Tornado IDS into the ECR configuration. Afterwards these aircraft were registered as EA-200B Tornado.

During the years **1993-96,** at CARL ZEISS-SONDERTECHNIK as the department for military optics of CARL ZEISS was named meanwhile, a further improved version comprising an additional narrow field of view was under development. Under the title "Stab-FLIR" the work was conducted at ZEISS's own expense and a demonstrator also was built. The line of sight of the narrow field of view was not fixed but could be shifted within the wide field of view. The "Stab-FLIR" was tested and delivered to MBB but was never qualified or installed in an aircraft. The space-limitations in relation to physical size of the Common Modules had resulted in significant losses in radiometric performance and had led to a "fine weather device". Perhaps this was the main reason for the failure of the chosen design concept.

On **27 February 1998,** the first Italian Tornado ECR was delivered.

In **1999,** German Tornado ECR aircraft took an active part in the NATO organized mission "Allied Force" in Kosovo: The army of Serbia should be forced into a retreat by heavy air raids. "On 24 March 1999, Federal Chancellor Gerhard SCHRÖDER announced in television the beginning of the NATO air raids: A war would be necessary to prevent a ‚humanitarian catastrophe'. Germany would not wage war however it was called upon to assert a peaceful solution … also by military means. To prevent a ‚humanitarian catastrophe', had to be sufficient as legitimation of aerial warfare as there was no UN-mandate for this mission." "Four RECCE as well as ten ECR Tornados of the Luftwaffe (German Airforce) from the Aufklärungsgeschwader 52 Immelmann (reconnaissance air force wing 52 Immelmann) and the Jagdbomber-Geschwader 32 (fighter bomber wing 32) together with 350 ground staff transferred to the north-Italian Piacenza; from here 390 missions were flown in the next months. According to the German Ministry of Defense the 14 aircraft fired in total 240 air-ground missiles of the HARM type (High Speed Anti-radiation Missile), without own losses. The missiles mainly were aimed at active and passive air defense, among this were airfields, radar installations as well as missile and anti-aircraft positions. What damage was done cannot be answered clearly. Because the Serbs were considered to be the masters of camouflage."[383]

The Italian Tornado ECR also did "patrol over the battle field with the task to suppress hostile air defense radar and thereby fired 115 AGM-88 HARM missiles."[384]

In **2011**, the Italian Tornado ECR aircraft were employed to keep the UN no-flight zone over Libya under surveillance.

In **2013**, all remaining 20 Tornado ECR of the German air force were given to the Taktische Luftwaffengeschwader TaktLwG 51 Immelmann in Jagel (tactical air force wing Immelmann). The Italian air force began a reconditioning of the 15 Tornado ECR which were still in service.

In **2017**, the actual state of German planning included to start measures in 2020 for a prolongation of the service life of the Tornado ECR until the year 2040.

To give a thermal imager in the specified quality to the Tornado ECR pilots further development work was necessary not only on the modules but on the design concept also. The optical reproduction of the thermal image had to be abandoned because in the Tornado aircraft a direct optical channel between ECR FLIR and cockpit was not possible. The thermal imager itself had to generate an electronical video signal that could be displayed on the head-up display in the cockpit.

Particularly, the temperature induced re-focusing had to work automatically. Previous tests had shown that large temperature differences could occur between the front lens near the dome and the lenses inside the fuselage. Therefore, a simple compensation of the thermal effects based on measuring temperature at a few points would not have worked properly. As a consequence, a new active control of the thermal effects had to be developed on which a patent was granted.[385] The functional principle was to measure the focus quality of a test beam which had passed the FLIR optic two times and use the result as a highly sensitive input for a controlled shifting of the focus lens.

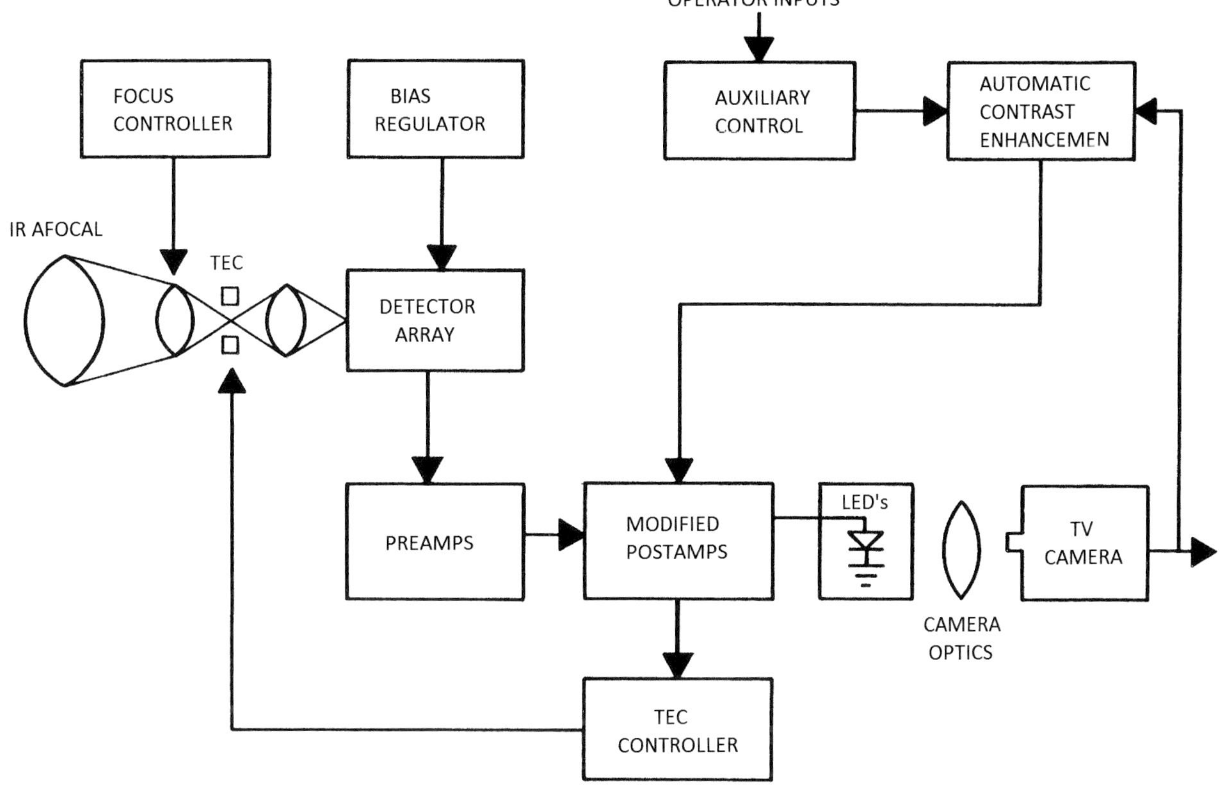


126 – *ECR FLIR signal flow for obtaining video images*

*The incoming thermal radiation reaches the IR detector via scanner and IR imager in the same way as in the WBG-X. The signal of each detector element is amplified and fed to an assigned LED: The higher the intensity of the thermal radiation, the brighter shines the LED. The red light emitted by the LED is reflected by the scanner back side and imaged by the camera optic onto the CCD-chip of the TV camera which generates a video signal for displaying the two-dimensional image on a monitor. An automatic adjustment of the contrast controls background and gain and therewith relieves the observer. The input for the contrast control comes from two thermo-electric cooler (TEC) elements which are arranged in the intermediate image plane of the afocal besides the used optical image. The temperature of the TECs is regulated by the TEC control to the average temperature of the scene. After each pass of the scanner, the detector looks onto one of the TECs for a short length of time and does a reference measurement which is used in the display of the next image.*

**127 – *Principle of thermal compensation of the ECR-FLIR***

*A small target (2) illuminated by a heat source (1) is coupled into the IR optic (6 & 7) closely besides the used field of view (2) by a small mirror (3). A second mirror near the window acts as retro-reflector. After having passed the optic two times, the light from the target is imaged onto a corner of the infrared image. The sharpness of the target image is evaluated by a special algorithm which generates a control signal for the servo-motor (15) to shift the focus lens (6b) in an appropriate manner to re-focus the system.*

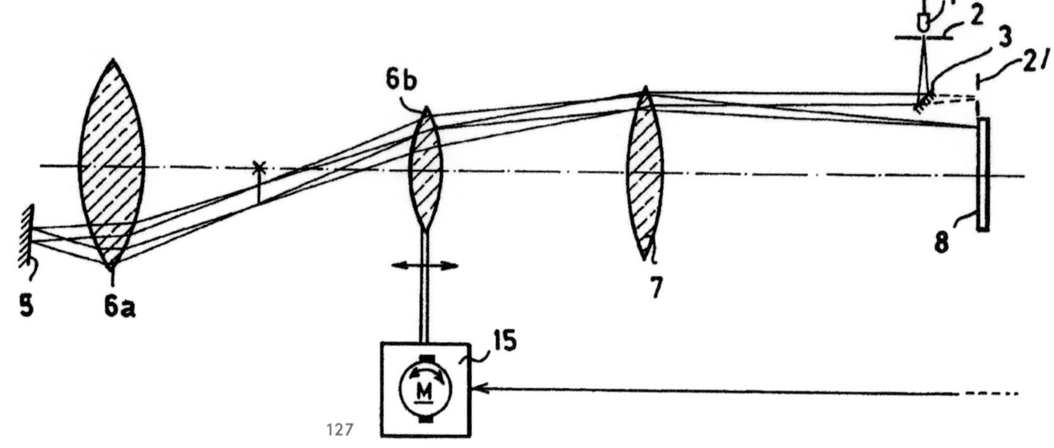


127

**128 – *Scanner with 4:1 interlace***

*For performing a 4:1 interlace the scanner needs two axes of rotation. For conventional horizontal scanning, the mirror is rotated around the vertical scan axis. To perform the interlace function the mirror is rotated a small amount around the inclined interlace axis during the reversal phase between two horizontal scans. Each time, the optical image is shifted by 1/4 of the detector pitch. After the 4th horizontal sweep, the scan mirror returns to the starting position.*

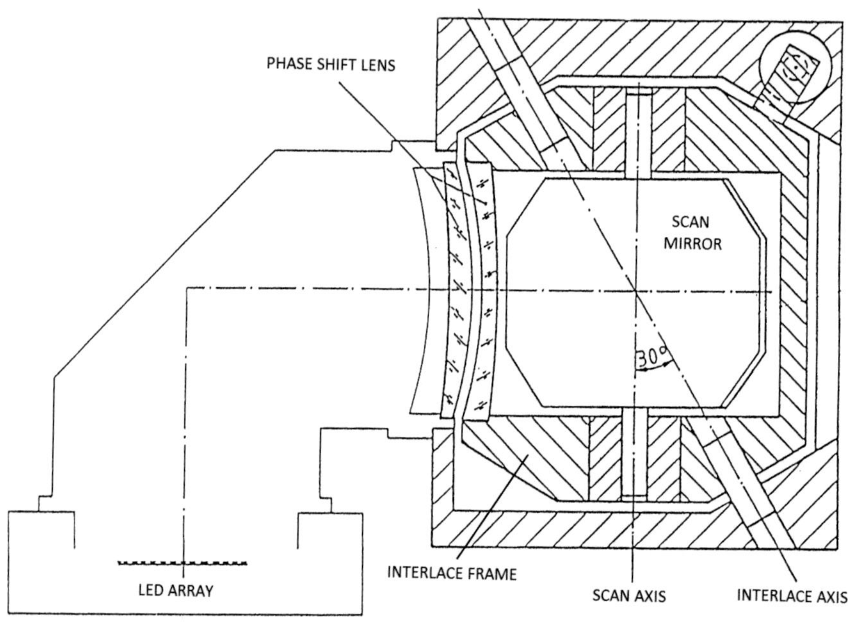


128

A significant contribution to the improvement of the ECR FLIR came from the modified scanner with a 4:1 interlace. Without any interlace the thermal image could only have the same number of lines as there were detector elements. With the usual 2:1 interlace of the Common Module thermal imagers the number of lines was doubled. By the new method the optical image was shifted four times, each time by a fraction of the size of a detector element. In total, an image was achieved which had four times as much lines as the detector had elements.

Some of the Common Modules could be kept unchanged (IR imager, pre-amplifier boards, visual collimator etc.) but the list of further or new developments was long:

- Detector array: Smaller dimensions of the detector elements, pitch unchanged, cold shields in two options for aperture F/1.2 and F/2.4, respectively; based on previous work at TEXAS INSTRUMENTS and at AEG TELEFUNKEN in Heilbronn.
- Dewar: The reliable DT-594 Common Module Dewar originally designed for 120 elements was adapted to 180 elements.
- Scanner: Development of a new scanners with 4:1 interlace on Common Module base; a previous development had been funded by the German BWB (Office for military equipment).
- Scan and interlace electronics: New development with closed-loop control of the scan process
- Infrared optic: New development of a long afocal system specially designed to fit into the available space in the Tornado aircraft including the dome section. The window made from germanium had to designed to withstand temperature, pressure, sand, rain and hail as well as the corrosive exhaust of the HARM missiles. The optical system had to fulfill the specifications for field of view and especially for the inclination of the line of sight with high precision.
- LED array: A further developed LED array with 180 elements with smaller dimensions was contributed by AEG TELEFUNKEN.
- Electro-optical multiplexer: New development of an optic for imaging of the red LED array onto a high-resolution CCD sensor of a commercial TV-camera, and a video interface to the electronic system of the aircraft.
- Automatic contrast control: Controlling of off-set and gain based on the reference signals of two thermos-electric cooler (TEC) elements which were regulated to average scene temperature.
- Thermal compensation control loop: De-focusing of a test beam as a reference for adjustment of the focus lens.
- Housings: Designed to fit into the assigned space in the Tornado fuselage with interfaces to the power supply and the air conditioning system of the aircraft.

129 – *ECR FLIR system and Head-up Display (HUD)*
*The left unit of the ECR system (with a dome that points downward when installed in the aircraft) is officially named FLIR sensor (LRU 1) and was developed and build by Carl Zeiss-Sondertechnik. The right unit is the Power Supply (LRU 2) build by Elettronica S.p.A. in Italy. The right-hand picture shows the Head-up Display (HUD) on which the real scene is superimposed by the thermal image.*

**Technical Data of FLIR for ECR Tornado**

| | |
|---|---|
| Detector | CMT (Hg-Cd-Te) |
| Wavelength range | 8 – 12 µm LWIR |
| Cooling | Stirling cooler with linear motor 77 K = -196 °C |
| Detector elements | 180 x 1 |
| Scanner | oscillating mirror with 4:1 interlace |
| Image lines | 720 |
| Frame rate | 42 Hz (60 Hz) |
| Number of image points | 691 200 |
| Infrared optics | IR lenses |
| Window | germanium substrate with DLC-coating |
| Aperture | F/1.6 |
| Field of view | 22.8° x 16.8° |
| Image reproduction | opto-electronically on head-up display |
| Power consumption | 350 W |
| Dimensions | 900 x 210 x 446 mm |
| Weight | 48 kg |

## Mobile Infrared Surveillance System

After the German reunification and the fall of the Iron Curtain in beginning of the 1990s Germany had an open eastern border which was not secured as a military front-line any longer. Therefore, the German BUNDESMINISTERIUM DES INNEREN (BMI) (Federal Ministry of Inner Affairs) had to look for novel technologies to protect this border at day-time and night.

In **1993/94**, the BMI published an invitation to bid for the procurement of in total 100 IR surveillance systems for the BUNDESGRENZSCHUTZ (Federal Border Security). 66 of these IR systems should be installed in VW T4 transport vans. The remaining 34 systems should be delivered in containers to be used later and as maintenance float. In addition, it was a constraint that at least 31% of the volume of orders had to be performed by a company located in the new eastern states. CARL ZEISS-SONDERTECHNIK submitted a quotation based on the reliable Common Module WBG-X thermal imager. At first it was intended to use the new "German Common Module" thermal imagers named Ophelios for this program. At the deadline for the quotation it was clear, however, that the ambitious delivery date of the BMI (all deliveries within

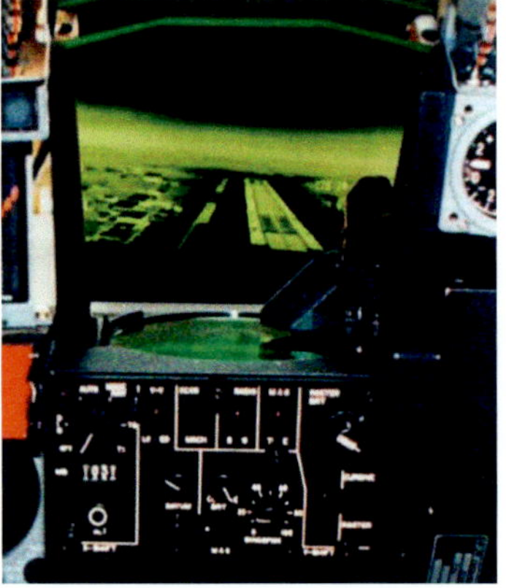

129

**Zeiss machte das Rennen:**

# Entspannung bei „Sondertechnik“

**Auftragsvolumen übersteigt 20 Millionen Mark**

**OBERKOCHEN (cz). Gegen starke nationale und internationale Konkurrenz hat die Firma Carl Zeiss, Oberkochen, als Generalunternehmer eine Ausschreibung des Bundesinnenministeriums über Infrarot-Beobachtungssysteme für den Bundesgrenzschutz gewonnen.**

Nach erfolgreicher Vorerprobung und ausführlichen Vergleichstests erhielt Carl Zeiss den Auftrag, Beobachtungssysteme sowohl für geländegängige Fahrzeuge als auch für den ortsfesten Einsatz zu liefern. Die Systeme dienen der Überwachung von Staatsgrenzen.

Die Realisierung solcher für Tag- und Nachteinsätze geeigneten Beobachtungssysteme basiert auf der Wärmebildtechnologie. Dabei werden Wärmebildgeräte, wie sie bereits in Tausender-Stückzahlen in Deutschland und anderen Staaten Europas in Anwendung sind, installiert.

Das Auftragsvolumen übersteigt 20 Millionen Mark. Ein bedeutender Anteil davon geht im Unterauftrag an Firmen in den neuen Bundesländern. Dadurch, daß es sich im wesentlichen um Fertigungsaufträge handelt, die bis Ende 1995 erfüllt werden, entspannt sich die schwierige Auslastungssituation in der Produktion des Geschäftsbereiches Sondertechnik.

130 – *Report about the border security program in the local press*

*“Zeiss won the race: Relaxation at “Sondertechnik“ Order income exceeds 20 million marks OBERKOCHEN (cz). Against strong national and international competition the company Carl Zeiss, Oberkochen, had won an invitation to tender of the Federal Ministry of the Interior for infrared surveillance systems for the Federal Boarder Police.”*

calendar year 1995) could only be fulfilled with the available and reliable WBG-X.

On **6 July 1994,** the confirmation of the order was given by BMI.

On **22 July 1994,** the production order was started divided into two parts (complete vehicle with thermal imager & stationary thermal imager system). To achieve the requested order volume in the eastern states only the assembly of the thermal imagers was done in Oberkochen while the system integration was performed by CARL ZEISS JENA. Moreover, the vehicles which were provided by the BMI were equipped by BINZ in Ilmenau who also did the acceptance testing. Altogether the program was a significant milestone for CARL ZEISS-SONDERTECHNIK in two aspects:

- The border security systems were the last big program based on the WBG-X and therewith marked the end of the Common Module era.
- It was the entry into a new para-military application. Border security was a field in which the next generations of thermal imagers Ophelios und Attica were successfully utilized over the coming decades until the present day.

Moreover, it was of peculiar significance for the business of CARL ZEISS-SONDERTECHNIK that it was the prime contractor of the BMI in this program and could act as a system house while in most of the other programs it was merely a sub-contractor, e.g. of the tank manufacturer KMW or the submarine shipyard TKMS[386].

In the years **1994-96** in total 120 IR surveillance systems installed in VW T4 synchro vans with four-wheel drive were delivered to the Bundesgrenzschutz (Federal border police). The vehicles were deployed mostly at the eastern border.

The advantage of using the WBG-X for the new application was, of course, that the proven WBG-X component could be utilized. Also, the required display of the thermal image on a monitor was already a solved problem. A new development was necessary for an afocal with the specified fields of view. Although this should not have been a problem it turned to be a real challenge because a narrow field of view of only 1.3° vertically was specified. As every amateur photographer knows

131 – *Sensor unit for the border security program*

*The housing comprises all important assemblies: sensor module, sensor electronics and electronic unit. The big cone on the right-hand side is the mounting of the large front lens made from germanium.*

132 – *VW-T4 frontier van with hoisted thermal imager*

from his telephoto lens a narrow field of view means a long focal length. In case of the WBG-X the large detector length of 12 mm made necessary a focal length of nearly 700 mm. This in turn resulted in a large IR germanium front lens with a diameter of 300 mm. This value was close to the limit of manufacturability and needed a lot of effort by the lens maker.

Moreover, it turned out that the internal lens needed for correcting the color error also needed to have a quite large diameter. It was made from the special IR glass IG 5 which was well-known but disliked for its high thermal expansion, although in earlier afocals it had worked quite well. Alas, there came a nice winter morning when an experienced development engineer carried a box with freshly produced IG 5 lenses over the yard of the Oberkochen facility and rested the it on his writing desk. After a short while he saw with great astonishment that one of the lenses formed cracks. Even small chips sprung off the material onto his table top and eventually the lens had decomposed itself. The problem could be solved together with the material manufacturer VITRON by optimizing the manufacturing process to diminish the internal stress in the lens substrates.

**Technical Data of WBG-X for Border Security**

| | |
|---|---|
| Detector | CMT (Hg-Cd-Te) |
| Wavelength range | 8 - 12 µm LWIR |
| Cooling | Stirling cooler with linear motor 77 K = -196°C |
| Detector elements | 120 x 1 |
| Scanner | oscillating mirror with 2:1 interlace |
| Image lines | 240 |
| Frame rate | 50 Hz |
| Number of image points | 76 800 |
| Infrared optics | IR lenses |
| Focal length | 695 mm |
| Fields of view | WFOV: 4° x 3° NFOV: 1.3° x 1° |
| Image reproduction | electronic display on monitor |

131

132

## 6.5 State of Thermal Imaging in 1992

At the end of the chapter about the 1st generation of Common Module thermal imagers a view on the entire field of thermal imaging seems to appropriate. The aspects of development shall be discussed as well as the different facets of application.

From very individual and genuine technical approaches which were followed well into the 1970s, the Common Module idea was developed, and by international acceptance a more homogenous technical landscape evolved from this idea. In a comprehensive study[387] conducted by JANE'S INFORMATION GROUP even a "typical modern thermal imager" of the time could be presented.

The given values are indeed in the range of the thermal imagers discussed in this chapter. Obviously, the chosen pool of devices was designed with more weight on higher temperature resolution whereby a lower spatial resolution was accepted. The thermal imagers of this chapter, however, were mainly designed for surveillance, i.e. weight was put more on spatial resolution. This can be seen also by the narrower fields of view. That is the kind of compromise, of course, that must be defined newly for every new application.

The JANE'S study also provided valuable insight into various fields of application. For our historic consideration, it is of special interest to see from the attached table how far Common Module thermal imagers have conquered virtually all branches of military service - land, sea and airborne forces.

For some applications, e.g. fire control, surveillance and reconnaissance, there was an amazingly huge supply of thermal imagers which contributes a lot to the high number of 250 imagers in total.

With respect to the high numbers of weapon systems equipped with thermal imagers, it is consistent that these were deployed in the First Gulf War in 1991 in multitude. During this greatest military conflict since the end of World War II,[389] thermal imagers could proof their capabilities in active service, and "Troops interviewed in the Gulf said that they very quickly came to regard thermal imaging as a 24 hour sight preferring it to the standard optics; some tank crews switched on the TOGS (Thermal Observation Gunner's Sight) at the start of the offensive and kept it running until ceasefire."

Also with respect to the still remaining competitive situation with the widely used sights with image intensifier tubes, significant experience could be gathered:

**Typical Modern Thermal Imager 1992**

| | |
|---|---|
| Cool down time | 10 min |
| Image format | 2 : 1 (horizontal : vertical) |
| Field of view | WFOV (wide) 30° x 15° |
| | NFOV (narrow) 7°x 4° |
| Output | Standard Video Format |
| Power | 150 W |
| Dimensions | 300 mm x 200 mm x 200 mm |
| Weight | 10 kg |
| Resolution | |
| spatial | WFOV: 0.25° |
| | NFOV: 0.03° |
| NETD | 0.1°C |

| | SYSTEM | APPLICATION | NUMBER |
|---|---|---|---|
| LAND | MISSILES | SEEKERS<br>ANTI-ARMOUR<br>AIR DEFENSE | 2<br>11<br>18 |
| | FIGHTING VEHICLES | TANK FIRE CONTROL<br>COMMANDERS SIGHTS<br>GUN SIGHTS<br>DRIVERS SIGHTS | 38<br>4<br>16<br>1 |
| | INFANTRY | WEAPON SIGHTS<br>FORWARD OBSERVATION<br>SURVEILLANCE | 3<br>5<br>28 |
| SEA | FIRE CONTROL | MISSILES<br>WEAPON DIRECTION | 2<br>32 |
| | SURVEILLANCE | MISSILE DETECTION<br>MARITIME SURVEILLANCE<br>PERISCOPES | -<br>20<br>11 |
| AIR | MISSILE SEEKERS | AIR-GROUND<br>AIR-TO-AIR | 5<br>30 |
| | MISSILE FIRE CONTROL | GROUND ATTACK | 31 |
| | RECONNAISSANCE | AIR RECCE<br>UNMANNED AIRCRAFT<br>MARITIME PATROL<br>SEARCH & RESCUE | 45<br>16<br>4<br>26 |
| | NIGHT FLYING | NAVIGATION<br>PILOT AIDS | 9<br>12 |

133 – *Market review of military thermal imagers*

*Assorted in land, sea and airborne forces for distinguished weapon systems various applications are listed together with the number of the dedicated thermal imaging systems.*[388]

"The superiority of thermal imaging over image intensification was further demonstrated by the fact that British tank troops were able to see out to beyond their own reconnaissance forces visual range."[390]

These insights for sure stimulated new ideas and it is not astonishing at all that in 1992 the "thermal imaging countries" (USA, Great Britain, France, Netherlands, Israel, Sweden, Germany) pushed forward thermal imaging with 27 technology programs on system level and further 21 programs on equipment level.

Even though the expectations about increasing quantities were muted somewhat after the end of the cold war, the application potential of thermal imaging was sufficiently proved – a good base for the 2nd generation of thermal imagers that was underway by that time.

7

# THERMAL IMAGERS OF 2ND GENERATION UNTIL 2005: SIMPLY DIGITAL

"Doubling the target identification range of current in-service IR sensors whilst maintaining ... the situational awareness (Field of View)."

*Objective of British STAIRS C program (early 1990s)*[391]

"One Principle. Several Versions. Many Fields of Application."

*Zeiss Optronik about the Ophelios program (1999)*[392]

The US-German Common Modules as well as their British and French counterparts had been of great success in technology and quantity as we have seen in the preceding chapter. Due to the availability of well performing thermal imagers in large numbers and to reasonable costs, many new applications could not only be tested but also brought to reality.

Nevertheless, already in the early 1980s the first ideas about further development or even replacement of the Common Modules were considered.

In **1981-82,** a study funded by the "Bundesamt für Wehrtechnik und Beschaffung" BWB (Federal office for military technology) was conducted by CARL ZEISS-SONDEROPTIK in which the development opportunities of all Common Modules were thoroughly investigated. At this time the expectation was that "the further development of the CM-Systems … in the midterm would be mainly in the electronic field and not so much in the context of the detector"[393]. "In this sense the parallel video signals injected by the CM-detector into the signal chain detector-preamplifier-main amplifier-LED will be digitized as close as possible to the detector (by means of a so-called Digital Scan Converters, DSC) and stored in digital form in a semiconductor memory to be converted by a so-called Digital Reformatter DRF in a serial standard video signal to be displayed on a commercial monitor."

In **1987,** a presentation of HUGHES AIRCRAFT COMPANY stated that the "2nd generation has taken longer than anticipated" and that the "present common modules are not fulfilling new system requirements". In consequence the study suggested to "implement a systematic planned upgrade to the common modules". To do this three stages of improvement of an "Advanced Modular FLIR" AMF I to III were presented which already comprised most of the measures (like polygon scanner, digital signal processing or linear cooler with remote cold finger) that were realized later in the 2nd generation. It was also proposed to put a part of the digital electronic in the Dewar of the detector. All plans, however, were still related to a photo-conductive CMT detector with a single column of elements.

In **1988,** CARL ZEISS-SONDEROPTIK submitted a definition study about a thermal imager named PC 288 to the BWB being the contracting entity. The layout of the PC 288 followed in principle the lessons learned in the previous study from 1981/82. The companies AEG in Heilbronn and Wedel, ELTRO in Heidelberg and MBB in Ottobrunn participated in the work – that was by now almost the same consortium which later would develop the German Common Modules in the 1990s. For the new PC 288 thermal imager, an impressive quantity of more than 339 devices were planned which should be installed in the gunner sight PERI-ZWFG and the commander's panoramic sight of the upcoming hunter tank Panther. A functional model should be ready until end of 1990 and a prototype built, tested and qualified until end of 1992. Series production should start in the mid of 1994.

From **1989-92,** a prototype of the PC 288 was developed, built and tested. Compared to the later thermal imagers of the 2nd generation the performance of this device was determined by the performance of the chosen detector AEG PC 288. With 288 detector elements and the commonly used 2:1 interlace the standard number of 576 image lines could be realized easily.

The AEG PC 288 detector like all Common Module detectors was of the photo-conductive (pc) type and therefore suffered from the typical limitations determined by its design:

- Very high number of wires
- Very high number of amplifiers which all had to be equalized
- Complicated Dewar
- Limited number of detector elements in only one column
- Direct connection to Si electronic circuits not possible

Therefore the PC 288 thermal imager had a kind of androgenic position between the 1st and 2nd generation thermal imagers. Effectively, it was a generation 1.5, and that perhaps was one of the reasons why it never made it to series production. Moreover, at the end of the 1980s it was already clear that great progress in the field of detector technology could be expected in near future.

Around **1990**, the first new linear detectors of the 2nd generation became available, namely the EURODIR IRCCD 48 x 4, the IRCCD 288 x 4 and the MULLARD 256 x 4.

For the new generation, several manufacturers offered so-called photo-voltaic (pv) detectors. Each detector element was composed p- and n-doted semiconductor material that formed a pn-Diode, similarly to what already was used in contemporary silicon electronics. These detector elements not only were more sensitive, their high electric impedance was suited perfectly for a direct connection to CCD registers in Si-technology. In this way, it was possible to perform a part of the digital signal processing already on the chip. With so-called on-chip multiplexers the signals of several detector elements could be fed into one electrical line and processed by the same amplifier chain in an identical manner.

**Technical Data of PC 288 (Basic Unit)**

| | |
|---|---|
| Detector | CMT (pc) |
| Wavelength range | 7.7 - 11.8 µm LWIR |
| Cooling | Linear-Split-Cooler, 77 K = -196°C |
| Detector elements | 288 x 1 |
| Element size | 25 x 25 µm |
| Scanner | oscillating mirror with 2:1 interlace & integrated Scan Position Sensor |
| Image lines | 576 |
| Image format | 768 x 576 |
| Frame rate | 25 images/sec |
| Number of image points | 442 368 |

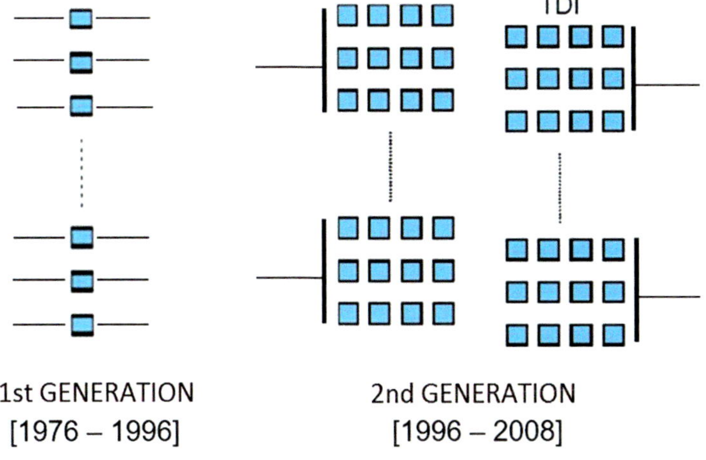


134

134 – *Detector geometry of 1st and 2nd generation*

*In a Common Module detector of the 1st generation all detector elements are arranged in a single column one beneath the other; each detector element is connected to two bond wires and consequently a high number of separate electrical conducting paths must be formed through the vacuum vessel to the outside connectors. Detectors of the 2nd generation have several elements side by side for the TDI process which are combined in groups via an on-chip multiplexer and connected to a lower number of read-out lines.*

135 – *Functional principle of 2nd generation thermal imagers*
*As in the Common Module imagers of the 1st generation, the thermal radiation emitted by the scenery is imaged via afocal, scan mirror and imager onto the detector surface. As there is no optical image reproduction any more, a Scan Position Sensor (SPS) is measuring the angular position of the scan mirror and supplies these signals for the synchronization of the read-our process of the detector. The detector signals are directed via multiplexer (MUX) and analog-digital converter (ADC) as digital values into an image memory. A standard video format (e. g. the CCIR standard) which can be displayed on a commercial TV monitor is created by a digital scan converter (DSC), also known as reformatter.*

Without the great many electrical wires, it was affordable to implement multiple detector elements, e.g. to arrange 4 elements in scan direction side-by-side. In that time 96 x 4 or 288 x 4 or 480 x 4 detectors became typical at companies like AIM, SOFRADIR or TEXAS INSTRUMENTS.

During the scan procedure, these neighboring detectors were read-out with a defined time delay, a method internationally known as "Time Delay and Integration" TDI. During the time when an image point is scanned across the 4 TDI-detector elements each element is read out at the right time so that the image point is measured once by each element. The 4 measured values are summed up to a total signal, and the signal/noise ratio of the sum is improved by a factor of 2, and so is the temperature resolution. As we shall see at the end of this chapter, this turned out to be the greatest advantage of the 2nd generation detectors.

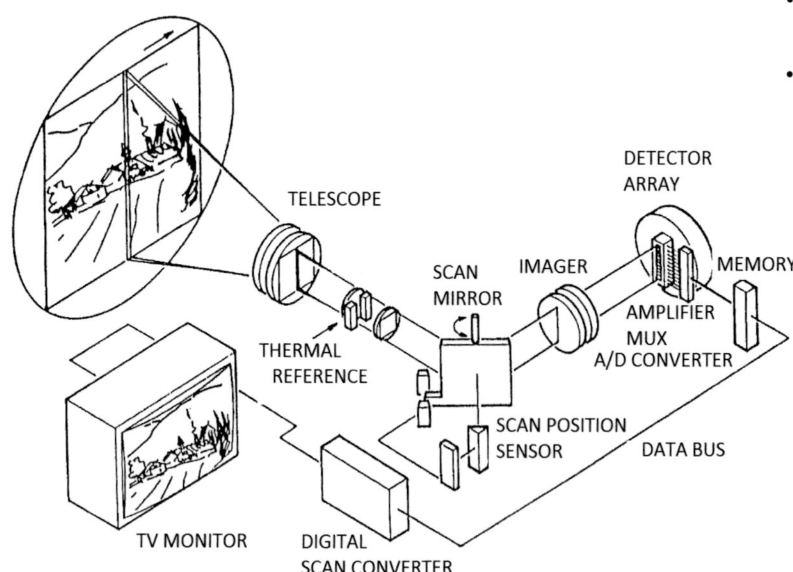


A further advantage of the TDI detector elements was that the defect of one element does not meant that an image line was missing as would have been the case with a Common Module detector. Only the signal quality was slightly decreased in this case.

To make the TDI-process work properly, the scanner motion must be timed exactly. To this purpose, the angular position of the scan mirror must be measured continuously with high precision. This was performed with a so-called "Scan Position Sensor" SPS.

For further processing of the digital output signals of the proximity electronics and conversion to a standard video signal, the algorithms for digital image enhancement offer a variety of options especially as the digital components become smaller and faster with every development step.

Very early, "second generation technology has split into two evolutionary paths:[394]

- "cheaper, smaller, lighter" like the German Ophelios and the French Synergi systems, and
- "higher performance in the same package" like the American HTI and British STAIRS C thermal imagers

## 7.1 Emergence of the 2nd Generation Common Modules

At the beginning of the **1990s**, the work on the next generation of thermal imagers started in all "Common Module countries" almost simultaneously. After the greatly successful Common Modules of the 1st generation it was obvious to follow a modular concept again.

The German industry engaged in thermal imaging, for example, conducted several studies on a new generation of thermal imagers and defined the following requirements which were phrased in similar words also in other countries:

- Reduction in volume, weight & power consumption
- Larger number of detector elements
- Improved thermal sensitivity
- Standard video output
- Cost reduction
- Longer lifetime

Around **1991**, the first 2nd generation detectors became available, "the French Sofradir 288 x 4 detector being the first detector to be widely available in series production"[395]. As a detector material, the established mercury-cadmium-telluride (HgCdTe or CMT) was used for a photo-voltaic (pv) type of detector. In this type, the incoming photons (light quanta) generated charge carriers which changed the voltage of a capacitor. Multiplexers in CCD technology were integrated on-chip to perform the TDI process and the serial read-out of the signals.

Based on this SOFRADIR detector two Common Module systems were created in France which were pushed by two groups of the domestic industry in parallel:

- The Synergi system of THOMSON-CSF OPTRONIQUE (later re-branded in TTD-THALES), of which also CARL ZEISS-SONDERTECHNIK and the British PILKINGTON OPTRONICS were a part: Synergi modules were used for Catherine-GP, FP or FC, the stabilized reconnaissance system Chlio-S and in the German LITENING II Pod.
- The Iris system of SAGEM: Iris modules were used in a visor of the Tigre helicopter, in a reconnaissance system for the Euroflir drone Sperwer (sparrow hawk) and in thermal imagers installed in the Mirage 2000 and in periscopes of the nuclear submarines SNLE-NG (Triomphant class).

For the performance of the French thermal imagers the following typical values for the detection range were anticipated whereby the values for recognition should be half of the value:

- Detection range against foot soldier 12 km
- Detection range against tank 15 km
- Detection range against fighter aircraft 25 km

In **1992**, a consortium of five companies began work on the development of German Common Modules (DCM):

- CARL ZEISS-SONDERTECHNIK in Oberkochen
- ELTRO, Gesellschaft für Strahlungstechnik in Heidelberg
- ATLAS ELEKTRONIK in Bremen
- AEG INFRAROT MODULE (AIM) in Heilbronn
- TELEFUNKEN MIKROELEKTRONIK (TEMIC) development center in Ulm

136 – *Consortium for the German Common Modules (DCM)*

*The company AIM in Heilbronn was in charge for detector and cooler, Eltro for scanner and multiplexer & analog-digital converter EMUX/ADC, TEMIC for the Proximity Electronics close to the detector, Atlas Elektronik for the system electronics und Carl Zeiss-Sondertechnik for IR imager, IR afocal and integration of the complete thermal imaging system. Carl Zeiss-Sondertechnik also acted as a speaker of the consortium towards the official authorities.*

To give name to the new module system, in the first time the acronym Helios was chosen. Because of trademark infringements the name was later changed to Ophelios which was the abbreviation for "Optronisches Passives Hoch Empfindliches Leichtes Infrarot-Optisches System" (Optronic high-sensitive light infrared-optical system).

The responsibility for the individual Ophelios modules was evenly distributed among the partners as is shown by the attached diagram.

In early **1993**, the optimistic planning aimed at a series ramp-up towards the end of 1994, alas it took somewhat longer in the end. The reason was perhaps that all components had to be newly developed. And this made necessary a lot of innovative solutions for almost any detail as is shown in the following. Furthermore, the chosen concept turned out to be a lot more demanding than the participants had anticipated based on their US-German Common Module experience.

At the beginning of the **1990s**, in Great Britain the DEFENCE RESEARCH AGENCY (DRA) and PILKINGTON THORN OPTRONICS jointly began a program called "Sensor Technology for Affordable IR Systems" (STAIRS C) to replace the first TICM generation. The predefined objection was "doubling the target identification range of current in-service IR sensors whilst maintaining … the situational awareness (Field of View)". The base of the planned high-performance system was a CMT detector with 768 x 8 detector elements that was used to record a HD (high-definition) image with 1250 x 768 image points together with a simple serial scanner. The signals of the 8 detector elements laying side-by-side were summed up using the "Time Delay and Integration" TDI method to improve the signal. It was common practice to screen out the worst performing detector elements and use only the best 6 of 8 elements for the TDI process.[396]

Midyear **1990s**, the US Army chose IRCMOS detectors with 480 x 4 and 480 x 6 detector elements for their new maximum-resolution thermal imager and integrated these into a "Standard Advanced Detector Assembly II" (SADA II). This unit became an essential part of the FLIR devices for the "Horizontal Technology Integration" (HTI) initiative. By enlarged detection ranges and broader horizontal fields of view the US Army tried to give their tank crews much better chances of survival.

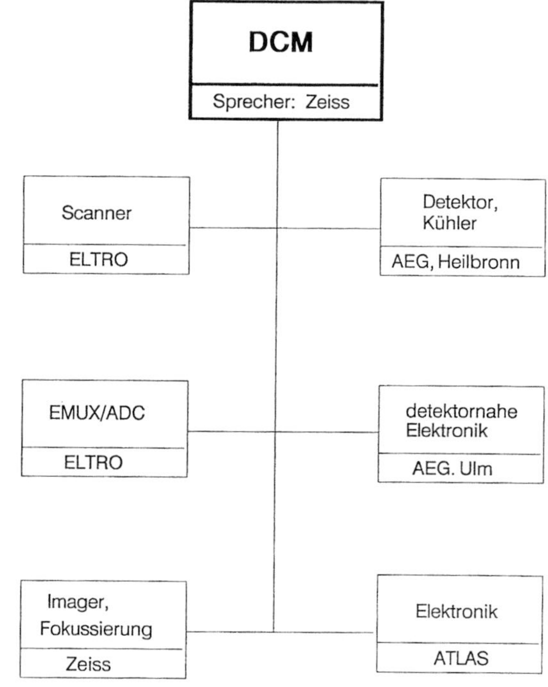

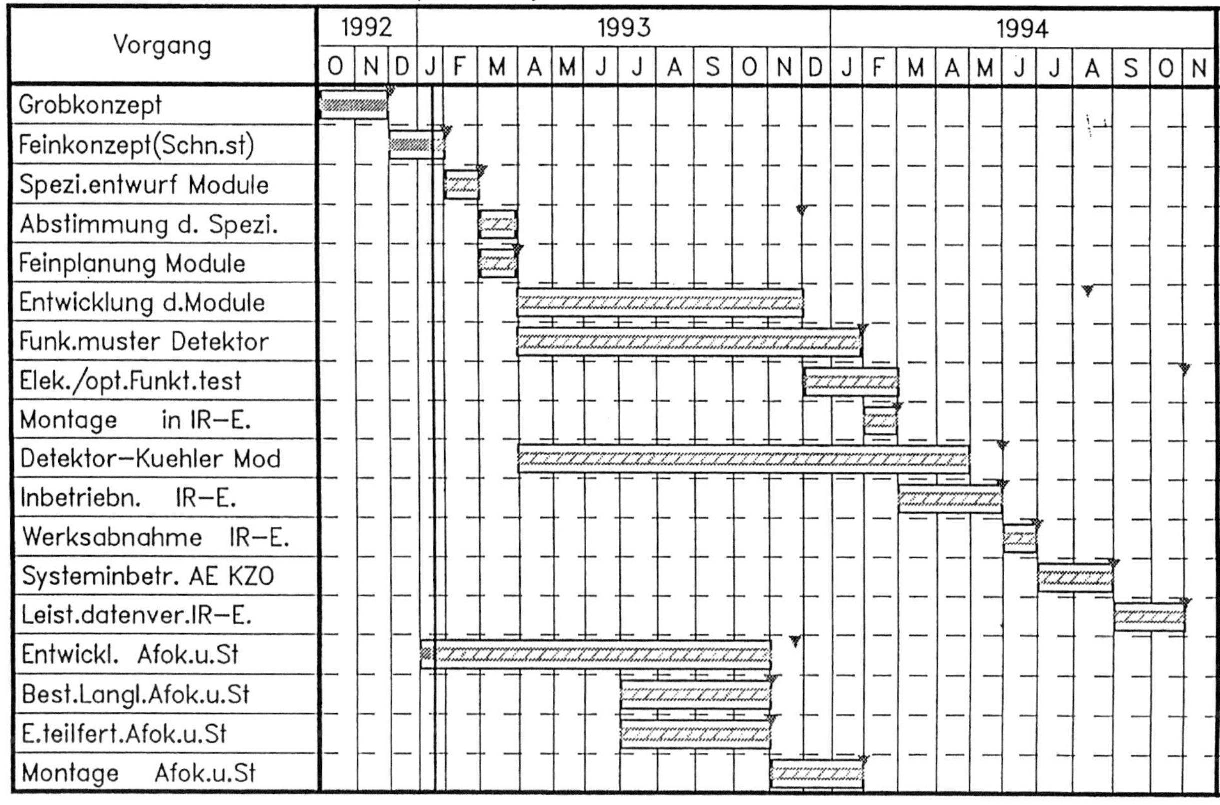


137 – *Development schedule for the German Common Modules*

*At the beginning of 1993, the planning of the KZO thermal imager was done which is given here as an example. First deliveries were anticipated optimistically to take place towards the end of the year 1994. At that time, the program was still known under the name Helios.*

In **1994**, the modules of the Synergi system went into production. At CARL ZEISS-SONDERTECHNIK the IR imager, the Scan Position Sensor SPS and some of the electronic boards were manufactured.

In **1995**, the company ELTRO GmbH in Heidelberg, a subsidiary of the DASA, was merged with CARL ZEISS-SONDERTECHNIK to become ZEISS-ELTRO OPTRONIK GmbH (ZEO) based in Oberkochen. By this merger ZEO became responsible for the largest workshare of the Ophelios modules.

In the same year work began at ZEO on the system concept and module definition of a "High Definition Infra Red" (HDIR) thermal imager. The objection was a further development of the 2$^{nd}$ generation:

- Image format 1920 x 1152 image points (h x v)
- Aspect ratio 16:9
- Video image compliant to the HDTV standard

This development was funded by the German "Bundesamt für Wehrtechnik und Beschaffung" BWB (federal office for military technology and procurement).

In **1996,** the series production of the Ophelios modules began with some delay compared to the original planning. Technical details and applications are discussed separately in Chap. 7.2 and 7.3.

Also in **1996,** the "Program Executive Office Intelligence and Electronic Warfare" in Ft. Belvoir (USA) sent first information about the American HTI thermal imager to ZEISS-ELTRO OPTRONIC after a "U.S.-German Research and Technology Projects Memorandum of Understanding" (MoU) had been signed by the governments of the United States and Germany in the year before on 17 March 1995.

On **19 June 1997**, an attachment to the MoU of 1996 titled "Project Agreement No. RTP-US-GE-A-97-0010" related to "Improved Forward Looking Infrared Devices" was agreed between the American DoD and the German BMVg. The objective of the work planned for a duration of 3 years (3/1997 to 1/2001) was to make the HTI thermal imager fit for installation into the German main battle tank Leopard 2. For this purpose, the Americans made available a so-called "B-Kit" as a loan that was "defined as the core components of the 2nd Generation FLIR which are common to multiple applications. These components are: the afocal assembly, imager assembly, detector/cooler assembly, scanner control CCA, digitizer CCA, Cooler control CCA, Video Processor CCA, Video Converter CCA, Interface Control CCA, Power Supplies, EMI filter and Electronics Unit housing."[397] (CCA = Circuit Card Assembly). The work of ZEISS-ELTRO OPTRONIC was to develop a completed and closed thermal imager based on these components that was compatible with the available space in the Leopard tank. This would then be the "A-Kit" according to American definition. The planned workflow was remarkable in the way that the German A-Kit had to be shipped to the United States to do the integration of the American B-Kit at TEXAS INSTRUMENTS. Afterwards the completed device came back to Germany again.

In this context, it is interesting that the managers at ZEISS-ELTRO OPTRONIC in this time obviously reckoned they still had to persuade the German officials of the benefits of the 2$^{nd}$ generation thermal imagers and they intended to use the HTI for this purpose. In the minutes of a meeting in 1997 was stated:

"Objectives of the Cooperation

- Superior to the program: Fostering the cooperation with the USA
- To convince the utility provider in D that 2$^{nd}$ generation FLIR is an improvement over Common Modules
- Comparison with Ophelios (installation of Ophelios can be justified by doing the comparison) (price-performance ratio) …"

At the same time, it should be avoided, of course, that the own German Ophelios thermal imager looked too bad.

The American-German HTI device was completed according to plan and tested in February 1998 and in January 1999 at the "Forschungsinstitut für Optik" FfO (research institute for optics) in Tübingen.

In 1997, the development work on an improved Synergi 2 system was started. At the company in Oberkochen by now branded as ZEISS OPTRONIK the thermally unstable IR-imager lens was further developed to an athermal (thermally compensated) objective.

In 2000, SOFRADIR followed the technical trend and switched their 288 x 4 IRCCD detector to IRCMOS technology. CARL ZEISS OPTRONICS assured themselves the availability of 80 wafers of the older version so that they could finalize the development work without design change.

In 2001, a HTI thermal imager was offered on the European market by CARL ZEISS OPTRONICS in cooperation with the American based company RAYTHEON who already had equipped the main battle tank M1A2 Abrams and the fighting vehicle M2A3 Bradley of the US Army as well as other platforms.

138 – *HTI modules (B-Kit) and German HTI-Sensor*

From **2002 to 2004**, a new development program "HTI for LEO II" was conducted at ZEISS OPTRONIK with the objective to bring the HTI thermal imager into the upcoming Leopard 2 program for Greece. The result of this effort became obvious later when all Greek Leopard tanks were equipped with Ophelios thermal imagers for the gunner and the commander as well. In the next chapter, this will be discussed in more detail.

Despite a higher performance compared to Ophelios and Synergi thermal imagers also in the following years CARL ZEISS OPTRONICS were unable to acquire any European customer for the HTI thermal imager.

RAYTHEON on the other hand could deliver more than 16,000 thermal imagers for the American HTI program in the following years.

From **2002 to 2007**, STAIRS C thermal imagers were produced by THALES for the air defense system Stormer SPHVM (Self Propelled High Velocity Missile) that was equipped with the Starstreak missile. In these thermal imagers, also a module from the Synergi system was used, namely the Scan Position Sensor (SPS) from which 135 devices were delivered by CARL ZEISS OPTRONICS. This did not happen without some teething problems because the own Synergi activities were virtually shut off at that time.

In **2002**, as a final development stage a prototype of the HDIR 2.2 with 576 x 6 CMT detector was finalized by CARL ZEISS OPTRONICS. Because of the limited thermal sensitivity, the large dimensions and the high weight this concept was not followed any further. The upcoming competition of the so-called staring thermal imagers with matrix detectors (e.g. with 320 x 256 detector elements) in that time as likely as not had

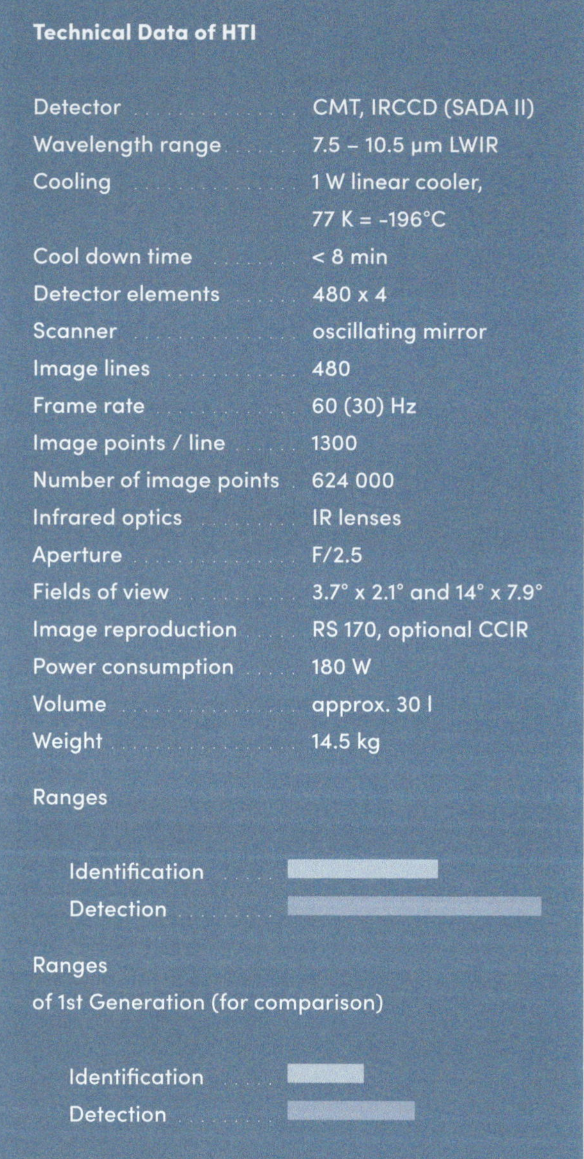

**Technical Data of HTI**

| | |
|---|---|
| Detector | CMT, IRCCD (SADA II) |
| Wavelength range | 7.5 – 10.5 µm LWIR |
| Cooling | 1 W linear cooler, 77 K = -196°C |
| Cool down time | < 8 min |
| Detector elements | 480 x 4 |
| Scanner | oscillating mirror |
| Image lines | 480 |
| Frame rate | 60 (30) Hz |
| Image points / line | 1300 |
| Number of image points | 624 000 |
| Infrared optics | IR lenses |
| Aperture | F/2.5 |
| Fields of view | 3.7° x 2.1° and 14° x 7.9° |
| Image reproduction | RS 170, optional CCIR |
| Power consumption | 180 W |
| Volume | approx. 30 l |
| Weight | 14.5 kg |

Ranges

Identification

Detection

Ranges of 1st Generation (for comparison)

Identification

Detection

**Technical Data of HDIR**

| | |
|---|---|
| Detector | CMT, IRCCD |
| Wavelength range | 7.5 - 10.5 µm LWIR |
| Cooling | 1 W linear cooler, 77 K = -196°C |
| Detector elements | 576 x 7 |
| Scanner | oscillating mirror with 2:1 interlace |
| Image lines | 1152 (HDTV) |
| Frame rate | 50 (25) Hz |
| Image points / line | 1920 |
| Number of image points | 2 211 840 |
| Infrared optics | IR lenses |
| Aperture | F/1.7 |
| Fields of view | 4.7° x 2.6° and 14.6° x 8° |
| Image reproduction | RS 170, optional CCIR |
| Power consumption | 140 W |
| Volume | approx. 40 l |
| Weight | 20 kg |
| Manufacturer | Zeiss Optronik GmbH |

contributed to the insight that the HDIR development path had been a blind alley.

In 2008, the series production of the Ophelios modules ended after more than 3000 German Common Module thermal imagers had been delivered.

139 – *HDIR thermal imager*

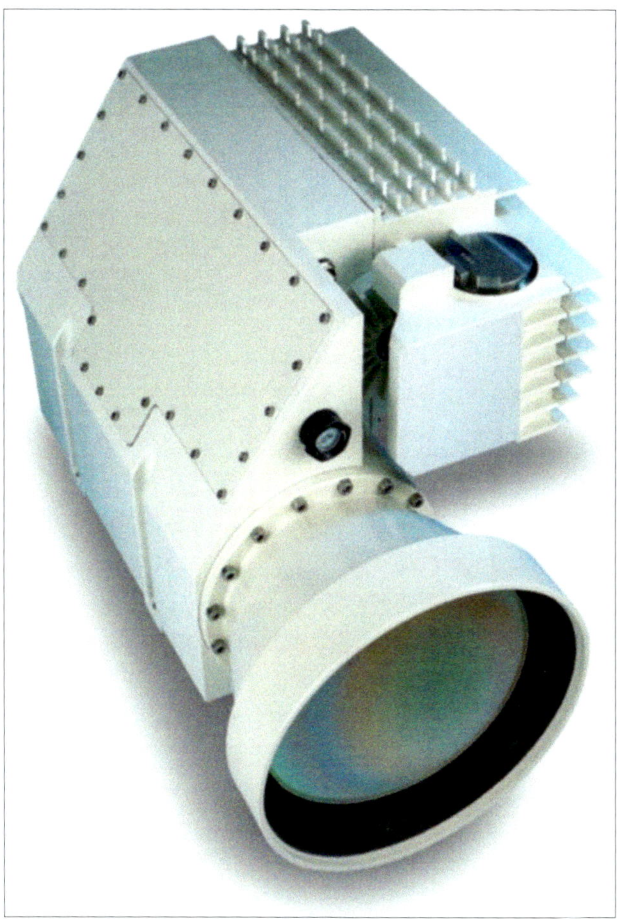

139

## 7.2 Technology of the German Ophelios Modules

The functional concept of Ophelios was based on a relatively small and therefore cost effective 96 x 4 detector in combination with a fast rotating polygon scanner. The scenery was scanned in 3 image strips stacked over each other, and which were combined to the total image in the image memory. By means of a 6-surface polygon the interlace function could be realized without further parts. In total 96 · 6 = 576 image lines were generated as required by the CCIR video standard. Due to the 4-fold TDI function (Time Delay and Integration) each image point was measured by 4 detector elements, and by this the signal quality was improved by a factor of 2.

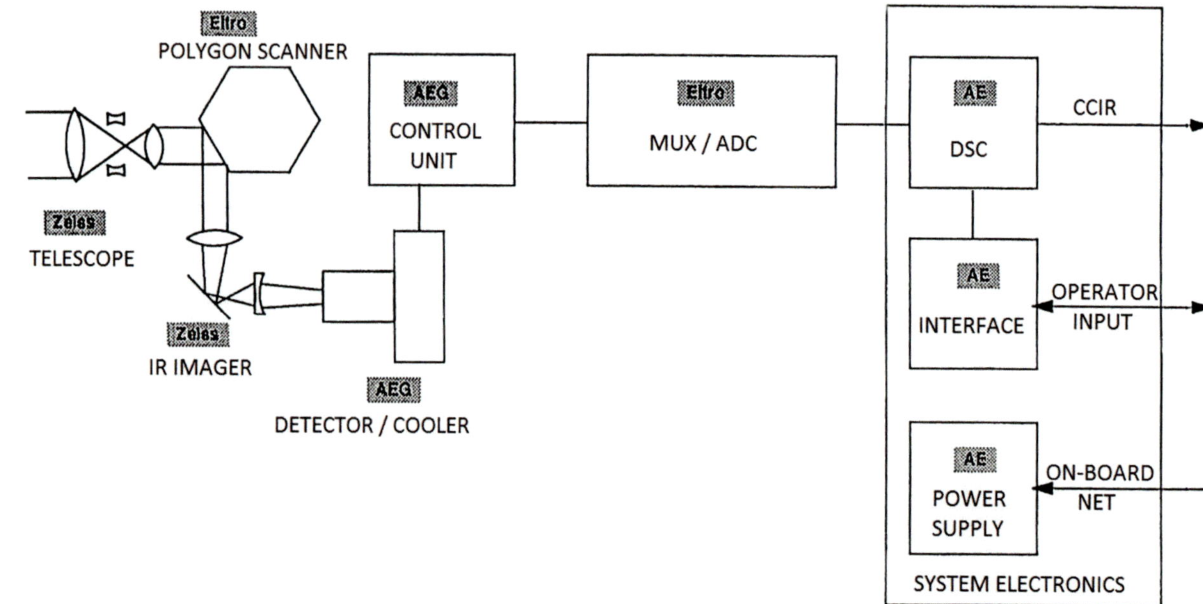


140 – *Functional principle of the DCM thermal imager Ophelios*

*The thermal radiation passes the afocal telescope by which the specified field of view is achieved and is then deviated by a surface of the rotating polygon into the imager. The scene is imaged onto the surface of the cooled detector which converts the irradiation into an electrical signal. The signals generated by the individual detector elements are directed via a multiplexer (MUX) into an analog-digital converter (ADC). The digital values are arranged in an image memory by a Digital Scan Converter (DSC) in a succession compliant with the chosen video standard. During each rotation, at one designated edge position of the polygon, the detector looks onto the surfaces of two thermo-electrical coolers (TEC). The temperature of the TECs is regulated to the average scene temperature. The signals measured in this manner are used as a reference for capturing and reproduction of the thermal images. (AIM = AEG Infrarot-Module, AE = Atlas Elektronik)*

141 – *Ophelios video image generation*

*With each revolution of the polygon, a full thermal image is recorded. Because the first 3 polygon faces are tilted against each other by 5°, the first 3 images strips recorded by means of this surfaces are shifted vertically by 10°. Each image strip is captured a second time by the remaining 3 polygon faces with a shift in height of one half of the line width. Therewith each half image comprises of 192 image lines and the full image has 576 lines in compliance with the CCIR standard. Once in every revolution at a defined edge position of the polygon the line of sight of the detector is directed onto two thermo-electrical elements to perform a reference measurement for the automatic calibration of the video images. The detector has 4 detector elements side-by-side in scan direction which are known as TDI detector elements: During image capturing every image point is scanned over his 4 TDI elements and is measured by each element. The total signal for the image point is the sum of the 4 single measurements, and this sum has a two times better signal-to-noise ratio. The detector signals are refined by the proximity electronics (PROXY) and passed through multiplexer and analog-digital-converter (MUX/ADC) into the memory of the system electronics.*

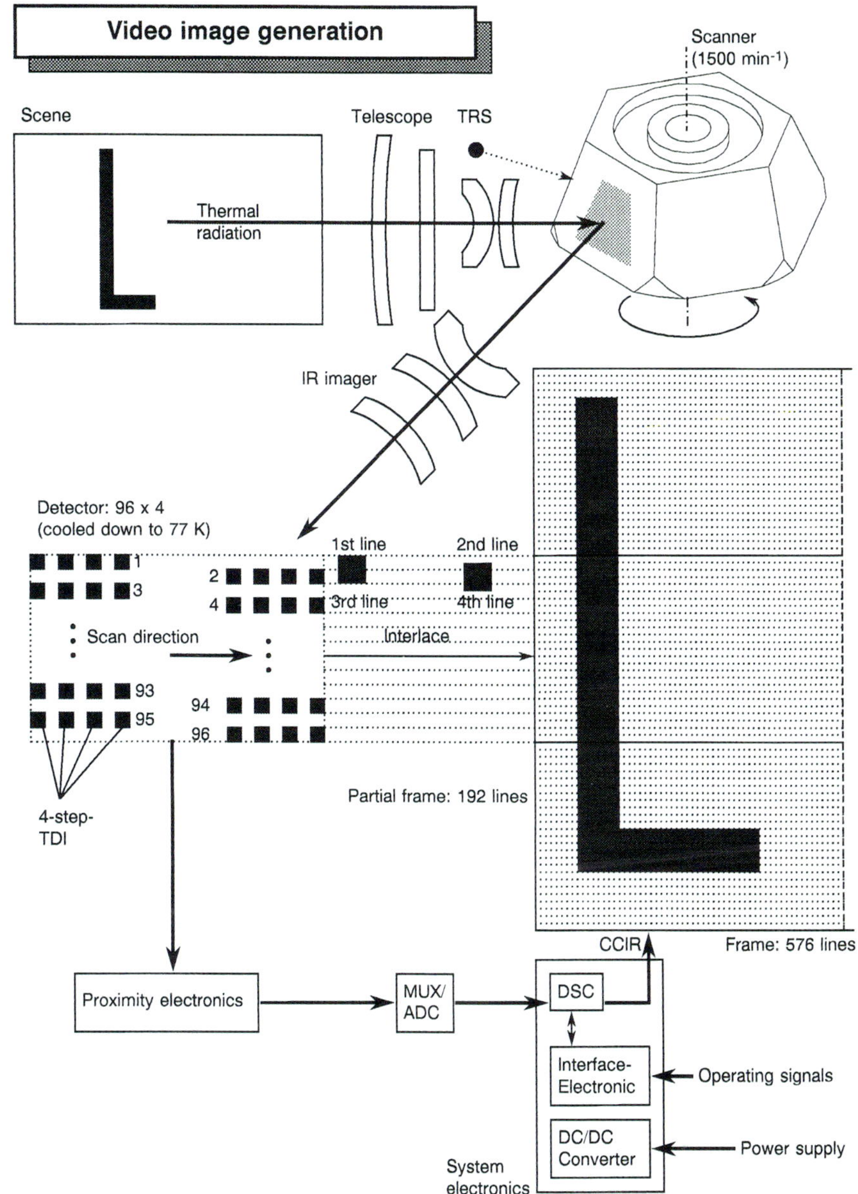

## Ophelios polygon scanner

As is depicted in the sketch of the image capturing the polygon scanner had great part in the generation of the video images. The polygon was driven by a small electro motor and rotated with a constant speed regulated to 1500 rev/min (or 25 revolutions per second). An incremental encoder was fixed on the axis which served as a so-called Scan Position Sensor (SPS) to measure precisely the angular position of the polygon. The rotation of the polygon happened with an exactly defined phase, and so could be synchronized with the read-out cycle of the detector.

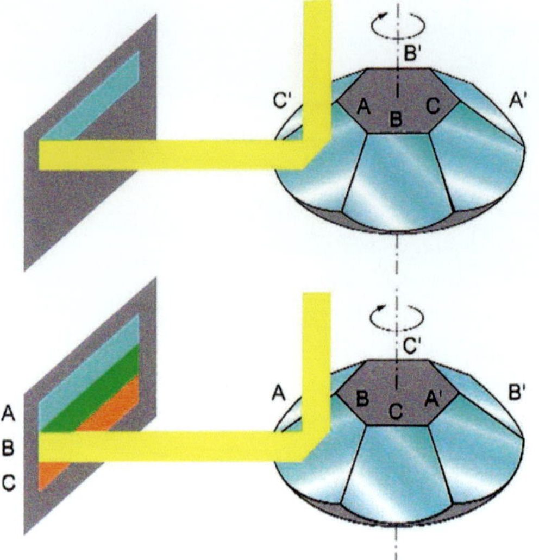


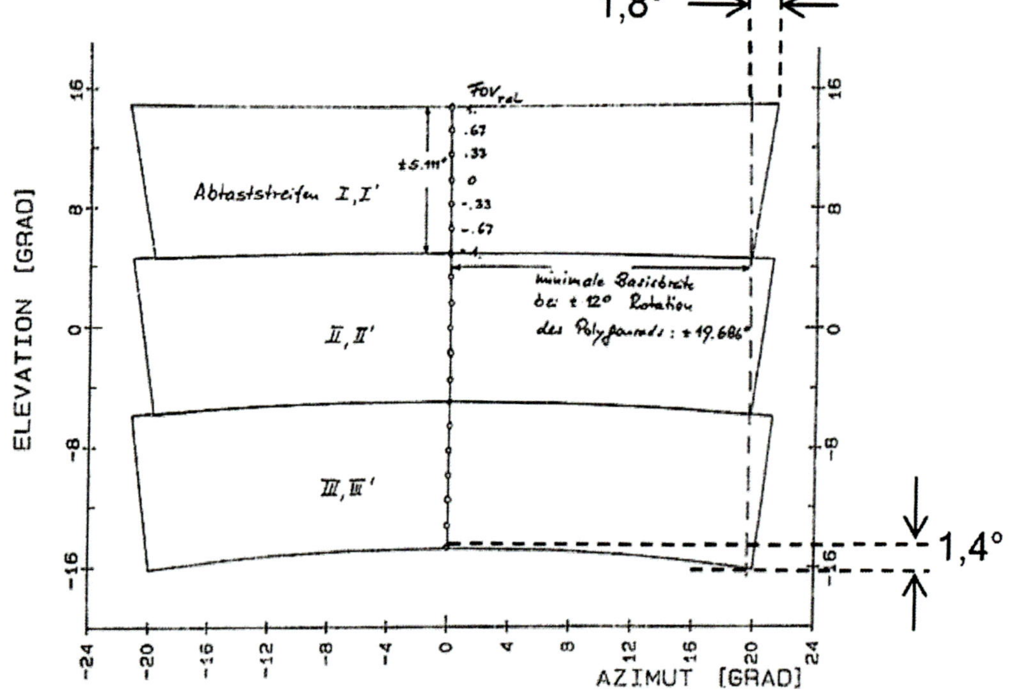


142 – *Ideal and real image stripes*

*In the sketch on the right the ideal function of the scanner is depicted: the image strips created by the faces A, B and C of 10° height are straight and attached perfectly to each other. Faces A', B' and C' create 3 image strips in an analog manner but with an interlace, i.e. the lines are shifted vertically by one half of the line width (≈ 5/100 angular degrees)*

*The diagram below shows the real shape of the image strips which are not oblong but show a significant distortion horizontally and vertically. The horizontal distortion amounts to up to 1.8° and is corrected by the system electronics. The vertical distortion (up to 1.4°) appears only towards the lower corners of the image and can therefore be accepted.*

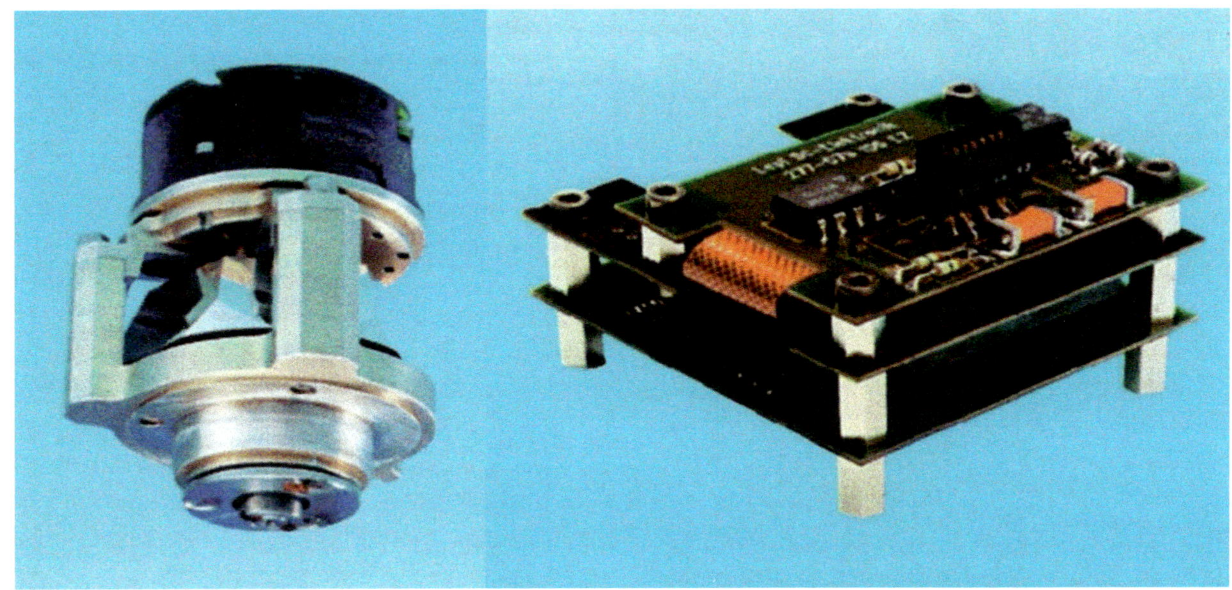

143

143 – *Polygon scanner (left) and scanner control board*

The 6 faces of the polygon were tilted by different angles. Therewith it could perform a combination of parallel and sequential scanning and capture a two-dimensional image. The faces A, B and C generated 3 connected image strips with a height of 10° which result in a full-format image of 40° x 30°. The second group of faces A‘, B‘ and C‘ performed the 2:1 interlace by generating a second image of the same size but shifted vertically by one half of the width of an image line.

Thus, using the principle of a polygon-scanner a full-format image including interlace could be captured with a very compact assembly. Practical experience showed, however, that the reflection of ray bundles at an inclined and rotating surface resulted in significant distortion of the image due to fundamental geometrical reasons. The solution presented here was the optimum

**Technical Data of Ophelios Scanner**

| | |
|---|---|
| Type | Polygon scanner with 6 faces |
| Revolutions | 1500 /min |
| Scan Position Sensor | incremental angular encoder |
| Image zones | 3 + 3 |
| Width of zone | 10° |
| Total field of view | 40° x 30° |
| Interlace | ½ image line, vertical |
| Designer | Eltro GmbH |
| Manufacturer | Zeiss-Eltro Optronics GmbH, later Carl Zeiss Optronics GmbH |

**144 – *Functional principle of a re-imager with intermediate image plane***

*The incoming parallel ray bundles are at first imaged by the front lens of the scanner into an intermediate image plane, and afterwards are imaged by the rear lens of the imager onto the detector surface. The re-imager has an accessible exit pupil behind the last lens in which the cold shield of the detector can be positioned easily. The cold shield, which protects the detector from disturbing thermal radiation emitted by the surroundings, is in turn imaged by the re-imager onto the entrance pupil where the scanner is located. The effect of this assembly is that the detector elements looking through the cold shield can only look into the scenery without seeing any unwanted radiation from the optical mounts or the inner housing.*

result after comprehensive modelling. The residual errors could only be compensated by electronic image processing which was done (at least in part) in the system electronic of the Ophelios thermal imager.

## Ophelios imager

Early in the 1990s a concept of the Helios imager, as it was called then, was derived from the Common Module imager.

In **1992**, a first specification was compiled in which a 3-lens imager with an intermediate image plane inside the imager was described, a so-called re-imager. As was correctly realized in the specification, an intermediate image plane is necessary for a correct optical linking of the optic to the cold shield of the detector.

However, the optical design of the Ophelios IR-imager proved to be more challenging compared to the Common Module imager for a number of reasons:

- A re-imager with intermediate image plane needed 3 times as much refractive power than a Common Module Imager, practically spoken this meant a larger number of lenses.
- To fit the three image strips generated by the polygon scanner together one to another without any gap, the nominal value of the focal length had to be adjusted with high accuracy.
- The focal length was not allowed to change with temperature as gaps between the image strips would have open otherwise or there would have been an equally disturbing overlap.
- Due to the small frame size the manufacturing tolerances were almost an order of magnitude tighter than that of the Common Modules.

In hindsight, it is not much astonishing that several development steps were necessary to fulfill all requirements with sufficient quality.

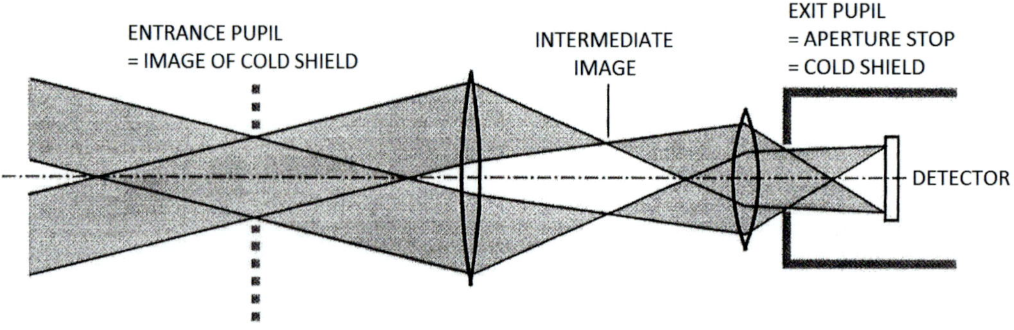

In **February 1993,** as a first optical design a 7-lens imager using the customary IR materials zinc selenide ZnSe, germanium Ge and the chalcogenide glass Amtir 3 was presented. The high number of lenses was necessary to correct color error and thermal error at the same time. Because of the number of lenses being (too) large compared to the specification this path was not followed any longer. To reduce the number of lenses new and innovative ideas were needed.

In **June 1993,** the optical design of a 4-lens imager with 2 aspheric and 2 diffractive surfaces was finalized. On a diffractive surface, a circular grating structure was formed which worked by diffraction of light and not by refraction. This opened an additional degree of freedom for correcting optical errors and helped to reduce the number of lenses.

As innovative paths always tend to be somewhat risky, an optical design of a 5-lens or 6-lens imager using the new infrared glass IG 6[398] was also performed. Because of the lower number of lenses, it was decided to build the 4-lens imager. Perhaps the decision was also influenced by the fact that a year before optical design and manufacturing of a modified Common Module imager with a diffractive surface were successfully accomplished.

In **1994,** the first pattern of the diffractive imager was optically measured on an infrared interferometer. It turned that the image quality in the designed image plane was satisfactory. However, the interferograms looked not as clean as anticipated because unwanted radiation was directed onto the detector by higher diffraction orders. As no corrective action was in sight this solution had to be abandoned and the design option with the IG 6 glass was revived.

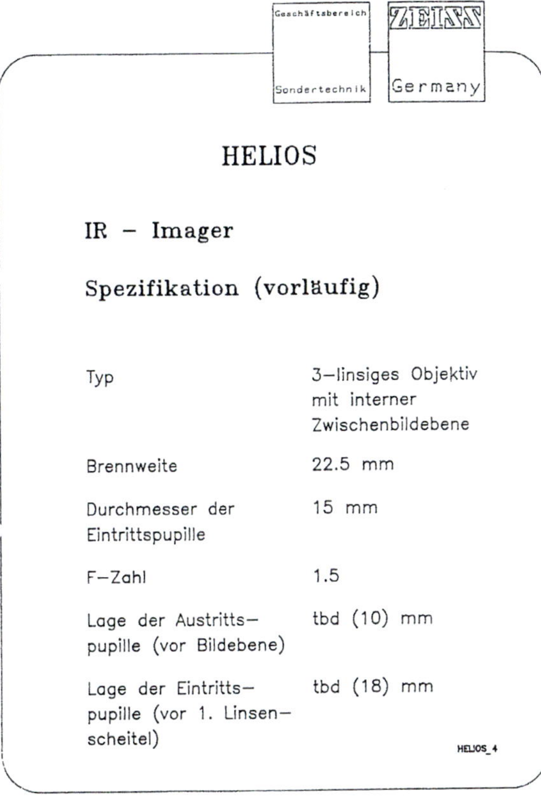

Geschäftsbereich Sondertechnik | ZEISS Germany

HELIOS

IR – Imager

Spezifikation (vorläufig)

| | |
|---|---|
| Typ | 3–linsiges Objektiv mit interner Zwischenbildebene |
| Brennweite | 22.5 mm |
| Durchmesser der Eintrittspupille | 15 mm |
| F–Zahl | 1.5 |
| Lage der Austritts–pupille (vor Bildebene) | tbd (10) mm |
| Lage der Eintritts–pupille (vor 1. Linsen–scheitel) | tbd (18) mm |

HELIOS_4

145

145 – *First specification of the Helios imager*

*Type: objective with 3 lenses and internal intermediate image plane; focal length: 22.5 mm; diameter of entrance pupil: 15 mm; F-number: 1.5; axial position of the exit pupil (in front of the image plane): tbd (10) mm; axial position of the entrance pupil (in front of the vertex of the first lens: tbd (18) mm. (tbd = to be defined)*

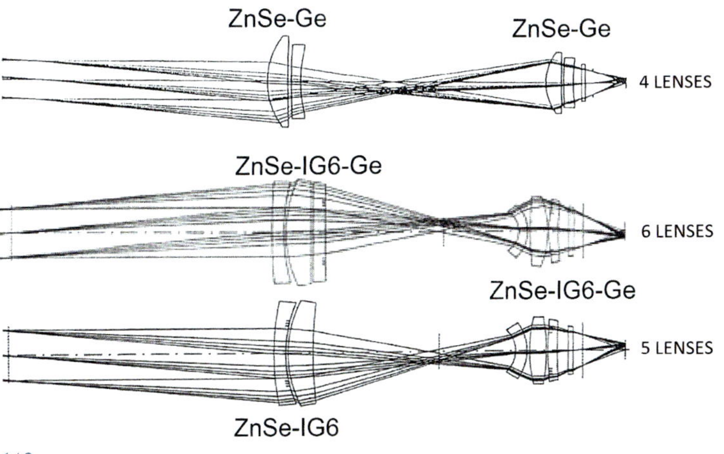


146

146 – *Designed optical layouts of the Ophelios imager*

*Top: 4-lens design with 2 diffractive surfaces on the Germanium lenses, respectively.*
*Center: 6-lens design with the then new IR glass IG 6.*
*Bottom: 5-lens design derived by optimizing of the 6-lens design.*

*The chosen lens materials are: Ge = germanium, ZnSe = zinc selenide and IG 6 = IR glass ($As_{40}Se_{60}$ mixture)*

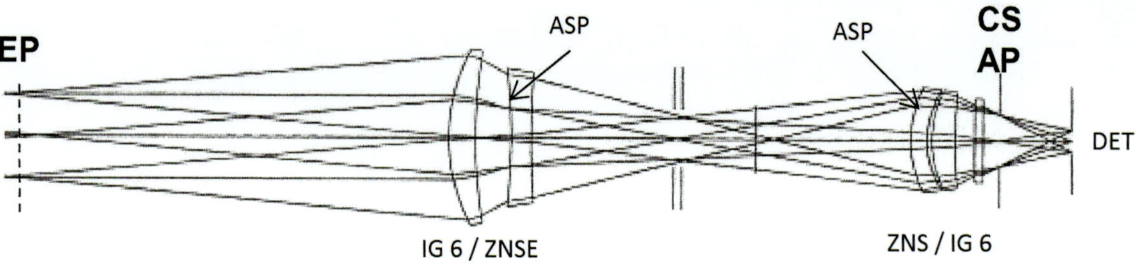


147

**147 – *Optical layout of the 4-lens Ophelios imager***

*The sketch shows the ray path with 4 lenses made from zinc selenide ZnSe, zinc sulfide ZnS and the IR glass IG 6. The aspheric surfaces (ASP) which contribute significantly to the reduction of the optical aberrations (errors) are indicated by arrows. It is characteristic for a re-imager that the exit pupil (AP) is freely accessible behind the last lens. Thus, the cold shield (CS) in front of the detector surface which acts as mechanical aperture stop can be easily mounted at this place. The cold shield is imaged by the re-imager optics onto the entrance pupil (EP) in front of the optics. This is the preferred position of the scanner because all beams are crossing here and the beam path is narrowest.*

148

**148 – *Diffractive surface and interferograms of the Ophelios imager***

*The upper picture shows the typical structure of a diffractive surface which acts similar as a diffraction grating. In general, it is designed in a way that the anticipated effect is achieved in the so called 1st order in which the optical wavelength difference between the zones is one wavelength (1λ). Unfortunately, radiation also is diffracted into higher orders which can cause disturbing ghost images. In the interferogram (1994) of the diffractive prototype imager from 1994 the dark straight fringes indicate the good optical quality of the anticipated image. However, there are also underlying artefacts due to an unwanted secondary image which is out-of-focus. In comparison, the interferogram (1996) of the series imager from 1996 exhibits fringes of a likewise, good quality but looks much cleaner.*

**Technical Data of Ophelios Imager**

| | |
|---|---|
| Type | Re-imager, 4 lenses with 2 aspheric surfaces |
| Corrections | achromatic & athermal |
| Focal length | 15 mm ±0.1mm |
| Field of view | 10° (vertical) |
| Aperture | F/1.5 |
| Transmission | > 88% |
| Entrance pupil | Ø 10 mm |
| Exit pupil | Ø 6.2mm (= Cold Shield Ø ) |
| Stop | stray light & anti narcissus stop (in intermediate image plane) |
| Manufacturer | Carl Zeiss Optronics GmbH |

In **May 1995,** the optical design of a 4-lens imager with the IR glass IG6 was successfully completed but without using any diffractive surface or any germanium lens. By avoiding germanium lenses a solution could be found to correct color and thermal error as well with only 4 lens elements. A European patent EP 0 783 121 "Achrathermer Reimager" was granted for this innovative solution where the adjective "achratherm" in the title is an acronym for "achromatic & athermal".

In **1996,** the 4-lens imager with color and thermal correction became the Ophelios series imager after some minor optimization work concerning the pupil positions.

In **October 1996,** a 2-lens imager with IG6 lenses and 2 diffractive surfaces was investigated as a further option, and a European patent EP 0 840 155 "Zweilinsiger, achromatischer athermalisierter Reimager" (Two-lens achromatic athermal re-imager) also was granted for this innovative solution. It was not put into effect because of the problems that had appeared with the diffractive surfaces in the first imager development.

149

149 – *Ophelios imager*

## Thermal reference

Different from visual CCD- or CMOS detectors in silicon technology that we use in our photo cameras or smartphones every day, infrared detectors in thermal imagers are made from exotic materials which are not at all cooperative in use. The individual detector elements typically differ from each other in the background signal (called "offset") and the sensitivity to radiation (called "gain") a great deal. Moreover, these properties also depend on the temperatures of the device and of the scenery, and they even change over time. To achieve a homogeneous image, at least a Nonuniformity Correction (NUC) must be performed. To do this, all detector elements typically are made to look simultaneously onto a temperature-controlled reference surface so that all elements see the same temperature, and then the individual signals of the elements is measured. As was stated above the measured signals are not as equal as they should be ideally. To make them equal, from the existing differences correction factors are calculated for each detector element which then are applied to all further images of the scenery.

At thermal imagers with scanner, it was advantageous to do the reference measurement in idle time, for example when an oscillating mirror was in the return phase. In case of a polygon scanner this was when an edge was pointing directly towards the detector. At the Ophelios scanner the edge between the surfaces #1 and #2 was chosen for this purpose.

The ideal method would have been to use the homogeneous distribution of the radiation in the pupil as a reference because the radiation emitted by the pupil is proportional to the average temperature of the scene. At the beginning of the Ophelios development the engineers at ZEISS-ELTRO OPTRONIK tried to realize this method but failed because the imaging errors of the winding reference beam path were too high. In a second attempt, the established method of using thermo-electric coolers (TEC) as homogeneous radiation sources was applied. Because of the splitting of the pupils by the polygon edge two TECs had to be implemented. For regulating the TECs to the average scene temperature the average signal level of the thermal images was evaluated and used as a control variable for the TEC temperature regulation.

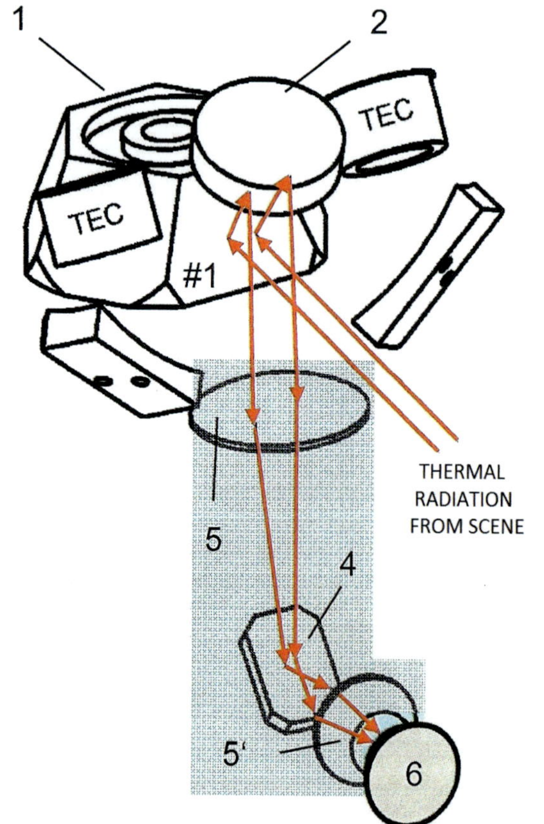


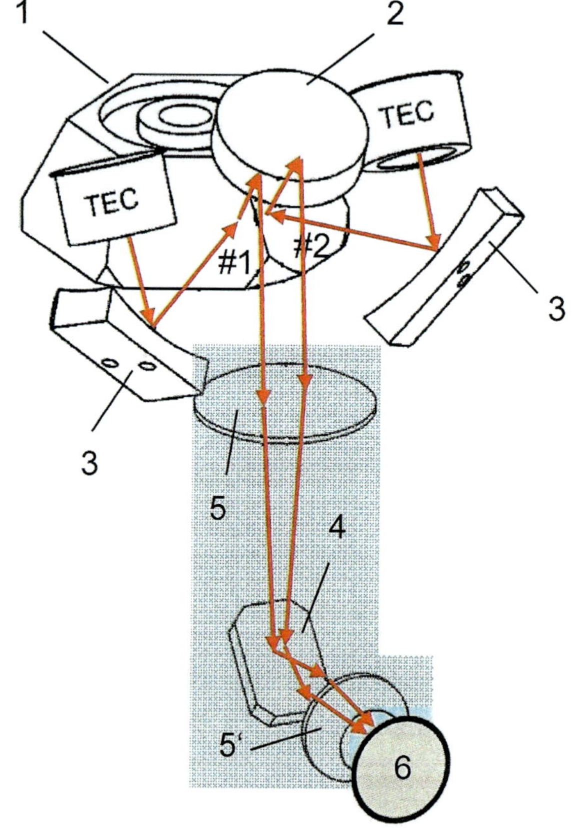


150 – *Image recording and reference measurement in the Ophelios system*

*Image recording: With the polygon (1) turning one of its faces to the scene (e.g. at an angular position of 0°±10°, 60°±10° etc.), the incoming thermal radiation of the scene is directed by this polygon surface and an auxiliary plane mirror (2) into the IR-imager formed by two lenses (5, 5') and a folding mirror (4). The IR-imager then creates an infrared image on the surface of the detector (6).*

*Reference measurement: With the polygon (1) rotated to an angular position of 30°, the thermal radiation emitted by two thermoelectric coolers (TEC) is directed by two toric mirrors (3), two polygon faces (#1 and #2) and an auxiliary plane mirror (2) into the IR-imager (5, 5'and 4). The imager focuses the thermal radiation emitted by the TECs onto the surface of the detector (6). The temperature of the two TECs is regulated to the average temperature of the scene.*

## Ophelios detector

The detectors of the 1st generation used mercury-cadmium-telluride CMT (HgCdTe) as sensitive material, and so did the Ophelios detectors of the 2nd generation. A great advantage of this semiconductor alloy is that its sensitivity can be optimized to the energy of the photons in the longwave region at 8 – 12 µm by means of the mixing ratio.

All CMT detectors used in the 1st generation thermal imagers were of the photo-conductive (pc) type, and suffered from the drawback that the electrical impedance (the frequency dependent resistance) was too small to connect the detector elements directly to a CCD register in silicon technology. Therefore, each detector element needed its own connecting wires and conducting paths through the Dewar to an external PC board. This resulted in a complicated Dewar, and therewith limited the possible number of detector elements.

To avoid this limitation, the Ophelios detector and all other detectors of the 2nd generation were designed as photo-voltaic (pv) detectors. In this new detector type, each detector element comprised of a n-doted and a p-doted area that behaved like a pn-junction of a conventional silicon diode. The incoming photons generated charge carriers which were accumulated in a capacitor and increased the voltage. Besides exhibiting an inherent low power consumption, the pv-detectors could be easily integrated in electronic circuits in silicon technology, and thus a reasonable part of the signal processing already could be done on-chip. Taken as a whole, a higher number of detector elements and a more complex detector architecture were possible, e.g. implementing several TDI detectors.

The IR chip made from CMT was connected directly to the electronic circuits on a standard silicon wafer by means of so-called indium-bumps. These bumps are small indium beads which melt at low temperature and establish a soldered connection to the silicon wafer without damaging the wafer or the IR chip by too high temperatures. This technology became known as "flip-chip" technology.

A direct connection between IR detector-chip and silicon circuit opened new possibilities of on-chip signal processing, for example to reduce the path length of the signals and avoid interfering effects.
On the Ophelios detector the following electronic functions could be realized:

- A 4-fold TDI function was implemented on-chip with CCD registers. By summing up 4 detector signals the temperature sensitivity was enhanced by a factor 2.

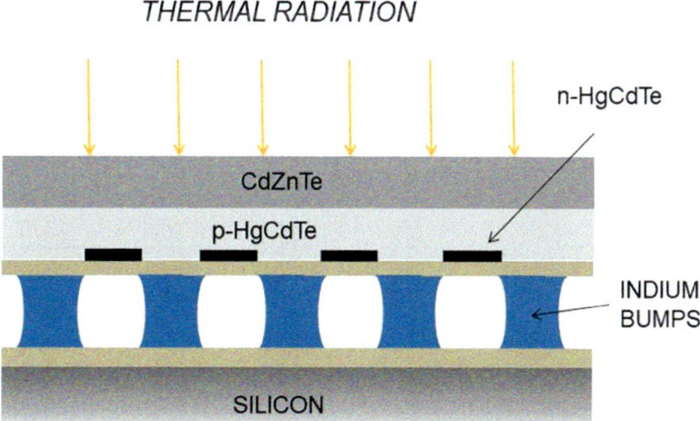


152

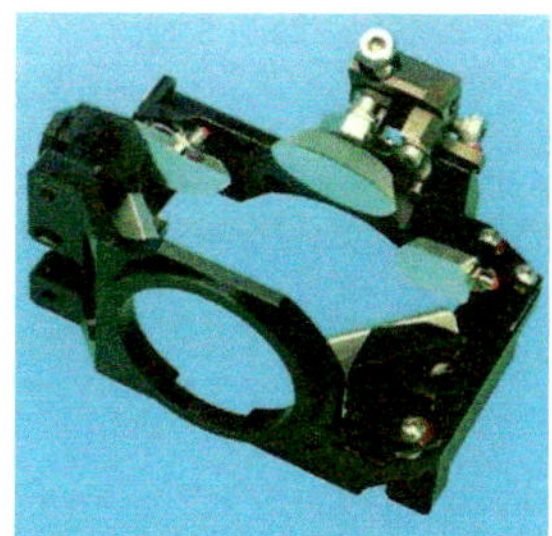

151

151 – *Ophelios thermal reference unit*

*The function of the auxiliary mirror and the toric mirrors is explained in Fig. 150. The reference unit also serves as mounting structure for polygon and imager. The imager is fixed in the large circular opening in the front. The polygon is mounted from beneath.*

152 – *Structure of an IRCCD detector*

*At the contact patches between n-doted n-HgCdTe and p-doted p-HgCdTe a pn-junction is formed which acts as a photodiode sensitive to thermal radiation. The indium bumps connect the IR detector elements with the electrical circuits on the silicon chip.*

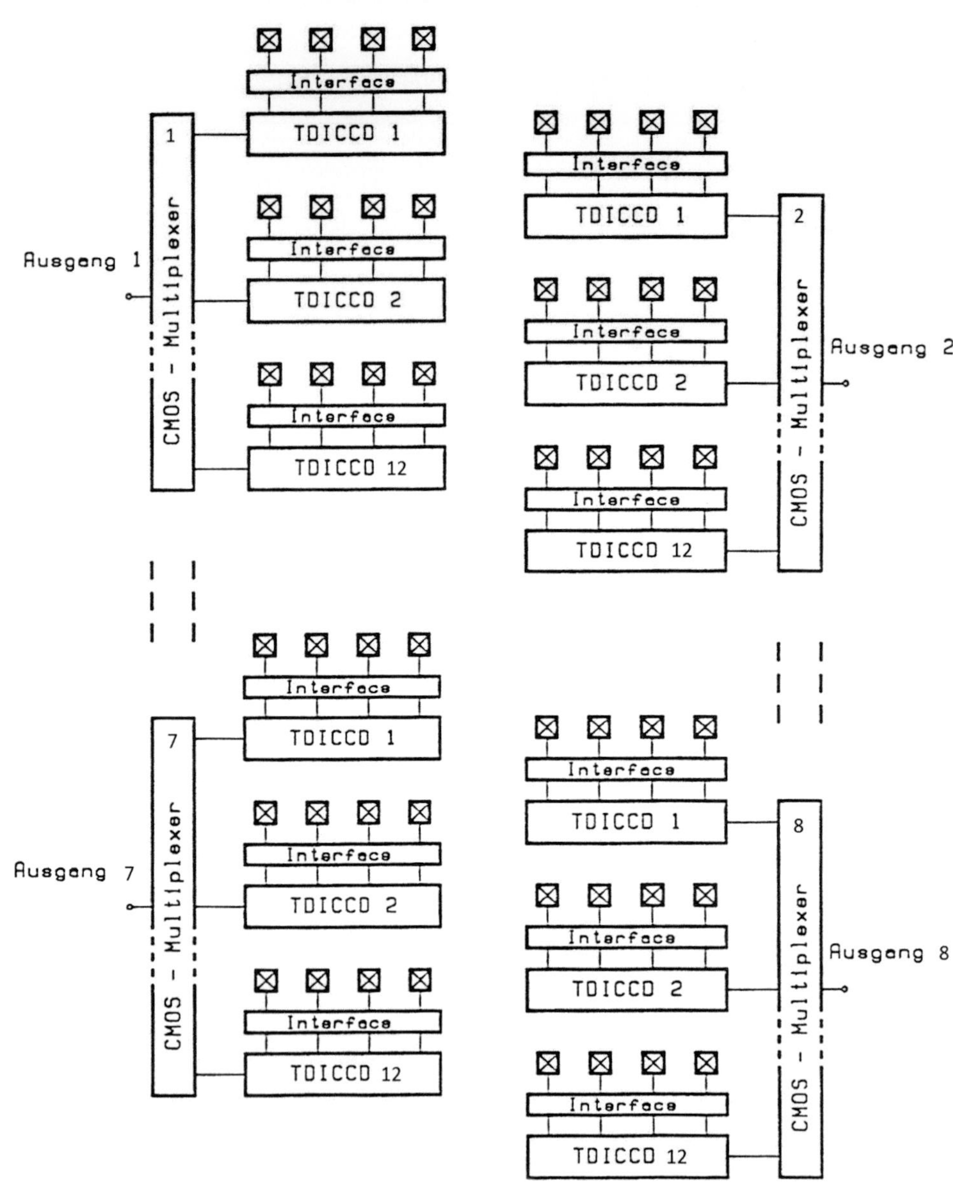


153 – ***Functional principle of the Ophelios IRCCD detector***
*The 4 detector elements used for the Time Delayed Integration (TDI) are connected by a CCD register (TDICCD). A 12:1 CMOS multiplexer connects 12 of the TDI groups to one of the 8 read-out lines, respectively.*

- With on-chip 12:1 multiplexers the number of analog video outputs was reduced to 8. This meant a lower number of exit lines and amplifiers and resulted in a more homogeneous thermal image.
- The integration time could be adjusted between 0.5 and 2.75 µs to find an optimum adjustment of the detector to a weak or strong level of thermal radiation.

For the necessary cooling detector, Dewar and cold finger were integrated in a single unit named "Integrated Detector Cooler Assembly" IDCA by the manufacturer. Besides a more compact construction also a better heat transfer between detector chip and cold finger could be achieved and that in turn contributed to low cooling power requirements of only 0.3 W.

As a temperature-regulated cooling machine a commonly called "Split-Stirling" cooler was installed. In this split-up configuration, the employed linear motor was connected to the cold finger only by a thin transfer pipe and could therefore be arranged at a remote location. This was a great ease to the design of a compact thermal imager.

**Technical Data of Ophelios Detector & Cooler**

| | |
|---|---|
| Detector | IRCCD: IR-Chip on Si-Wafer, 77 K |
| Material | CMT Cadmium-Mercury-Telluride |
| Principle | photo-voltaic (pv) |
| Wavelength range | LWIR 7,5 – 10,5 µm |
| Elements | 96 x 4 |
| Element size | h x v: 25 µm x 28 µm |
| Detector size | h x v: 333 µm x 2688 µm |
| Thermal load | 0.3 W |
| Readout circuit | CCD & CMOS circuit on Si |
| Cooling | Split-Stirling cooler with linear motor |
| Cool down time | approx. 7 min |
| Manufacturer | AEG Infrarot-Module GmbH (AIM) |

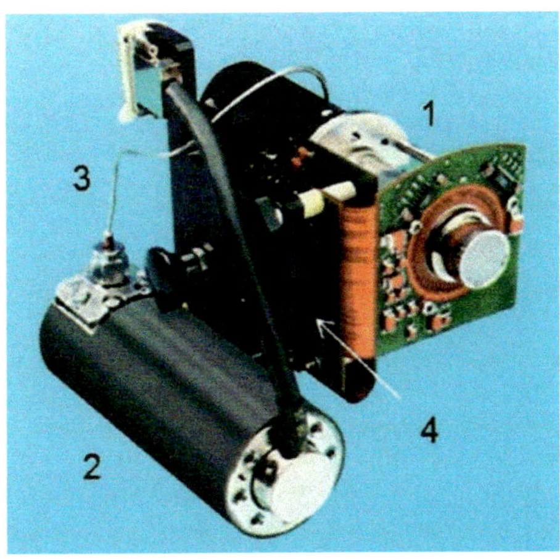

154

154 – *Ophelios detector-cooler assembly*

*The cold finger (1) is integrated in the Dewar (vacuum vessel) and connected to the linear motor-compressor unit (2) via a small transmission pipe (3) only. The proximity electronics (4) is mounted close besides the detector.*

## Ophelios afocals (telescopes)

From Common Module experience it was clear that a 40° x 30° field of view of the Ophelios basic unit would be too large for many applications and that an afocal telescope was necessary to adapt to various applications. Unlike the US-German Common Modules, the Ophelios system used a Keplerian telescope because an exit pupil laying far behind the last lens allowed free access to mount the scanner at this favorable location. Therewith, a proper pupil imaging could be achieved. The cold shield of the detector (= exit pupil) was first imaged onto the scanner and then into the entrance pupil of the Keplerian telescope. This prevented unwanted thermal radiation from the interior to reach the detector elements.

For some reasons only germanium lenses were used in the new afocal:

- Germanium has the highest refractive index n = 4 of all IR materials and thus can generate high refractive power with a low curvature of the lens surfaces. This assists to reduce optical aberrations.
- By using the method of diamond-turning, which was mastered on germanium lenses already at that time, aspheric surfaces could be manufactured in high quality. This reduced the necessary number of lenses.
- Germanium lenses had only a low color error that could be corrected by a diffractive structure. Such a structure could be formed on an aspheric surface also by diamond-turning in the same run of the machine.
- The thermal change of the refractive power of a germanium lens could be accepted, because the resulting de-focusing could be actively compensated. For thermal compensation, correction values were evaluated from the signals of two temperature sensors. Based on these values the focusing lens was shifted to preserve a sharp image.

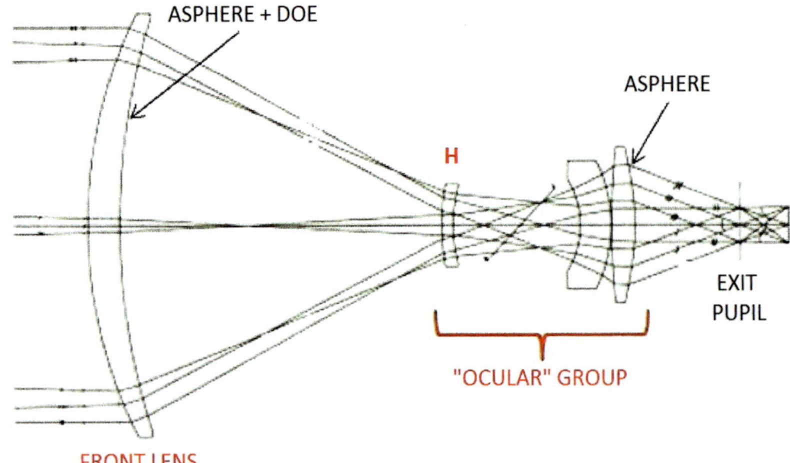


155 – *Ophelios afocal with modular ocular group*

*The front lens has a combined aspheric and diffractive (DOE) surface on the inner side. Because this surface is far enough from the detector the negative effects which were observed with the diffractive imager cannot occur. The rear lens (H) is used for focusing to close distance and for thermal correction. The common 'ocular' group was optically and mechanically identical for all series afocals. On the right-hand side, the free accessible exit pupil is clearly to be seen.*

The following time line of the development shows that also in this case several iteration steps were necessary to bring the concept into reality:

In the years **1993-94,** at CARL ZEISS-SONDER-TECHNIK a basic optical design was established for the application of Ophelios thermal imagers in various military vehicles (Leopard 2, M48, Zobel) and for border security purposes as well. The practical trials were not satisfying because disturbing image artefact appeared again and again under distinct operating conditions.

In **November 1995,** a system analysis showed that there were mainly two reasons for the appearing artefacts:

- The scanner distortion was not appropriately taken into consideration during the optical design process.
- At the edges of the optical beam path many errant rays existed, and propagated outside the nominal ray bundles.

The result of both effects was that the detector elements not only could look to the scene but also onto parts of the lens mountings, and thus could see thermal radiation emitted by these parts. As a first remedial action the diameters of the critical lenses were enlarged by up to 3 mm.
In **1996,** further work was done during the preparations for series production to achieve a better optical correction and to reduce the temperature depending changes of the field of view.

In **December 1996,** the idea of a modular system for the afocals was born. After it had become obvious that the troublesome optimization work that was just finished had to be repeated with every new afocal, it was decided to design a unified "ocular-group". This group comprised a sliding rear lens and the actual eye piece. Optically and mechanically of identical construction this module was used in all afocals designed thereafter. By this modular concept development cost could be saved because only the front lens had to be changed to adapt the magnification to a new application. Furthermore, production costs could be reduced due to the larger quantities of the ocular-group. A European Patent EP 0 843 190 "Modulares Infrarot-Kepler-Fernrohr" (modular infrared Keplerian telescope) also was granted for the modular ocular-group which was used at CARL ZEISS OPTRONICS until the end of the Ophelios production time in 2008.

| DEVICE | FIELD OF VIEW | Γ | EP [mm] | RELEASED |
|---|---|---|---|---|
| MOLF | 2.8° x 3.7°<br>9° x 12° | 10.6<br>3.3 | 106<br>33 | 7/1994<br>9/1995 |
| Zobel "Afokal 2" | 2.66° x 3.54°<br>9.0° x 12.0° | 11.7<br>3.46 | 116.7<br>35 | 8/1995<br>2/1996<br>12/1996 |
| Brevel Zoom | 3.6° x 4.8°<br>10.1° x 13.5°<br>28.8° x 38.4° | 8.53<br>3.02<br>1.03 | 85.9<br>30.2<br>10.4 | 2/1995<br>3/1999 |
| Leo 2 "Afokal 3" | 3.6° x 4.8°<br>12.3° x 16.4° | 8.5<br>2.56 | 85.3<br>25.6 | 7/1993<br>2/1996<br>12/1996 |
| BMI "Afokal 1" | 1.54° x 2.04°<br>5.24° x 7.00° | 20<br>6 | 200<br>60 | 4/1994<br>12/1996 |
| OMS | 3.1° x 4.1°<br>9.3° x 12.4° | 9.6<br>3.3 | 96<br>32 | 1994 |
| SERO 14 | 3° x 4°<br>10° x 13.3° | 10.5<br>3.14 | 105<br>31.4 | 6/1998 |
| Exp. FLIR III | 2.0° x 2.7°<br>8.0° x 10.8°<br>22.6° x 30.5° | 14.8<br>3.8<br>1.3 | 148<br>38<br>13 | 6/1995 |

156 – *Ophelios afocals for various applications*

*In the column Γ the magnification is given for each field of view. EP is the diameter of the entrance pupil. The value for the highest magnification is corresponding approximately to the diameter of the front lens. The column RELEASED gives the date when a new configuration was readily developed. (Γ = magnification, EP = entrance pupil diameter)*

157 – *Proximity electronics (left) and MUX/ADC board of the Ophelios system*

158 – *System electronics of the Ophelios system*

## Ophelios electronics

The electronic boards of the Ophelios system were supplied by several companies.

The electronic board close to the detector called Proximity Electronic was developed by TEMIC and served the following functions:

- Generation of clock-pulses
- Buffer for detector signals
- Read-out of the detector signals of 8 analog video outputs with matched impedance
- Power supply for the detector

The MUX / ADC board from ELTRO comprised the following functional units:

- 8 analog video inputs
- Analog amplifier with offset correction for each image line
- 8 analog/digital converters with 10-bit resolution
- 8:2 - multiplexer for the digital video signals
- 12-bit video data interface

The system electronic from ATLAS ELEKTRONIK was equipped with a digital scan converter, a video-processor and the power supply for all voltages to provide the following functions:

- Correction of the image data
- Correction of the scanner distortion
- Controlling of the CAN-Bus[399] interface
- Controlling of the telescope (thermal compensation, field of view change or zoom function)
- Generation of a target mark
- Operation of a system BITE[400]

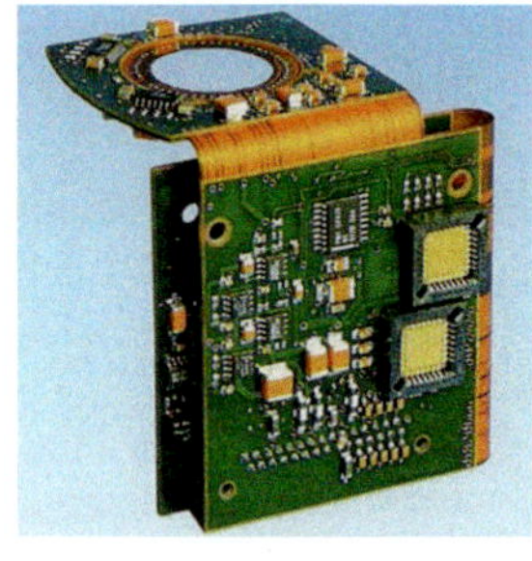

157

158

## Typical Ophelios Thermal Imager

At the close of this technology chapter, a typical Ophelios thermal imager shall be presented which had the highest production number and was installed in many vehicles and systems for border security.

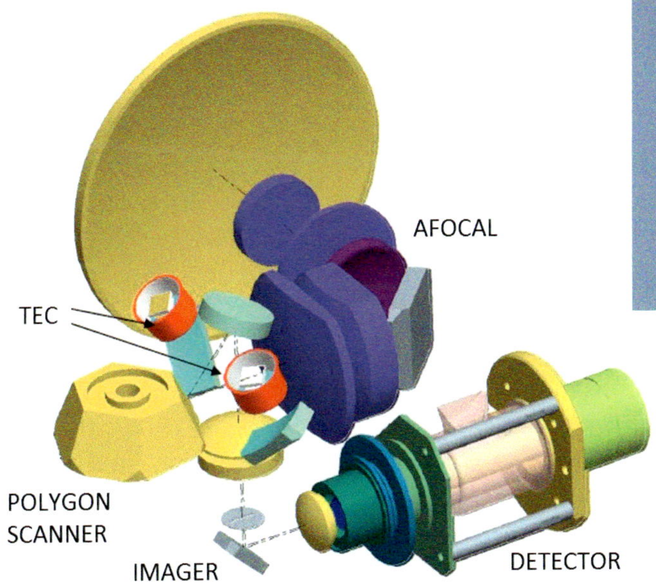


159

**Technical Data of Ophelios**

| | |
|---|---|
| Detector | CMT, IRCCD |
| Wavelength range | 7.5 - 10.5 µm LWIR |
| Cooling | linear cooler, 77 K = -196°C |
| Detector elements | 96 x 4 |
| Scanner | rotating polygon with 6 faces |
| Image lines | 576 |
| Frame rate | 25 Hz |
| Image points / line | 768 |
| Number of image points | 442 368 |
| Infrared optics | IR lenses |
| Aperture | F/1.5 |
| Fields of view | 4.7° x 2.6° and 14.6° x 8° (example) |
| Image reproduction | CCIR |
| Power consumption | 60 W |
| Dimensions | 157 x 230 x 150 mm |
| Weight | 7 kg |
| Thermal resolution | < 120 mK |
| Manufacturer | Zeiss Optronik GmbH |

159 – *Picture and 3D-model of a typical Ophelios thermal imager*

*The 3D-model shows that the optical system is folded four times. Without these folds, an assembly as compact as shown in the sketch would not be possible. However, folding an optical path does not make the adjustment any easier.*

## 7.3 German Ophelios Modules in Active Service

The large list of different afocals shown in the previous chapter indicates that Ophelios thermal imagers were employed in a lot of programs during its 10 years of production. The most interesting programs for land, maritime and airborne systems will be discussed in this chapter.

### Land systems

160 – *Overview of the most important land applications*

*Beginning at top left: remote BAA system, scouting vehicle Fennek, main battle tank Leopard 2, border surveillance van and air defence system LeFlaSys.*

### Light scouting vehicles

Mid of the 1980s the armored reconnaissance brigade of the Bundeswehr (federal army of Germany) demanded a new vehicle that should replace the reconnaissance tank Luchs and that could also be deployed by the armored infantry battalions.

In **1989,** the German Heeresamt (Army Office), after having investigated the market and conducted several studies, could begin with the comparison of two vehicles available on the market.

In **January 1990,** the assessment was finalized. The experimental vehicle Zobel developed by the GESELLSCHAFT FÜR SYSTEMTECHNIK GST (corporation for system technology) and built by the INDUSTRIEWERKE SAAR turned out to be superior in all criteria to its French competitor PANHARD VBL.

The political revolutions of 1989/90 had the effect that the whole planning of scouting vehicles was revisited.

In **October 1992,** the Heeresamt issued a request for quotation which demanded for "retractable observation contrivance which secures the observation out of the vehicle at day, night and under poor visibility"[401]. In a first lot 336 vehicles should be built and 4 experimental troop models of these should be delivered in advance. In response to this announcement 5 quotations were received.

In **June 1993,** cooperation with the Netherlands surprisingly materialized where similar conceptual and operational framework conditions in relation of a new scouting vehicle were discussed. Despite of an undisputed high technical potential the Zobel program came to a full stop – a political knock-out after the second successful (technical) round.

It is also interesting to see how the planned quantities were changing during a few years:

| Year | Quantity |
|---|---|
| 1985 | 1714 |
| 1988 | 800 |
| 1991 | 336 (+ optional 380 after 2001) |
| 1992 | 4 (+ series later) |
| 1993 | 0 |

This was the hard fate shared by many contemporary procurement programs of military equipment.

End of **March 1994**, the consortium DAF/WEGMANN won the changed call for bids issued end of 1993 with their vehicle Fennek.

In **August 1995**, the optical design of a first Zobel-afocal for the Ophelios thermal imager was finalized at ZEISS-ELTRO OPTRONIK. After the introduction of the modular ocular-group the optical design was reworked end of 1996.

To fulfill the requirements of the Fennek specification, the "**B**eobachtungs- und **A**ufklärungs-**A**ustattung" BAA (observation and reconnaissance equipment) was developed at CARL ZEISS OPTRONICS comprising the following sensors:

- Eye safe laser rangefinder MOLEM
- Visual monochrome CCD camera
- Thermal imager Ophelios (identical construction as the last Zobel configuration)

The BAA could be elevated by 1.5 m up to a total height of 3.29 m. Furthermore, it could be easily dislodged from the vehicle, mounted on a tripod and operated on a remote location up to 40 m away from the Fennek vehicle.

**Technical Data of Zobel Afokal 2**

**Principle**
revolving magnification changer, modular ocular group

| Fields of view | H x V | Magnification | Entrance pupil |
|---|---|---|---|
| NFOV | 3.5° x 2.7° | 11.7x | 117 mm |
| WFOV | 12.0° x 9.0° | 3.5x | 35 mm |

From **1997** on, the first 4 experimental models were submitted, and intensive technical testing was performed followed by field trials at the troops.

In **May 2002,** ZEISS OPTRONIK got an order for series production of 364 BAA sensor platforms which were to be delivered between 9/2002 and 7/2007 with a rate of up to 13 devices per month.

The flow of information and hardware was typical for procurement programs of the time (and still is today): From the ordering countries Germany and the Netherlands it went via the system house KRAUSS-MAFFEI-WEGMANN and STN ATLAS to ZEISS OPTRONIK and the Dutch company NEDISCO in Venlo an erstwhile subsidiary founded by Zeiss in 1920.

On **10 December 2003,** the first series vehicle came to the troops at the Panzertruppenschule (school of the armored corpse) in Munster. As early as in 2004 in the course of a so-called "einsatzbedingten Sofortbedarfs" (immediate demand induced by deployment) 4 vehicles were prepared as scouting vehicles and deployed by German field artillery observers in Afghanistan from autumn 2004 on.

161 – *Scouting vehicle Fennek with hoisted BAA*

**Technical Data of Afokal 2 & Afokal 3**

**Principle**
revolving magnification changer, modular ocular group

**Afokal 2 for EMES 15**

| Fields of view | H x V | Magnification | Entrance pupil |
|---|---|---|---|
| NFOV | 3.6° x 2.7° | 11.7x | 117 mm |
| WFOV | 12.0° x 9.0° | 3.5x | 35 mm |

**Afokal 3 for Peri R17 A2**

| Fields of view | H x V | Magnification | Entrance pupil |
|---|---|---|---|
| NFOV | 4.8° x 3.6° | 11.7x | 85 mm |
| WFOV | 16.4° x 12.3° | 3.5x | 26 mm |

In **2011**, the series production came to an end and the two manufacturers KRAUSS-MAFFEI-WEGMANN (KMW) and their Dutch subsidiary DUTCH DEFENCE VEHICLE SYSTEMS (DDVS) had delivered in total 410 vehicles to the Netherlands and 222 to Germany.

## Main battle tank Leopard 2

When in the beginning of the 1990s the Kampfwertsteigerung KWS (upgrade) of the main battle tank Leopard 2 A5 was started the panoramic periscope Peri R17 A2 of the commander was equipped with a thermal imager. However, not the Zeiss WBG-X was chosen. This would have been sensible for logistic reasons because the WBG-X was already installed in the gunner sight EMES 15. Instead, the TIM thermal imager supplied by the Israel based company EL-OP ELECTRO-OPTICS INDUSTRIES was chosen, perhaps due to political reasons.

From **1993 to 1996**, a compatible thermal imager based on the new Ophelios modules was developed by ZEISS-ELTRO OPTRONICS as an alternative solution for the Peri R17 A2 and the EMES 15 as well. The two newly developed afocal telescopes utilized the modular ocular-group and were internally named Afokal 2 and Afokal 3.

From **2006 to 2008**, Greece purchased in total 170 battle tanks of the type Leopard 2 A6 HEL equipped with the gunner sight EMES 15 and the panoramic periscope Peri R17 A2 for the commander. In both devices, thermal imagers of CARL ZEISS OPTRONICS were installed which were developed from the Ophelios module set. The thermal imagers were in part also assembled by a Greek company to comply with the offset-regulations in the contract demanding that the domestic industry should participate in the program.

From **2008 to 2011**, an order from Switzerland followed to install the new Peri R17 A2 with Ophelios thermal imagers into the Panzer 87 Leopard WE (Swiss version of the Leopard 2 main battle tank). In total 134 tank periscopes were supplied and 50 additional systems for training and as maintenance float.

Altogether, in this 5 years 514 thermal imagers of the Ophelios system were delivered just for the Leopard programs.

An interesting aspect specially of the Greek Leopard program was that the two different thermal imagers for EMES 15 and Peri R17, respectively, showed impressively how distinct two devices could be which both were designed from the Ophelios module set.

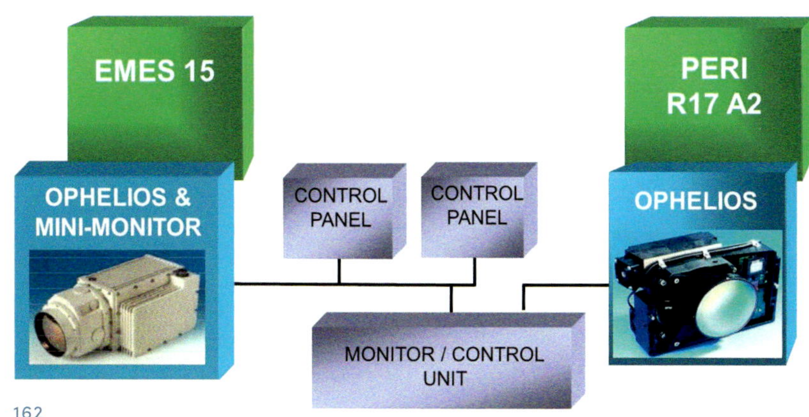


162

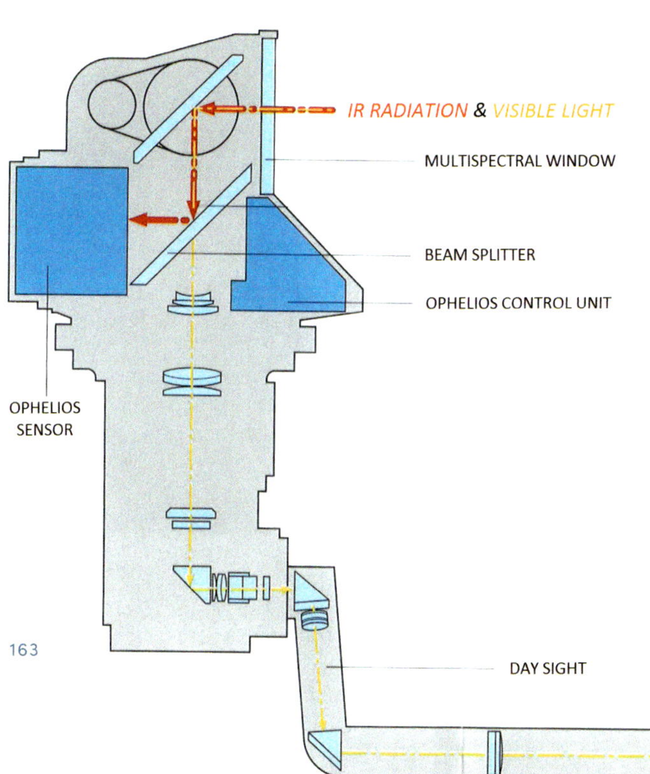


163

162 – *Block diagram of the Leopard 2 A6 HEL sensor system*

*EMES and Peri are connected to a common control unit. Commander and gunner have an own control panel each.*

163 – *Sketch and picture of a R17 A2 with Ophelios thermal imager*

*The thermal radiation and the visible light enter the periscope through a multispectral window made from zinc sulfide and are directed downwards by the elevation mirror. A beam splitter reflects the thermal radiation into the Ophelios thermal imager. The visible light is transmitted and reaches the eye piece via the day sight channel.*

164 – *Ophelios thermal imagers for Peri R17 (left) und EMES 15*

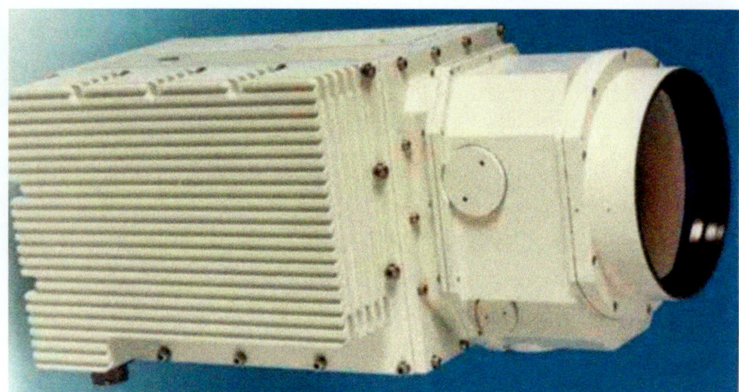

165 – *Mercedes G-models equipped for border surveillance*

*The picture taken in 2006 is a historic document in so far that it shows the transition from Ophelios modules to the subsequent Attica thermal imagers. The last car in the row of vehicles ordered by Latvia is s till equipped with an Ophelios thermal imager which can be easily recognized by the larger front lens. The two G-models in the front already have the new sensor units with the modern, staring Attica thermal imagers which came in general use from that time on.*

*(Photo: Karl-Heinz Renner)*

## Border surveillance

As was already mentioned in Chap. 6 about the US-German Common Modules, the first order of vans for border surveillance of the German border police in the years 1995-96 had to be fulfilled with the Zeiss WBG-X because the Ophelios series production had not been started then.

In **1998**, however, the next following order from Spain was served with the Ophelios thermal imagers. As can be seen in the attached table, many more programs followed in the subsequent 10 years. In total, more than 170 Ophelios thermal imagers were supplied to border surveillance programs all over Europe.

| YEAR | SYSTEM | COUNTRY | QUANTITY |
|---|---|---|---|
| 1998-99 | VW T4 | Spain | 4 |
| 1999 | MORS | (Thales) | 19 |
| 1999 | with SNK | Austria | 2 |
| 2000 | VW T4 / for mast | Estonia | 3 / 1 |
| 2000 | VW T4 | Poland | 9 |
| 2000 | VW T4 | Spain | 4 |
| 2001 | VW T4 | Latvia | 3 |
| 2001 | tripods | UAE | 4 |
| 2002 | systems | Austria | 2 |
| 2002 | VW T4 | Hungary | 10 |
| 2002 | VW T4 | Spain | 2 |
| 2002 | tripods / VW T4 | Algeria | 20 / 5 |
| 2003 | VW T4 | Spain | 2 |
| 2004 | VW T4 / for mast | Hungary | 7 /8 |
| 2004 | VW T4 | Poland | 10 |
| 2005 | Sprinter | Poland | 49 |
| 2006 | VW T5 | Spain | 1 |
| 2006 | G-Modell | Latvia | 1 (11) |
| 2008 | Ophelios | UAE | 10 |
| | | Total | 176 |

166 – *Border surveillance programs using Ophelios thermal imagers*

167

168

## Maritime and airborne systems

Also in these application fields, many important sensor systems were equipped with thermal imagers. Again, the most interesting are presented in greater detail on the next pages.

Two different platforms for the service on surface ships were developed at CARL ZEISS OPTRONICS for paramilitary and military applications, respectively.

The simpler "Maritime Electro Optical System" MEOS was delivered in low quantities to customs and coast guard.

From **1994,** on a first MEOS version existed which still was based on the US-German Common Module WBG-X.

From **1996** on, the MEOS platform was offered with an Ophelios thermal imager.

In **1999,** MEOS systems were installed on 7 custom vessels.

In **2000,** a MEOS system was experimentally installed on a HSC ferry. It was hoped that the argument of enhanced safety would open a civil market but the operation was not carried on.

In **2002,** a contract for equipping the coast guard ships of the Bundesgrenzschutz (federal border police) could be won. In total 6 guard ships were supplied with MEOS systems. Spectators of the German TV series

167 – *Overview of important maritime and airborne applications*

*Beginning at top left: army drone KZO, Breguet Atlantic, helicopter Sea King, Class 212 submarine, frigate and patrol boat with MEOS*

168 – *Patrol boats of the German coast guard*

*The MEOS sensor platform can be spotted on top of the bridge above the bright red rail.*

"Die Küstenwache" (coast guard) could see thermal images in several episodes which had been recorded by the MEOS platform.

All Ophelios thermal imagers of the MEOS platforms were equipped with the Afokal 2 which utilized the modular ocular-group.

## Multi-sensor platform MSP

In the **1990s,** RHEINMETAL DEFENCE ELECTRONICS (RDE) developed the "Multi-Sensor-Plattform MSP 500", a high-precision stabilized electro-optical sensor system with long observation ranges against sea-, ground- and airborne targets, and equipped for automatic tracking and fire control as well. The sensor equipment of the MSP 500 was supplied by CARL ZEISS OPTRONICS and comprised

- Ophelios thermal imager with Afokal 1 (FOV 2,0° x 1,5° and 6,7° x 5,1°)
- Laser rangefinder (6 Hz, range up to 40 km)
- TV camera with 10x-zoom

From **1995** on, the installation on 10 speed boats of the 143A Gepard-class and 8 frigates of the F122 Bremen-class was carried out.
From **2001** on, the German Navy participated in the operation "Enduring Freedom" in the Indian Ocean at the Horn of Africa. All deployed units got the MSP 500 to spot fast small boats and drifting mines at night or under adverse weather conditions.

In **2004-05,** the "Norwegian Advanced Surface to Air Missile System" NASAMS, a modern Norwegian air defense system, was supplied with 13 MSP 500 systems.

From **2004 to 2006,** the MSP 500 systems were installed in 3 frigates of the F124 Sachsen-class.

169

169 – *Multi-sensor platform MSP 500*

**Technical Data of Afokal 2**

**Principle**
revolving magnification changer, modular ocular group

| Fields of view | H x V | Magnification | Entrance pupil |
|---|---|---|---|
| NFOV | 3.5° x 2.7° | 11.7x | 116.7 mm |
| WFOV | 12.0° x 9.0° | 3.46x | 34.6 mm |

170 – *Pictures of IR unit and IR sensor of the KZO system*

*The complete IR unit (left) is fixed in a cardanic suspension in the drone and can be panned and tilted. The IR sensor also can rotate within the black painted structure. In the picture of the IR sensor (right) the elements for adjusting the zoom lenses can be easily recognized by their L-shape.*

## Reconnaissance drone Brevel or KZO

In **1983,** MESSERSCHMIDT-BÖLKOW-BLOHM (MBB) in Bremen and the French company MATRA DÉFENSE in VÉLIZY developed a common reconnaissance drone which was named "Brevel" after the cities **Bre**men and **Vél**izy where the companies were located, respectively. The most important sensor of the drone was the thermal imager.

In **1988,** the Bundeswehr (federal German army) decided to use the Brevel drone for their program "Kleinfluggerät Zielortung" KZO (small aircraft for target spotting). In the following time, the Brevel drone was mostly called KZO in Germany.

From **1992 to 1998,** a special IR unit based on the Ophelios modules was developed and built for the KZO drone by ZEISS-ELTRO OPTRONICS. The KZO afocal was the first to be equipped with an 8x zoom lens which changed the fields of view by axially sliding two lenses. For the company, this was the first IR zoom lens of many more to come. In the following years, the older afocal systems with swivel-type field-of-view changers were increasingly replaced by zoom systems. From 1998 on, all KZO zoom afocals were built by PILKINGTON in St. Asaph in Wales.

In **1996,** the German ministry of defense BMVg began measures to extend the usage of 18 German reconnaissance aircrafts of the type Breguet 1150 Atlantique MPA (Maritime Patrol Aircraft) until 2010. These had been in service as of 1966. The thermal imaging unit of KZO was chosen for this program.

In **1998,** the first prototype of the KZO drone was delivered.

In **1999,** the first Sea King helicopter was equipped with a KZO IR-unit. The Bundesmarine (German navy) had used this type of helicopter ever since 1975.

In **2002,** the delivery of 1+14 KZO IR-units for Breguet Atlantic was started.

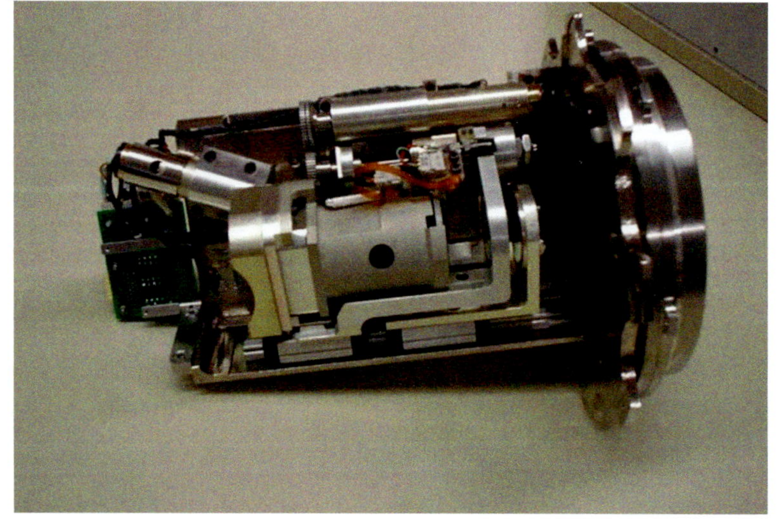

From **2005 to 2014,** in total 61 KZO systems were supplied to the Bundeswehr (German forces).

From **Juli 2009** on, the KZO reconnaissance drone was deployed in Afghanistan. In total 18 KZO drones were lost: 12 drones were destroyed and 6 were reported missing.

In the years **2002-2003,** more KZO IR-units were delivered to equip further 12 Sea King helicopters.

A technical feature of the KZO systems was that the whole IR-unit could be swiveled in two axes and additionally be rotated around the axis of the cylindrical IR-unit. The rotation was needed for erecting the image: this was necessary to secure an upright thermal image also during rolling of the drone or aircraft. Therefore, the IR-unit of KZO was performing a full 3-axis stabilization.

The afocal system again is of the Keplerian type comprising an intermediate image plane and an accessible exit pupil in which the polygon scanner could be mounted. Because of the newly introduced zoom function with two sliding lenses a new ocular group optimized to the zoom concept also became necessary; the old modular ocular-group was not compatible any more.

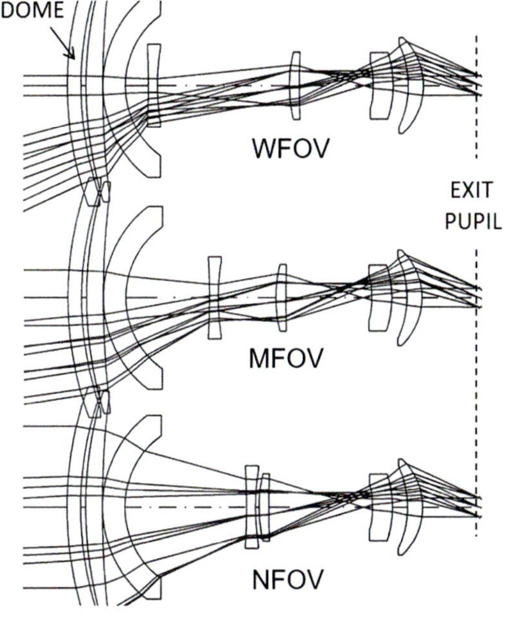


171

171 – *Functional principle of the KZO zoom telescope*

*The sketch shows clearly the freely accessible exit pupil of the Keplerian telescope. From top to bottom the positions of the two movable lenses are depicted for the wide field of view (WFOV), the medium field of view(MFOV) and the narrow field of view (NFOV). The lenses are all made from the IR materials germanium and zinc selenide. In total 5 lens surfaces have an aspheric shape to create more degrees of freedom for improving the image quality.*

**Technical Data of KZO Afocal**

**Principle**
3-position zoom lens (ocular group not modular)

| Fields of view | H x V | Magnification | Entrance pupil |
|---|---|---|---|
| NFOV | 4.8° x 3.6° | 8.53 x | 85.9 mm |
| MFOV | 13.5° x 10.1° | 3.02 x | 30.2 mm |
| WFOV | 38.4° x 28.8° | 1.03 x | 10.4 mm |

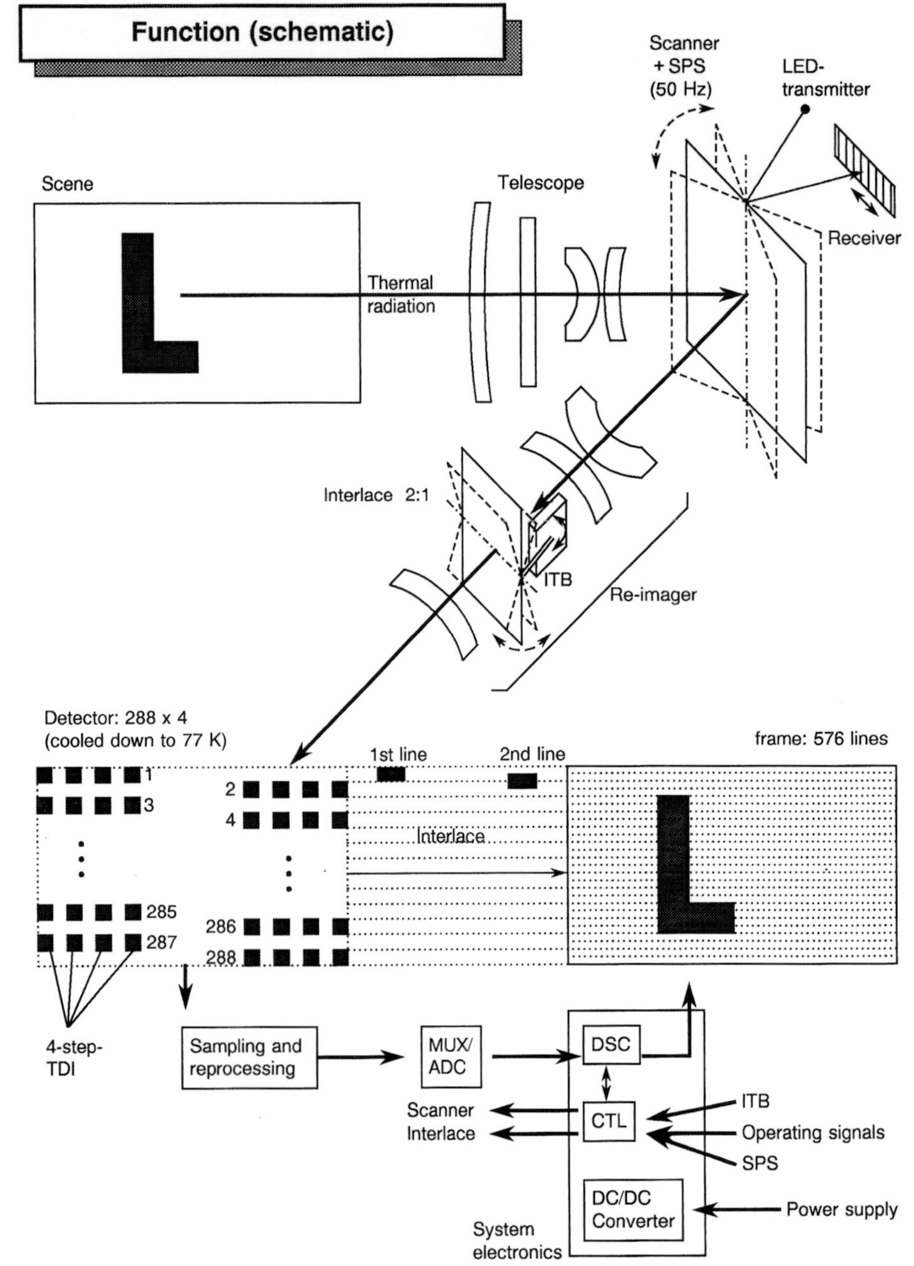


172 – *Functional principle of the Synergi thermal imager*

*The thermal radiation emitted by the scene reaches the oscillating scan mirror though the telescope. It is focused then onto the detector surface by the imager. The imager comprises a plane parallel plane which can be tilted between two angular positions to shift the optical image vertically by half of the width of an image line. This creates the interlace necessary for doubling the number of images lines. With the method of a 4-fold TDI the signals of 4 detector elements lying side-by-side are added, and therewith, the signal-to-noise ratio of the summed signal is improved by a factor of 2. To synchronize the scanner motion with the detector read-out the angular position of the scan mirror is measured continuously with high precision. This is done by reflecting the light bundle of a LED on the backside of the scan mirror. The position of the reflected beam is measured by a Position Sensitive Device (receiver). All detector signals are processed in the Proximity Electronics close to the detector. By means of a Digital Scan Converter (DSC) the signal values are converted into a standard video signal which can be displayed on a commercial monitor.*

## 7.4 Technology of the French-German Synergi Modules

The functional principle followed the pattern of the 2nd generation devices already discussed: A photo-voltaic 288 x 4 CMT-detector with 4 TDI-detector elements and a "Read Out Integrated Circuit" ROIC consisting of CCD- and/or CMOS-registers on-chip made up the basic concept. Like all fellow detectors of the 2nd generation the Synergi detector had a cold shield in front of the sensitive area to protect the detector elements from disturbing thermal radiation emitted by the surroundings.

For the image generation, a simple oscillating scan mirror was used, and an additional tiltable plane plate was performing the interlace function. By this, a 576 x 768 video image was generated compliant with standard CCIR image format.

The assembly and the function of the individual modules is discussed in more detail in this chapter.

### Synergi detector

In the beginning, the detector was built in IRCCD technology but was changed to IRCMOS technology in the time of the Synergi development. Compared to the Ophelios detector, it had three times as much detector elements while the number of 4 TDI elements was the same. This made possible a simple oscillating scanner mirror and an enhanced scan efficiency. Due to the larger size of the detector chip the necessary cooling power of 0.5 W was a little higher than that of the Ophelios detector.

288 x 4 DETECTOR
1
2
3
4
8,1 mm
0,4 mm
285
286
287
288

173 – *Detector geometry of Synergi detector*

| Technical Data of Synergi Detector & Cooler | |
|---|---|
| Detector | IRCCD, CMT, 77K |
| Wavelength range | 7.5 – 10.5 µm |
| Ave. sensitivity | $1.5 \cdot 10^{11}$ cm Hz$^{1/2}$ W$^{-1}$ |
| Detector elements | 288 x 4 |
| Read out procedure | 16 groups of elements |
| TDI | 4-fold |
| Element size | 25 µm x 28 µm |
| Thermal load | 0.5 W |
| Cooler | Stirling cooler with linear motor |
| Cooldown time | 7 min typ |
| MTBF | 3500 h |
| Manufacturer | Sofradir |

## Synergi imager

Because of the cold shield well in front of the detector also the Synergi imager – like the Ophelios imager – had to be designed as re-imager with an intermediate image plane. The interlace unit as part of the imager was a thin plane plate made from germanium that was switched between two angular positions. However, this plate could also be tilted continuously to cover two further functions:

- For basic adjustment, the error of the line of sight of the thermal imager could be corrected.
- By sensitively controlled tilting of the plate a fine stabilization of the imager's line of sight could be performed in vertical direction. Due to the small mass of the plate the angular corrections could be done with high speed.

To allow for a very compact structure of the thermal imager the optical axis was folded two times. Additionally, a fixed plane plate was mounted on the detector side for an optional spectral filtering.
The development was done in two steps. In the first step - later called Synergi 1 - a 3-lens imager was developed which had a sliding front lens because of two reasons:

- The thermal imager could be imaged to close laying objects.

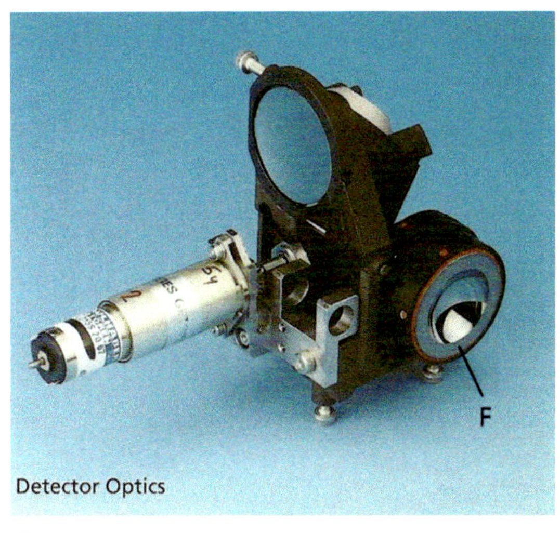


174

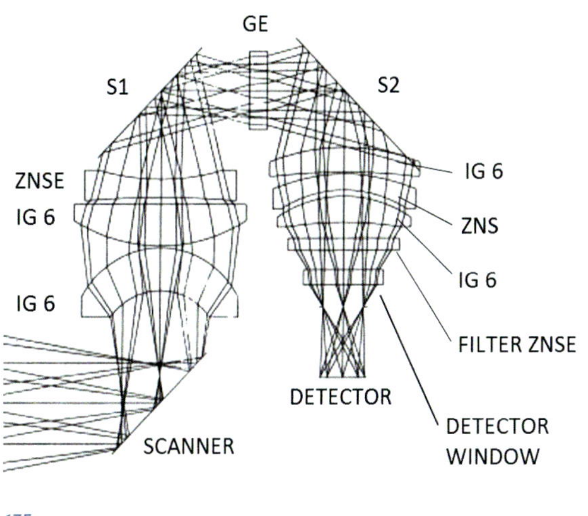


175

**174 – *Synergi 1 re-imager***

*For focusing, a lens group (F) is axially shifted. In the final assembly this lens is facing the detector.*

**175 – *Optical design of the Synergi 2 re-imager***

*The lenses are made from zinc selenide (ZnSe), zinc sulfide (ZnS) and the IR glass IG 6. The tiltable interlace plate is from germanium (Ge). S1 and S2 are 2 folding mirrors which are not only folding the optical system in the paper plane but in three dimensions.*

- The de-focusing caused by temperature changes could be compensated, i.e. the image could be sharpened again

The imager assembly comprised also a control board and an electric motor for tilting of the interlace plate.

In the second development step to Synergi 2 it was decided to place the focusing into the afocal and to design the imager as a fixed focus lens. The consequence of this decision was that the imager necessarily had to be temperature compensated and, of course, still had to fit exactly into the same space as the first imager so that the old Synergi 1 structure could be used further on.

The solution of the problem eventually was to use the same concept that was invented and patented at ZEISS-ELTRO OPTRONICS already for the Ophelios imager. By using the infrared materials well tried by now (zinc sulfide (ZnS), zinc selenide (ZnSe) and IR glass IG 6) also the new Synergi imager successfully was temperature compensated.

The limited space which was further reduced by the two folding mirrors turned out to be a real issue. In the end, it was very tight and from the IG 6 front lens even a part of the rim had to be cut off to prevent that the scan mirror would hit the lens at the slightest irregularity.

176 – *Comparison of Synergi 1 & Synergi 2 imagers*
*The most demanding requirements during the advancement from Synergi 1 to Synergi 2 are indicated by white script.*

| | SYNERGI 1 | SYNERGI 2 |
|---|---|---|
| Wavelength range | 7.5 – 10.5 µm | |
| F-number / pupil | 1.7 / 13.4 mm | |
| Field of view | 20° vertical | |
| Line of sight tolerance | ±1 mrad | |
| Focal length 20°C<br>tolerance | 22.7 mm<br>none | 22.6<br>± 0.5% |
| Back focal length<br>tolerance | adjustable<br>none | fixed<br>< 0.2 mm |
| Therm. stability<br>focal length<br>back focal length | refocussed<br>± 0.6%<br>± 0.6 mm | athermal<br>± 0.6%<br>± 100 µm |
| Image quality<br>20 Lp/mm, 20°C | MTF > 0.45 | MTF > 0.53 |
| Dimensions | exactly equal | |
| Weight | 500 g | 570 g |

## Scanner & Scan Position Sensor

The Synergi scanner used an oscillating mirror which was rotated by a mechanical angle of 13.5° at a frequency of 50 Hz. Because the oscillation followed a sawtooth profile the scan efficiency was higher than 70%.

The Scan Position Sensor SPS was designed as an optical sensor and followed basically the old concept of optical image reproduction used in the 1st generation. However, a laser test beam instead of the visual image was now reflected by the back side of the scan mirror. The angular deviation of the laser beam was measured

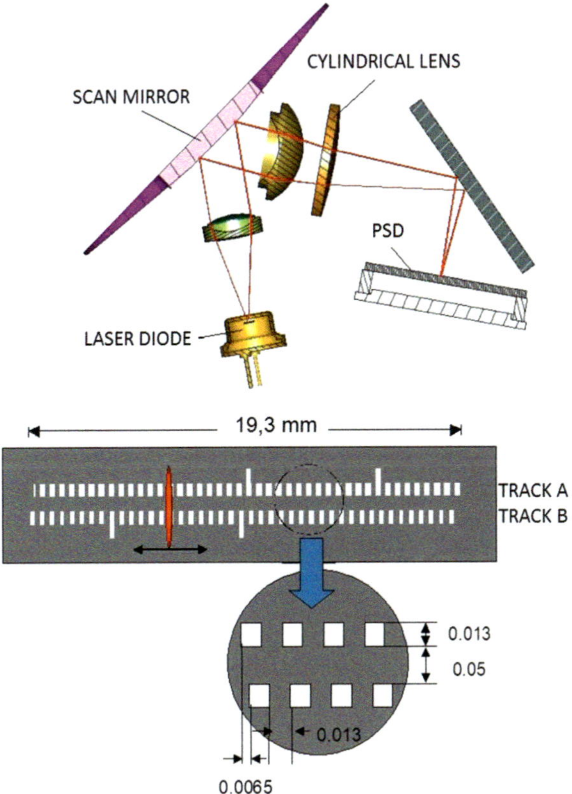


177

178

*177 – Functional principle of the Scan Position Sensor*

*A laser diode emits light with wavelength 780 nm that is collimated and reflected at the backside of the scan mirror. An objective with a cylindrical lens element is imaging the laser beam onto the surface of a position sensitive device (PSD) which generates an electrical signal depending on the location of the light spot on the PSD surface. Because of the cylindrical lens the light spot has the shape of a strip and can reliably cover both tracks of the PSD. The two tracks are staggered by half of the element width and therewith can detect the running direction of the light spot.*

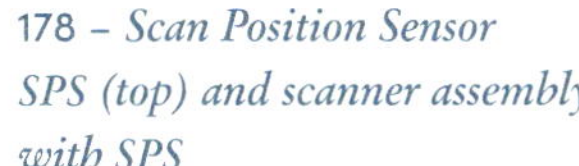

*178 – Scan Position Sensor SPS (top) and scanner assembly with SPS*

| Technical Data of Synergi Scan Position Sensor | |
|---|---|
| Principle | scanning of laser beam |
| Transmitter | laser diode λ = 780 nm |
| Collimator | 3.5 / 13 mm |
| Scan lens | F-theta design |
| Focal length | 26.3mm |
| Light spot | 13 μm x 200 μm (h x v) |
| Receiver | Position Sensitive Device PSD 2 tracks with 742 elements each |
| Manufacturer | Carl Zeiss Optronics GmbH |

with high precision by a "Position Sensitive Device" PSD and the measured values were used for controlling the scanner motion and for synchronizing the detector read-out cycle.

This SPS unit could be used universally for all oscillating scanners, e.g. also in ZEISS-ELTRO OPTRONICS's own HDIR thermal imager. Beginning in 2002 also THALES used this SPS in their STAIRS C thermal imagers for the air defense system Stormer SPHVM.

### Synergi 1 Thermal Imager

To allow comparisons to other thermal imagers the data of a Synergi 1 thermal imager for general purpose are given here. To provide as much flexibility as possible for various applications the device could be assembled in one housing ('monoblock') or in two housings ('bi-block'). In the latter case, the thermal imager is split into a sensor unit and an electronic unit.

Assuming similar fields of view, the Ophelios thermal imager had advantages in size, weight and power consumption while the Synergi thermal imager had a better temperature resolution.

| Technical Data of Synergi 1 | |
|---|---|
| Detector | CMT, IRCCD |
| Wavelength range | 7.5 - 10.5 μm LWIR |
| Cooling | 0.5 W linear cooler, 77 K = -196°C |
| Detector elements | 288 x 4 |
| Scanner | oscillating mirror with 2:1 interlace |
| Image lines | 576 |
| Frame rate | 50 Hz |
| Image points / line | 768 |
| Number of image points | 442 368 |
| Infrared optics | IR lenses |
| Aperture | F/1.7 |
| Field of view | 4° x 3° (example) |
| Image reproduction | CCIR; optional RS 170 digital |
| Power consumption | 110 W |
| Volume | 8.5l + 5l |
| Weight | 11.2 kg |
| Thermal resolution | < 90 mK |
| Manufacturer | Zeiss Optronik GmbH |

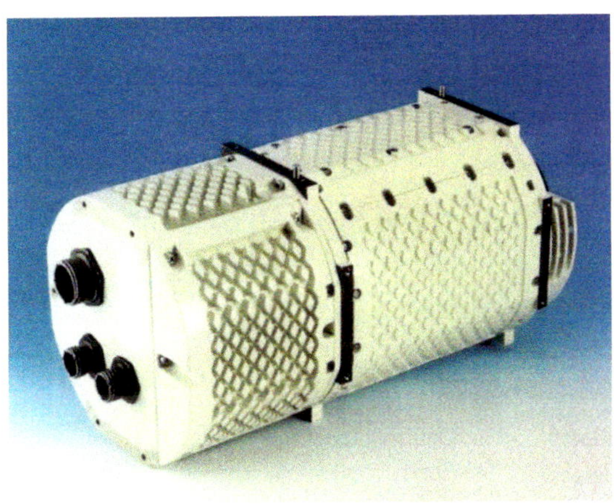

179 – *Two-part structure of Synergi 1 thermal imager ('biblock')*

*In both pictures the electronic unit is shown on the left side attached to the rear of the sensor unit.*

## 7.5 Applications of German Synergi Thermal Imagers

In France, the Synergi modules were employed in several national programs, e.g. in the Catherine thermal imagers. Besides the supply of Synergi modules to other programs, at CARL ZEISS OPTRONICS in Germany complete thermal imagers only were developed for two programs. The modular idea, however, was based so firmly at that time that the two afocals to be developed were conceptualized with identical ocular groups already in the first optical design.

### BAMSE Air Defence Missile System

The Swedish air defense system BAMSE comprised a mobile launching platform equipped with 6 ground-air missiles. To support the fire control a combination of radar and thermal imager was planned.

In **1992**, the Swedish office started for the BAMSE development program.

In **1993**, corresponding development orders were given to BOFORS AB and ERICSSON.

End of **1995**, ZEISS-ELTRO OPTRONIK began the development of a special BAMSE afocal based on the "Development Specification BAMSE-CIDS-343".

In **1997**, the optical design of ZEISS-ELTRO OPTRONIK was presented to BOFORS.

From **10/1999 to 5/2001**, a prototype of the new BAMSE thermal imager had to undergo very detailed qualification tests which were finalized successfully by a qualification review at the company in Karlskoga that meanwhile had been renamed to SAAB BOFORS.

In **Mai 2008**, the first series BAMSE systems were delivered, however, equipped with the in-house thermal imagers that had become available due to the merging with SAAB. The German BAMSE thermal imager based on the Synergi modules did not get beyond the delivered single prototype - or the unpopular "one twelfth of a dozen" as cynics at ZEISS-ELTRO OPTRONIK tended to dub this quantity.

The specially developed afocal for the BAMSE program had two fields of view compliant to the specification:

- Narrow field of view NFOV: 2.3° x 1.7° and magnification 11.6x
- Wide field of view WFOV: 6.9° x 5.1° and magnification 3.9x

The total weight of the BAMSE thermal imager was 19.5 kg. An optical peculiarity of the afocal was a special beam path for thermal reference which was woven into the imaging beam path.

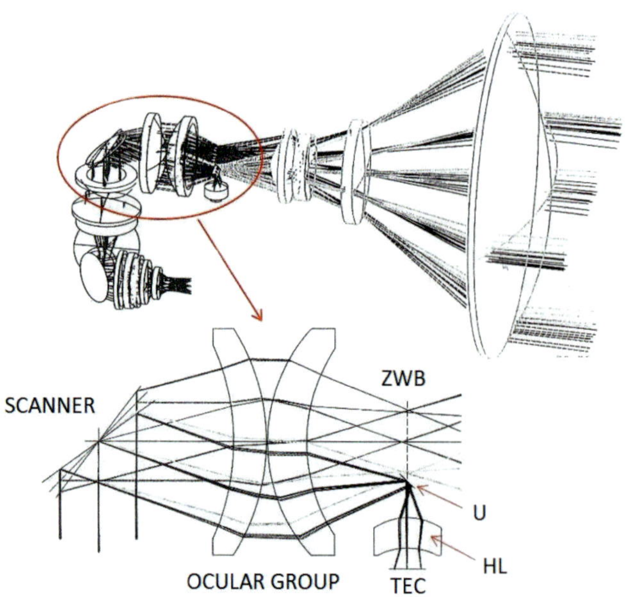


180 – *Optical beam path of BAMSE afocal with TEC beam path*

*The figure shows the position of the lenses in the narrow field of view NFOV. Beneath the intermediate image (ZwB) a small folding mirror (U) is fixed which directs the line of sight of the detector to the thermal reference for a short length of time at the end of each sweep of the scanner. By means of the auxiliary lens (HL) a great part of the TEC area can emit radiation onto the detector elements to establish a very homogeneous reference source.*

## LITENING Targeting and Navigation Pod

According to the words of its conceptual description, the pod combined "several sensors in a single pod of maximum flexibility at low expenses" and should be suitable "for all modern fighter aircraft". The so-called pods were mostly mounted underneath the fuselage of fast and agile fighter aircraft for reconnaissance of hostile territories and designating of significant targets if need would be.

The functionality of the LITENING pod was generated by the following devices:

- a FLIR system
- a TV-camera
- a Laser-designator
- a Laser-spot tracker
- an electro-optical point- and inertial tracker

In **1992**, the Israel company RAFAEL started the LITENING development program. In the 1st generation a FLIR of the contemporary 1st FLIR-generation was installed.

In **1995**, a further developed 2nd LITENING generation was introduced at the Israel air force.

In the same year a Joint Venture was agreed between RAFAEL and ZEISS-ELTRO OPTRONIK (ZEO) for the equipping the German Tornado IDS until 2002 which had an order volume of 20 pods (worth approx. 80 Mio DM). ZEO's work share was

- the development and delivery of a FLIR thermal imager of the 2nd -generation based on Synergi modules
- the build-up of an own assembly line for the LITENING pods of the German air force

181

181 – *Tornado Recce of German Airforce with LITENING Pod*

*The Tornado Recce is a version of the Tornado IDS with a reconnaissance pod mounted on the fuselage station plane (yellow arrow) to be applied for image recording airborne reconnaissance. The LITENING Pod is 2.2 m long, has a diameter of 40.6 cm and weighs 220 kg.*

**Technical Data of LITENING Afocal**

**Principle** 3-position zoom lens

| Fields of view | H x V | Magnification | Entrance pupil |
|---|---|---|---|
| NFOV | 2.4° x 1.8° | 11.4x | 152 mm |
| MFOV | 8.6° x 6.5° | 3.2x | 42 mm |
| WFOV | 29.1° x 22.1° | 0.92x | 12 mm |

In the end, 13 systems of the planned volume were built and a few additional FLIR thermal imagers.

From 2001 on, RAFAEL and NORTHROP GRUMMAN were cooperating at the further LITENING generations without German participation.

For the new FLIR thermal imager named LITENING TI (LITENING Thermal Imager) a new afocal with the specified fields of view had to be developed. To achieve a compact construction that fitted in the available space the afocal was designed as a zoom lens with distinct positions of the sliding lens elements optimized for the specified fields of view. In accordance with the idea of the best possible modularity the new LITENING afocal was designed to be compatible with the ocular-group already used in the BAMSE afocal.

182 – *LITENING Pod (top) and LITENING TI*

## 7.6 Comparison of 2nd Generation Thermal Imagers

In the previous parts of this chapter about the thermal imagers of the 2nd generation we became acquainted with several systems of different thermal imagers. Although being widely international, each system was based to some extend on the technical traditions of the country in which it was developed. And of course, the gripping question at the end of this chapter is: Which was the best device?

Before starting the discussion of the differences, let us first have a look at the obvious commonalities. All development teams had learned the advantages of a "Common Module" set from the 1st generation and had even internalized that knowledge. For this there are at least two indicators.

Not only the systems were consequently designed in a modular manner, even within the modules, i.e. on a sub-level beneath the system level, carry-over parts were used wherever possible. A good example are the identical ocular-groups used in the afocals Ophelios system and the German Synergi thermal imagers, respectively.

The second indicator is that useful modules for standard functions necessary in every system were utilized for different systems, even if they were competing on the market. The Scan Position Sensor (SPS) developed and build at ZEISS-ELTRO OPTRONICS for the German Synergi imagers is a good example for this. It was used in identical shape also for the in-house HDIR thermal imager and later even was delivered to THALES for the STAIRS C system.

We find more commonalities in the detector technology. All systems uniformly used a CMT detector in the LWIR wavelength range perhaps for the following reasons:

- The power of the thermal radiation at room temperature is highest in the LWIR range
- The atmosphere has a broad transparent window in the LWIR range and the stray light of mist or smoke is relatively low due to the long wavelengths around 10 μm
- The CMT detectors work close to the theoretical optimum in the LWIR range

Obviously, the capabilities of the manufacturers SOFRADIR, BAE und AIM in production technology also were comparably high. All companies used

- photo-voltaic (pv) detectors
- cooling of the detectors to 77 K
- on-chip multiplexers and registers in IRCCD and/or IRCMOS-technology
- 4 or more TDI detector elements
- virtually the same staggered detector geometry
- a cold shield with nearly the same f-number
- a linear motor as cooling machine

After all there are some differences in the functional concepts of the thermal imagers indeed. Each system might be characterized best in a short verdict:

| | OPHELIOS | SYNERGI | HTI | HDIR |
|---|---|---|---|---|
| Detector | CMT (pv), LWIR 7.5 to 10.5 μm | | | |
| Elements | 96 x 4 | 288 x 4 | 480 x 4 | 576 x 6 |
| Image format | 576 x 768 | 576 x 768 | 480 x 1300 | 1152 x 1920 |
| Frame rate | 25 Hz | 50 Hz | 60 / 30 Hz | 50 / 25 Hz |
| Aperture | F/1.5 | F/1.7 | F/2.5 | F/1.7 |
| NETD | 100-120 mK | 80-90 mK | 90 mK* | 120 mK |
| Video standard | CCIR | CCIR | RS 170 | CCIR, HDTV |
| Power consumption | 70 W | 110 W | 180 W | 130 W |
| Volume | 6 Liter | 13 Liter | 29 Liter | 39 Liter |
| Weight | 7-8 kg | 11 kg | 14.5 kg | 20 kg |

183 – *Comparison of 2nd Generation thermal imagers*

*Most of the values are taken from data sheets of the manufacturers. The values earmarked with (*) are best guess.*

### Ophelios

The emphasis lay on a small and cost-effective detector needing low cooling power, to build a compact and light device with low power consumption. The technical penalty for this was an inferior temperature sensitivity and significant image artefacts in case of a moving scenery due to the three-part division of the image.

### Synergi

The emphasis lay on high temperature-sensitivity combined with a high frame rate. Due to the larger detector, the cooling needs were higher, and the device was slightly larger and heavier, and had a higher power consumption.

### HTI

The emphasis lay on best possible temperature sensitivity of an image with standard number of image lines, and largest possible horizontal field as was already indicated be the naming "horizontal technology initiative". The technical drawback was a relatively large volume and a higher power consumption. The weight was surprisingly low due to a distinguishing design of the mechanical structure.

### HDIR

The emphasis lay, before anything else, on a thermal image in full HDTV format. The necessarily large detector and a scanner with interlace brought losses in temperature sensitivity, and made the device big, heavy and hungry for electrical power.

### STAIRS C

Although not as many details are available to the author as of the other systems, the emphasis seemed to be broadly similar to the HTI: By means of a large 768 x 8 detector a "high definition" image with 768 x 1250 image points and a temperature resolution as high as that of the HTI thermal imager was realized. The resulting consequences with respect to size, weight and power consumption also were similar.

## 7.7 Generation Gap: 1st Generation vs. 2nd Generation

At the end of a historical treatment of the scanning thermal imagers it is interesting to raise the question for the technical progress over the generations.

Did the 2nd generation fulfill the high hopes stated at the beginning?

Where has the anticipated technical progress taken place?

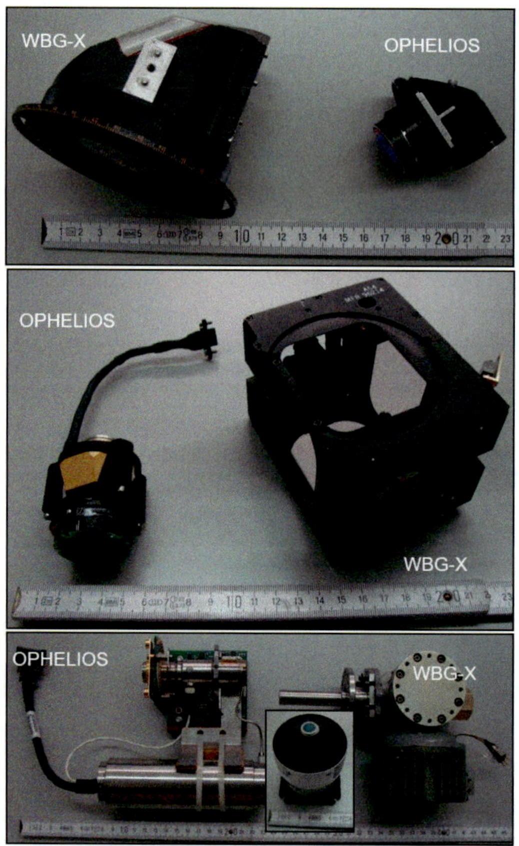


184 – *Comparison of 1st and 2nd Generation modules*

Volume and weight: The differences in size are more than obvious. Due to a smaller detector for the same field of view, the Ophelios thermal imager needed a shorter focal length than the WBG-X which in turn resulted in a shorter length. The optical aperture which defines the quantity of radiation reaching the detector is the ratio of pupil diameter and focal length. Therefore, a shorter focal length results in smaller lens diameters and therewith in reduced height and width of the thermal imager. A comparison of the most important modules (IR imager, scanner and detector-cooler assembly) shows striking differences. Due to the smaller optic, also the mechanical structure and the housing could be made smaller and lighter.

Power consumption: Due to the smaller detector and the integrated IDCA assembly, the need for cooling of the Ophelios was significantly lower and thus also the power consumption of the cooling compressor. Furthermore, the dramatically reduced number of electronic amplifiers also contributed to the reduction.

Thermal resolution: The achieved temperature resolution, e.g. the measured NETD was improved by the 4-fold TDI process applied to each image point. The resulting improvement factor of 2 also can be found in the comparison chart.

Video output: Due to the digital buffer memory, the image data could be arranged easily to comply with any established video standard.

Cost reduction: Smaller lenses, a smaller housing, a lower number of electronic boards and the omission of the optical reproduction channel saved material expenses. Lower production costs also resulted due to the fact, that the installation times were shorter. At the US-German Common Modules each detector element had its own chain of amplifiers which had to be balanced thoroughly regarding to offset and gain to make the image lines equal between themselves and therewith prevent a streaky image.

In conclusion, it can be stated that the change to the 2nd generation of thermal imagers yielded reasonable progress. Besides all number crunching this can be proven with impressive thermal images as well.

185 – *Comparison of 1st and 2nd Generation thermal images*

*The pictures show the spire of the Catholic Church in Oberkochen. The left one was recorded with the WBG-X, a common module thermal imager of the 1st generation. The right picture was recorded with the HDIR, one of the most ambitious thermal imagers of the 2nd generation.*

186 – *Comparison of 1st and 2nd Generation imager data*

*The WBG-X is chosen as representative for the US-German Common Modules and the Ophelios thermal imager for the 2nd generation Common Modules.*

| | 1st GENERATION WBG-X | 2nd GENERATION OPHELIOS |
|---|---|---|
| Detector material | CMT (HgCdTe) | CMT (HgCdTe) |
| Detector type | photo-conductive (pc) | photo-voltaic (pv) |
| Wavelength range | LWIR 7.6 – 11.8µm | LWIR 7.5 – 10.5µm |
| Detector elements | 120 x 1 | 96 x 4 |
| Time delay & integration | none | 4-fold TDI |
| Thermal load | 1 W | 0.3 W |
| Image format | 240 x 900 | 576 x 768 |
| Aperture | ~ F/2 | F/1.5 |
| Imager entrance pupil | ~ Ø 34 mm | Ø 10 mm |
| Thermal reference | none | yes |
| Thermal compensation | none | yes |
| NETD | 200 mK | 100 - 120 mK |
| Electronics | analogue | digital, CCD & CMOS |
| Image reproduction | optical, CCIR | CCIR |
| Power consumption | 180 W | 60 - 80 W |
| Volume (typ.) | 10.4 l | 5.5 l |
| Weight (typ.) | 18 kg +13 kg | 7 - 8 kg |

186

8

# REVIEW AND OUTLOOK: THE CHANGING FACE OF THERMAL IMAGING

"If I have seen further it is by standing on y$^{e}$ shoulders of giants."

*Isaac Newton (1676)*[402]

In the preceding chapters, we followed the historical development of thermal imaging, and we have seen how innovative scientists and engineers have pushed the borders of knowledge further and further over a period of more than 2500 years. Many great ideas and sometimes a small but important impetus have led to surprising and impressive results and devices even according to today's measures.

In my personal review of the most outstanding feats, nevertheless, nature as an architect is my favourite once more.

More than 20 million years ago, nature equipped the rattle snake with genuine thermal imagers in the form of pit organs on the front of their head. These devices were not at all for decoration but were important sensors for detecting the warm-blooded prey and were triggering the final strike via a regular "fire control system" – even after the death of the animal.

Hippocrates was a Greek physician probably known to most of us by the Hippocratic Oath named after him. It is rarely known, however, that as early as in 430 AD he used a moist and clay soaked cloth in the diagnosis of lung disease to locate the centre of the disease. Doing so, for sure he was the first human being who set eyes on a thermal image.

Sir William Herschel was a multi-talented musician, astronomer and research scientist who had to flee from the French troops, which were to occupy his hometown Hannover in northern Germany, and went to England where he discovered the existence of thermal radiation in 1800 and initiated the research of this phenomenon.

Sir John Herschel, as an astronomer and scientist not less gifted than his father Sir William, 2300 years after Hippocrates probably was the second human who came to see a thermal image. With a method later called "Evaporography" he managed to make a thermal image visible on a thin metal plate by applying a thin oil film on the surface of the plate. A thermal image was formed on the surface of the plate by a collecting mirror. At an image point where the incident thermal radiation was high the oil was evaporating faster, and a visible image was generated shimmering like an oil film on a puddle. This was a thermal imager that did not need electric power.

Max Planck, a highly talented German physicist, was told by the experts of his time to let hands off the "old physics". But 60 years after the work of the Herschel family, he was able to derive a radiation law of general validity which is named after him until the present day. He could only succeed by using the revolutionary assumption that a light particle must exist. He called that an act of despair and he never was completely happy about this for the rest of his life. Light particles or photons, as we call them today, and discrete quantized energy states of mater led right into the theory of quantum mechanics, one of the greatest revolutions of scientific thinking that ever happened.

Samuel Langley, an American aviator, astronomer and inventor, parallel to Planck's theoretical work developed a so-called bolometer detector based on the fast heat up of a thin plate by thermal radiation. The device was more sensitive than the measuring equipment in the time of the Herschels by three orders of magnitude. Langley's principle is still in use today in the uncooled micro-bolometer thermal imagers.

The WPG Donau 60, a German thermal pointer using an improved version of Langley's bolometer, was the first thermal device in series production that was put into service before the end of World War II, and was used for reconnaissance in night time as well as for target tracking.

The Pyroscan thermal imager was built with very simple means, and was initially dedicated to British merchant shipping which, however, was not interested. Instead the Pyroscan initiated a 10-year flourishing of medical thermography.

The US-German Common Modules lifted thermal imaging to a new professional level of cost-performance ratio and enable a lot of new applications with considerable width. The experience gathered with the Common Modules influenced the following generations of thermal imagers in a decisive way.

This choice can only be a subjective one, of course, and the reader may celebrate his own favorites.

Today's state of thermal imaging is characterized by two-dimensional matrix arrays which have changed the conception of the thermal imagers drastically.

Because the two-dimensional scenery now can be directly recorded by a two-dimensional detector the optical-mechanical set up can be made much simpler:

- There is no need for a scanner any more
- Without a scanner, the scan position sensor is not necessary either
- Without a scanner, the optics need not have an intermediate pupil for the scanner and the usual partition of the optics into afocal telescope and imager is obsolete; there is only one objective which does all the imaging of the thermal radiation.

The biggest advantage of a two-dimensional detector with respect to signal capturing is that all image points are measured simultaneously by the related detector elements over the whole integration time of the frame. Due to this so-called "multiplex advantage", the integration time for every single image point is much larger than that of an imager with scanner resulting in a much better signal-to-noise (S/N) ratio.

We have already seen that the thermal imagers of the 2nd generation with 4 TDI detector elements and a 4-fold measurement of each image point exhibit a signal-to-noise ratio improved by a factor of 2. A 4-fold measurement means that the effective integration time is also times longer. Because the S/N improvement is proportional to the square-root of the time factor, a 4-fold integration time results in an improved S/N ratio by a factor of 2.

Let us now consider, for example, a CCIR image with 768 image points per line which are measured simultaneously in a staring imager. This means that every image point is measured 768 times longer than in a comparable scanning thermal imager and the resulting multiplex advantage is almost a factor 28.

The table gives an overview how superior today's 3rd generation thermal imagers are with respect to the achievable temperature resolution.

Because the better signal-to-noise ratio has direct influence on the achievable temperature resolution, thermal imagers with two-dimensional arrays, often called "staring imagers", typically show extreme good values of the "Noise Equivalent Temperature Difference" NETD.

In practical applications, however, it is the total package that counts and a fraction of the NETD advantage is often sacrificed to improve other important properties as the following examples may illustrate:

| | 1st GENERATION WBG-X | 2nd GENERATION OPHELIOS | 3rd GENERATION ATTICA | UNCOOLED UCM | UNCOOLED UCM |
|---|---|---|---|---|---|
| Detector Temperature | CMT (pc), 77 K | CMT (pv), 77 K | CMT (pv), 80 K | a-Si μ-Bolo. 295 K | a-Si μ-Bolo. 295 K |
| Wavelength | LWIR 7.7 – 11.8 μm | LWIR 7.5 – 10.5 μm | LWIR 7.7 – 9.3 μm | LWIR 8 – 14 μm | LWIR 8 – 14 μm |
| Elements | 120 x 1 | 96 x 4 | 640 x 512 | 640 x 480 | 1024 x 768 |
| Format | 240 x 900 | 576 x 768 | 640 x 512 | 640 x 480 | 1024 x 768 |
| Image rate | 42 Hz | 25 Hz | 50 Hz | 50 Hz | 50 Hz |
| Effective detector area | 45.8 mm x 12.2 mm | 10.8 mm x 8.1 mm | 9.6 mm x 7.68 mm | 10.88 mm x 8.16 mm | 17.41 mm x 13.06 mm |
| Aperture | F/2 | F/1.5 | F/3 | < F/2 | < F/2 |
| NETD | 200 mK | 100-120 mK | 30 mK | 50 mK (at F/1) | 50 mK (at F/1) |
| Power consumption | 180 W | 70 W | 20 W | < 3 W | 4 W |
| Volume | 10 + 20 Liter | 6 Liter | 4 Liter | 0.05 Liter | 0.07 Liter |
| Weight | 18 + 13 kg | 7-8 kg | 4 kg | 40 g | 50 g |

187

187 – *Comparison including modern thermal imagers*

*The listed thermal imagers are belonging to different generations: WBG-X (Common Modules of 1st generation), Ophelios (German Common Modules of 2nd Generation) and Attica (staring thermal imager of 3rd generation). All these devices have cooled detectors that make use of Mercury-Cadmium-Telluride (CMT) as photo-sensitive material. The UCM (= Uncooled Modules) are thermal imagers based on micro-bolometer detectors made from amorphous silicon which do not need cooling. (Manufacturer of all devices is or was the company which today is named HENSOLDT Optronics)*

- Enlarging the F/# (= smaller aperture) makes the NETD worse (even by the factor in square) but reduces the required diameters of the lenses, and makes the imager assembly more compact.
- Making the integration time of an image shorter makes the NETD worse (with the square-root of the factor) but reduces the smearing of the images remarkably. This is important in case there are fast moving objects in the scenery or the imager itself is moved or rotated.

What can we hope for in future?

Staring thermal imagers can be equipped with mega-pixel detectors already today. The way ahead will for sure lead to ever more detector elements. However, there is a limitation of element size due to the unavoidable diffraction of light by the edges of the aperture stop. This effect prevents that an image point can become smaller than a certain limit, the so-called diffraction limit. The minimal diameter d is determined by the wavelength λ and the F-number F/# and can be estimated by the equation $d = 2 \cdot \lambda \cdot F/\#$. LWIR thermal imagers reach this limit quite early. Even with an extremely good F-number F/1.0 the diffraction diameter is 22 µm. This is larger than the element size of 15 µm which is widely used today. Making the detector elements much smaller thus will not bring major gain in spatial resolution.

Because the diameter of the diffraction spot in the MWIR range from 3 to 5 µm is smaller due to the smaller wavelength, and because indium antimonide (InSb) is available as an excellent detector material for this wavelength region, MWIR thermal imagers probably will gain more importance in future.

The table also shows that there is a second promising option besides the cooled thermal imagers, namely the uncooled micro-bolometer detectors. These are used in civil applications for a long time already but in the military field they suffer a shadowy existence as driver sights and hand-held sights with short observation range. A look at the thermal performance compared to the 2nd generation thermal imagers reveals that with today's micro-bolometer imagers many applications for sure can be covered which had needed cooled imagers with scanners in former times. Moreover, the uncooled thermal imagers are significantly more compact and cost effective and have a lower power consumption by far.

In the frame of a European program for protecting the maritime borders[403] the imaging performance of an uncooled thermal imager together with an objective of long focal length could be demonstrated by CARL ZEISS OPTRONICS several years ago. The technical evolution towards smaller detector elements in higher numbers is already under way also in this field as is shown by the last column of the table.

Notwithstanding the trends in hardware development, the software implemented in an imager gains more and more importance already today. In past times, it was sufficient for a scanning thermal imager to perform a non-uniformity correction NUC of the detector elements. Today defect pixel correction, edge enhancement, image homogenisation and local adaptive dynamic compression are customary. In future, this trend will grow stronger for sure.

This guess is supported by the fact that a nice thermal image that enjoys the observer simply is not enough anymore. Increasingly the task is given to derive quantitative information from the thermal images by means

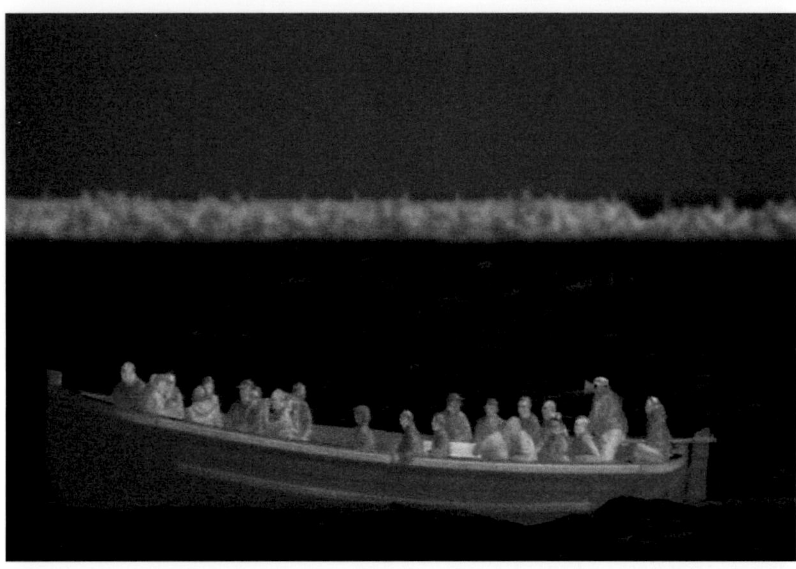

188

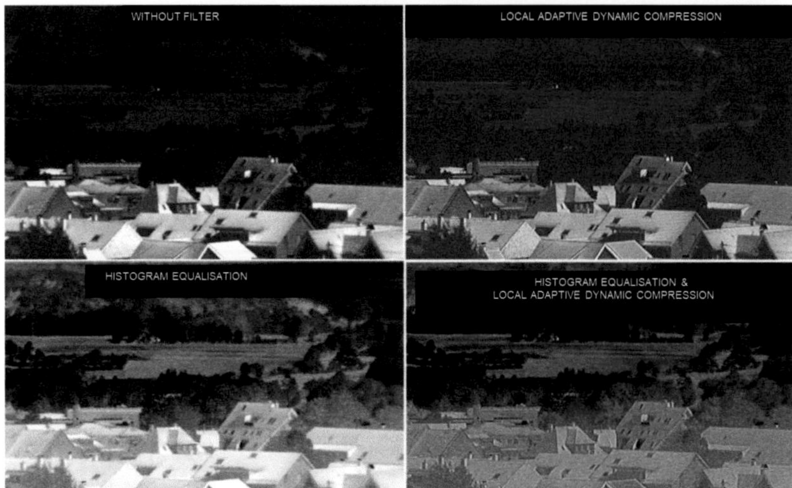


189

of software tools for image evaluation. In a thermal imager applied to detect approaching missiles, for example, the radiometric measurement of the infrared target signature with high accuracy is of crucial importance.

Moreover, the state-of-the-art of thermal imaging is such high today that the atmospheric conditions during observation have become increasingly important. For long range thermal imaging for border surveillance in a desert area, for example, the achievable observation range is limited by the strong air turbulence caused by the high radiation level of the sun. The thermal imager itself could look farther due to its high thermal and spatial resolution but the details of the scenery are smeared by the turbulence. Nevertheless, also in this field first approaches have been launched to restore the original image content using turbulence modeling. So we can be sure, thermal imaging will keep changing.

188 – *Thermal image captured by an uncooled thermal imager*

*The image was obtained with an uncooled UCM640-210M thermal imager that was developed during the European AMASS program (focal length 210 mm, aperture F/1.4, horizontal field of view approx. 3°).*

189 – *Modern image processing algorithms*

Who knows where the limits may be?

If we can learn something useful from the history of thermal imaging then it is probably from a small episode out of the life of the great German physicist Max PLANCK: As a young student, he was very much interested in physics. But even the Munich physics professor Philipp von Jolly advised Planck against going into physics, saying, "in this field, almost everything is already discovered, and all that remains is to fill a few holes."[404] And what happened? The greatest revolution of the physical conception of the world - the introduction of quantum physics - was standing right in front of the door and PLANCK played a great part in that development.

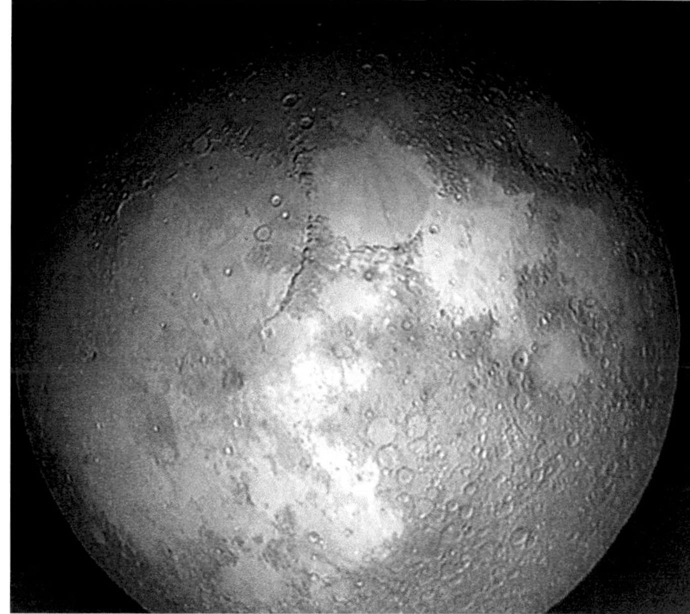

190

190 – *Thermal image of the moon's surface*

# APPENDIX

## Timeline: General History of Thermal Imaging

The milestones of thermal imaging history are given together with some pioneering inventions in related fields of technology.

**20 million years ago**
– the evolution equipped pit vipers with pit organs, by which the snakes could see a thermal image of their prey

**430 BC** – HIPPOCRATES of Kos (GR) made visible the temperature distribution at the body of a patient by means of a moist, clay soaked tissue; he used the temperature dependent drying rate

**By 1660** – at the ACADEMIA DEL CIMENTO in Florence (I) experiments with thermal radiation were conducted

**1672** – Isaac NEWTON (GB) found that white light is composed of variant colors; thus, variant types of light must exist

**1679** – Edme MARIOTTE (F) experimented with thermal radiation

**1682** – Robert BOYLE (GB) experimented with thermal radiation too

**1737** – Émilie de CHÂTELET (F) postulated that thermal radiation does exist, and must be of the same nature as light

**1752** – Thomas MELVILL (GB) made flames shine by means of salts and found infrared radiation

**1777** – Carl Wilhelm SCHEELE (D) investigated thermal conduction and could distinct between convection and thermal radiation
– Marsilio LANDRIANI (I) repeated NEWTON's experiment and measured the visible spectral colors with a thermometer

**1791** – Marc-Auguste PICTET (CH) tried to measure the speed of thermal radiation

**1800** – Sir William HERSCHEL (D/GB) discovered infrared radiation in sunlight and rose the question about the relationship to light

**1804** – Thomas YOUNG (GB) for the first time measured the wavelength of light in a double-slit experiment

**1814** – Joseph von FRAUNHOFER (D) discovered infrared lines in the solar spectrum besides the widely known "Fraunhofer-Lines"

**1821** – Thomas Johann SEEBECK (Baltic/D) discovered the thermoelectric effect ("SEEBECK effect")

**1829** – Leopoldo NOBILI (I) used the "SEEBECK effect" to build an electrical thermometer and radiation-meter

**1833** – Leopoldo NOBILI and Macedonio MELLONI (I) built a first thermopile and their "thermomultiplier", a combination of thermopile and galvanometer

**1835** – André-Marie AMPÈRE (F) proved the similarity between light and thermal radiation

**1839** – Macedonio MELLONI (I) identified the absorption lines of water vapor in the solar spectrum

**1840** – Sir John HERSCHEL (GB) discovered the atmospheric transmission windows by means of an "actinometer" developed by himself
– John HERSCHEL also was the first to make a thermal image visible to the human eye; he used the "evaporography" method based on differential evaporation of a fluid film

**1847** – Leon FOUCAULT and Hippolyte FIZEAU (F) measured the wavelengths of infrared radiation up to 1.5 μm
– Hermann KNOBLOCH (D) confirmed the equivalence of light and thermal radiation

**1858** – Gustav KIRCHHOFF (D) formulated the radiation law named after him today

**1862** – Gustav KIRCHHOFF (D) coined the term "black body"

**1872** – John TYNDALL (GB) published quantitative measurements of infrared absorption bands of the gases of the atmosphere, and became a pioneer of infrared spectroscopy

**1873** – Willoughby SMITH (GB) discovered the light sensitivity of the electrical resistance of selenium

1878 – Samuel Pierpont LANGLEY (USA) invented the "bolometer" that used a thermally induced change of the electrical resistance to measure radiation

1879 – Joseph STEFAN (A) formulated the "Stefan-Boltzmann Law" about the temperature dependence of the emitted thermal radiation which was confirmed theoretically by Ludwig BOLTZMANN in 1884

1880 – Paul-Quentin DESAINS and Pierre CURIE (F) measured the wavelengths of infrared radiation up to 7 µm
– The term "infrared" denoting invisible radiation with long wavelength became more and more popular

1893 – Wilhelm WIEN (D) formulated his "Displacement Law" about the spectral shift of the maximum of radiation over temperature

1896 – Wilhelm WIEN (D) formulated his radiation law that for the first time provided a full formal description of the spectral distribution of thermal radiation; at long wavelengths, however, the formula deviated from reality

1897 – Heinrich RUBENS (D) measured the wavelengths of infrared radiation up to 20 µm, and determined quantitatively the spectral radiation distribution with high accuracy

1898 – Heinrich RUBENS (D) and Ernest Fox NICHOLS (USA) determined the wavelengths of infrared radiation up to 150 µm

1900 – John William STRUTT, Lord RAYLEIGH (GB) formulated a new radiation law based on classical electrodynamics; with the corrections of James JEANS in 1905, it is known as "Rayleigh-Jeans Law" today; at short wavelengths, however, the formula was far from reality ("ultraviolet catastrophe")
– On 19 October, Max PLANCK (D) presented a new radiation formula which was fully proved by the measurements of RUBENS
– On 14 December, PLANCK (D) officially presented his theoretically proved radiation law; this was the "birthday of quantum physics"

1917 – Theodore W. CASE (USA) developed the "thallofide cell"
– In Great Britain, the first IR Search & Track (IRST) device was developed; achievable range was 1 mile against aircraft

1926 – An active IR pointing system for detection of aircraft was developed in the United States

1929 – Marianus CZERNY (D) began the development of a thermal imaging device based on evaporography; John HERSCHEL had used the same principle in 1840
Kalman TIHANYI (H) invented an infrared sensitive electronic TV camera

1930s – Edgar W. KUTZSCHER, B. GUDDEN and P. GÖRLICH (D) developed photo-conductive lead sulfide (PbS) cells

1932 – Herman WILLENBERG (D) improved CZERNY's method
– An active IR pointing system could track an airship up to 2100 m altitude (USA)

1935 – A passive thermal pointer achieves ranges of up to 17 miles against large ocean liners (USA)
– At CARL ZEISS (D) the development of thermal pointers with bolometers began

1937 – Marianus CZERNY (D) presented a practical thermal imager and coined the term "Evaporography"

1939 – At CARL ZEISS (D) a thermal pointer based on bolometer for continuous radiation demonstrated ranges of 5 km against driving cars (D)

1940 – At CARL ZEISS (D) thermal pointer based on bolometer for continuous radiation and 150-cm mirror demonstrated 30 km range against ships

1941 – The first German Naval Artillery units at the Atlantic coast were equipped with thermal pointers (D)
– Rudolph KOOPS produced the new IR-materials KRS 5 and KRS 6 at CARL ZEISS (D)
– Robert J. CASHMAN (USA) was ordered by the NDRC to continue the work at the "thallofide cell"
– At CARL ZEISS (D) the development of thermal pointers with bolometers based on modulated radiation began

**1942** – In Great Britain and Germany, the photoconductivity of lead selenide (PbSe) was discovered

**1943** – ELAC in Kiel (D) produced lead sulfide (PbS) detectors as well as ZEISS-IKON in Dresden (D)

– ELAC in Kiel (D) built the "Wärmepeilgerät S2" with PbS-cell for the German navy

– CARL ZEISS in Jena (D) built "Wärmepeilgerät S1" mit modulated light bolometer for the German navy

– Lab models of the target seeker head "Madrid" (D) for air-to-air missiles used PbS-cells

**1941-44** – Approx. 50 devices of the ZEISS (D) thermal pointer "Donau 60" were deployed at the Naval Artillery

**1944** – CASHMAN (USA) also developed PbS detectors

– At E. LEYBOLD'S NACHFOLGER (D) the thermal imager "Potsdam" was developed which used a common scanner for obtaining and reproducing the thermal image

– At ELAC (D) the PbSe cells reached production status

**1944-45** – CARL ZEISS (D) built the infrared search device FuG 280 "Kiel IV" to be deployed in a night hunter aircraft (Ju 88 G)

**1953** – H. GOBRECHT and W. WEISS (D) presented a reworked version of CZERNY's thermal imaging sensor

– The BAIRD Ass. company (USA) filed a patent application for a thermal imaging sensor (US2,855,522)

**1953-54** – Development of the first IR-seeking missile "Firestreak" (GB)

**1954** – A group around W. D. LAWSON developed detectors made from indium antimonide (InSb) at the RRE (GB)

– First approaches (USA) towards modern imaging IR-cameras with thermopiles (20 min per image) and bolometer (4 min per image)

**1955** – At KELVIN HUGHES (GB) the development of the IR-camera "Pyroscan" began

**1956** – BAIRD Ass. (USA) put the (first?) commercial evaporograph models "JZ-1" and "KR-1" on the market (USA)

– TEXAS INSTR. (USA) got a development contract for line-scanner cameras

**1957** – R. LAWSON (CAN) conducted first examinations of the female breast with an "Evaparograph" from BAIRD

– BARNES Eng. (USA) produced the electro-optical "FAR-IR Camera" with high temperature resolution

**1959** – A group around W. D. LAWSON at the RRE (GB) developed mercury-cadmium-telluride (HgCdTe, also known as CMT or MCT)

– R. LAWSON (CAN) introduced quantitative breast examinations with the IR-camera of BARNES

**1961** – S. SMITH & Sons (GB) produced the "Pyroscan Mk. I" thermal imager for the 5 µm-region

**1960s** – Significant advances in the field of photolithography and substantial improvement of the reliability of the detector cooling equipment

**1963-64** – TEXAS INSTR. (USA) developed the first "Forward Looking Infra-Red" (FLIR) System that could deliver 20 images per second

– S. SMITH & Sons (GB) presented the "Pyroscan Mk. II" specially designed for medical applications

**1965** – The "Thermographic System 660" of AGA (S) was regarded as the first civil IR-camera in series production

– A lab model of a FLIR of TEXAS Instr. (USA) was installed in an aircraft that was deployed as a "Gunship" in Vietnam

**1966** – At TEXAS Instr. (USA) the production of FLIR-systems began

**by 1968** – Beginning of series production of linear IR-sensor arrays by means of photolithography (1st Generation)

**1969** – TEXAS instr. (USA) began the series delivery of the "AN/AAD-FLIR" for the American gunship aircrafts in Vietnam

**by 1970** – World-wide approx. 20 different commercial thermography devices were offered for medical application

**1970** – CARL ZEISS (D) built "WBG 1/25" and "WBG 1/50" thermal imagers with linear detector arrays comprising 25 and 50 elements, respectively

1971 – CARL ZEISS (D) presented the "PST", a combination of image intensifier and infrared thermal pointer
– ELTRO (D) built the infrared tracker "ELTRO D 14-4" CARL ZEISS (D) developed and tested the high-resolution thermal imager "WBG 1/100" with a linear detector array of 100 elements

1972 – TEXAS Instr. (USA) had already built more than 385 FLIR devices, however, in 55 different configurations; to reduce effort and cost a "Common Module" concept for FLIR systems was proposed

1973 – GRETAG (CH) put the "PANICON 7510" thermal imager on the market that was based on thermoplastic deformation of an oil film instead of differential evaporation
– CARL ZEISS (D) offered the thermal pointer "AWO 10" to be attached to the "ORION 110" image intensifier device

1974 – The Department of Defense (USA) officially presented the "Common Module" concept that had been proposed by TEXAS Instr. and the NIGHT VISION LAB
– C. Thomas 'Tom' ELLIOTT (GB) developed the SPRITE technology for CMT-detectors, also called TED (Tom Elliott's Device); this became the basis of the British "Thermal Imaging Common Modules" (TICM)
– In France, an own "Système Modulaire Thermographique" (SMT) was developed

1975 – CARL ZEISS (D) presented the "ORIOTHERM", a night vision device with integrated thermal pointer
– CARL ZEISS (D) conducted a study about the integration of the combination of night vision device & thermal pointer in their "PERI R 12" commander periscope (Peri-Passiv)
– The German Ministry of Defense (BMVg) settled a decision for the "Common Modules"; analog decisions were made in Taiwan, South Korea and Denmark

1976 CARL ZEISS (D) offered the thermography device "IKOTHERM"
– TEXAS Instr. (USA) started the production of Common Modules
– CARL ZEISS (D) became sole licensee for the sale of FLIR Common- Modules of TI on the territory of the BRD

1977 – CARL ZEISS (D) presented the "WBG 1/200-10", a thermal imager of the FLIR-type

1979 – Cooperation between TEXAS Instr. (USA) and CARL ZEISS (D) in German programs, especially in the German Tank Thermal Sight (GTTS) program
– TEXAS Instr. (USA) installed a factory for Common Module detectors and coolers at AEG Telefunken in Heilbronn
– CARL ZEISS (D) got a series contract for equipping the main battle tank Leopard 2 with thermal imagers

1980 – TEXAS Instr. (USA) signed 'second-source' agreements in USA and in Germany as well: AEG-TELEFUNKEN in Heilbronn (detector & cooler, pre-amplifier, LED-array), ELTRO in Heidelberg (scanner) and CARL ZEISS in Oberkochen (visual & IR optics)

1982 – The specifications of the Night Vision Electro Optical Laboratory NVEOL (USA) of the "Common Modules" became official MIL standards

1984 – The cooperation of TEXAS Instr. (USA) and CARL ZEISS (D) had delivered more than 2300 systems; TI alone had produced more than 13,000 FLIR systems

1990er – The German Thermal imaging industry conducted studies about a new generation of thermal imagers (D)

1992-96 – The German consortium (CARL ZEISS, ELTRO, ATLAS ELEKTRONIK, AEG & TELEFUNKEN) developed the new German Common Modules (DCM) named "Ophelios"
– At the same time, in Great Britain the "STAIRS C" program is pursued by the DEFENCE RESEARCH AGENCY (DRA) and PILKINGTON THORN OPTRONICS; in France SAGEM worked on the "Iris" modules, and the "Synergi" modules were developed in French-British-German cooperation by THOMSON-CSF OPTRONIQUE (later TTD-THALES), PILKINGTON OPTRONICS and CARL ZEISS-SONDERTECHNIK

1996 – The thermal imagers of the 2nd generation reach production status in France ("Synergi" & "IRIS"), Great Britain ("STAIRS C", "Synergi"), USA ("HTI") and Germany ("Ophelios" & "Synergi")

1998 – TEXAS Instr. (USA) had produced already more than 30,000 FLIR systems according to their own records
– CARL ZEISS (D) had produced more than 10,000 Common Module thermal imagers for the equipment of Leopard 1 and Leopard 2 battle tanks, Marder and Luchs fighting vehicles, submarine periscopes SERO 14, ECR Tornado reconnaissance aircrafts and mobile infrared surveillance systems

by 2005 – The transition from Common Modules of the 2nd generation to staring thermal imagers with matrix detectors, which were regarded as 3rd generation, was under way

2008 – The "Ophelios" series production was discontinued after more than 3000 devices (D)

## Glossary

**A** – Ampere: Unit of electrical current

**A/D, ADC** – Analog/Digital, Analog/Digital Converter

**Afocal** – Designation derived from Latin for telescope (without focal point)

**AfOM** – Attica for Optronic mast; special version of the Attica thermal imager

**Adaption** – Adjustment of the human eye to changing brightness by changing the pupil diameter (fast) or the sensitivity of the receptors (slow)

**Accommodation** – Adjustment of the human eye to changing distance by changing the refractive power of the eye lens

**Atmospherical windows** – The atmosphere transmits radiation but not in all wavelength bands. There are two distinct infrared wavelength bands (windows) with very high transmission: the mid wave IR window between 3 – 5 µm (MWIR) and the long wave IR window between 8 – 12 µm (LWIR)

**ATTICA** – Name of a staring cooled thermal imager of HENSOLDT

**BITE** – Built-In Test Equipment

**Black body** – Object that absorbs 100% of the radiation of all wavelength and emits thermal radiation according to Planck's law

**BMVg** – Bundesministerium für Verteidigung: Federal Ministry of Defense of the Federal Republic of Germany

**Bolometer** – Device for measuring radiation based on thermal variation of the electrical resistance

**Bundeswehr** – Military forces of the Federal Republic of Germany

**Camera lucida** – Optical device to overlay two images (e.g. scene and drawing plane) in the eye of the observer (Latin for bright chamber)

**Camera obscura** – Pinhole camera to generate real images without using a lens (Latin for dark chamber)

**CAN Bus** – Controller Area Network: Digital data bus for controlling several electronic devices (e.g. cameras)

**Cassegrain** – Mirror telescope widely used in astronomy: a large primary mirror with a hole in the center collects the incoming light which is then directed through the central hole by a smaller secondary mirror which is also curved.

**catadioptric** – An assembly of optical elements comprising lenses and mirrors

**CCIR standard** – Comité Consultatif International des Radio-communications: widely used video standard for thermal images (corresponds to black/white video standard)

**CM** – Abbreviation of Common Module

**CMT** – Cadmium-Mercury-Telluride: widely used material for infrared detectors

**Common Modules** – US-Common Modules or US-German Common Modules: Set of modules for thermal imaging developed by TEXAS INSTRUMENTS (USA); also manufactured and used in Germany

**DCM** – Deutsche Common Modules: German Common Modules developed and manufactured by a German industry group

**Descanner** – Opto-mechanical device to compensate the image shift of a moving camera

**DLC coating** – Diamond-Like Carbon: very hard and environ-mentally resistant anti-reflection coating of infrared windows and lenses

**DoD** – Department of Defense of the USA

**Dome** – Window with the shape of a hollow sphere; also known as calotte

**Donau 60** – German thermal pointer with bolometer detector and 60 cm mirror that was in service in WW II

**Donau 150** – German thermal pointer with PbS detector and 150 cm mirror

**DRA** – British Defence Research Agency

**DSC** – Digital Scan Converter: converts the sequence of detector signals into an image format

**ECR** – Electronic Combat Reconnaissance: special version of the Tornado fighter aircraft for Germany and Italy

**Entrance pupil EP** – In the entrance pupil, all bundles are crossing that enter the optic and contribute to the imaging. The optic images the entrance pupil onto the exit pupil.

**Evaporography** – Generation of thermal images based on local heating of a fluid film and resulting changes of the film thickness; can be made visible by means of interference effects

**Exit pupil** – In the exit pupil, all bundles are crossing that leave the optic. The optic images the exit pupil onto the entrance pupil.

**Fabry-Perot interferometer** – Consists of two plates with high reflection; narrow bandpass filter due to the interference of the reflected beams

**FLIR** – Forward Looking Infra-Red: thermal imager capturing a two-dimensional infrared image

**Focal length** – Distance between the lens and the focus point in which a parallel light bundle (e.g. from the sun) is concentrated

**Focus** – The spot in which a collecting lens concentrates the incoming light

**focusing** – The process to adjust a detector or screen in the image plane to get a sharp image

**FOV** – Field of View

**Galilean telescope** – Telescope comprising a collecting and a diverging lens invented by the astronomer Galilei

**Generations of thermal imagers** – 1st generation: Thermal imagers with linear detector array, scanner, analog electronics and optical image reproduction
– 2nd generation: Thermal imagers with linear detector array with TDI, scanner, digital electronics and standard video images
– 3rd generation: Thermal imagers with two-dimensional detector array, digital signal processing & image reproduction

**Germanium (Ge)** – Semi-conductor material with large transmission range in the infrared; most commonly used in infrared optics

**Pit organ** – Thermal imager of rattle snakes that works like a modern uncooled bolometer

**GTTS** – German Tank Thermal Sight program to equip German tanks with thermal imagers

**HARM** – High Speed Antiradiation Missile to combat radar systems

**HELIOS** – Hoch-Empfindliches Leichtes Infrarot-Optisches System: High-sensitive light infrared optical system; early name of a scanning thermal imaging system developed by a German industry group in the 1990s; later renamed to Ophelios

**HTI** – Horizontal Technology Integration program for FLIR cameras in USA

**Hz** – Hertz: Unit of frequency 1 Hz = 1/sec

**IDCA** – Integrated Detector Cooler Assembly

**IDS** – Interdiction, Strike: Standard version of Tornado-fighter aircraft

**InSb** – Indium antimonide: widely used detector material for the mid IR range 3 – 5 µm

**IR** – Infra-red: this radiation cannot be seen by the human eye because of the long wavelength

**IRCCD** – Infra-red Charge Coupled Device: shift register in infrared detectors using the CCD process

**IRCMOS** – Infra-red Charge MOS: shift register in infrared detectors using the CMOS process

**IR radiation** – Infra-red radiation: not visible to the human eye because of the long wavelength

**ITB** – Interlace with Transducer Board: control board for interlace process

**IRST** – Infra-Red Search and Track

**K** – Kelvin: temperature unit [K] = [°C] – 273,13°C

**Keplerian telescope** – Telescope comprising two collecting lenses invented by the astronomer Johannes Kepler

**Line of sight LOS** – Direction in which an optical instrument is pointing

**LRU** – Line Replaceable Unit

**LWIR** – Long Wave Infra-Red; infrared wavelength range between 8 - 12 µm

**Magnification** – Increasing or decreasing the angle under which an object is perceived; typically done by means of a telescope

**MCT** – Another term for CMT (Hg-Cd-Te)

**Mrad, milliradian** – Milliradian: angular unit, 1 mrad = 1/1000 rad
**MRTD** – Minimum Resolvable Temperature Difference
**MUX** – Multiplexer: electronic switch between different signals
**MWIR** – Mid Wave Infra-Red; infrared wavelength range between 3 - 5 µm
**NETD** – Noise Equivalent Temperature Difference
**Newtonian telescope** – Mirror telescope with a large collecting mirror and small plane mirror in the center of the beam path to direct the light out of the entrance aperture
**NUC** – Non-Uniformity Correction: used for infrared images
**Objective** – Lens facing the object
**Ocular** – Lens facing the eye of the observer
**OPHELIOS** – Optronisches Passives Hoch-Empfindliches Leichtes Infrarot-Optisches System: Optronic passive high-sensitive light infrared optical system, a scanning thermal imaging system developed by a German industry group in the 1990s
**Optronics** – Integration of optical and electronical components in an imaging system
**PANICON** – Swiss thermal imager of GRETAG based on thermos-plastic changes of the surface of a film
**PE** – Proximity Electronics: very close to the detector
**Periscope** – Optical instrument in which entrance and exit are laterally displaced
**PSD** – Position Sensitive Device: Sensor for detecting the position of a light spot
**Pupil** – In a pupil of an optical system all light bundles are crossing
**Rad, radian** – Radian: angular unit, $2\pi$ rad = 360°
**Radiation law** – Gives the power density of the radiation as a function of wavelength; parameter is the temperature of the radiator
**ROIC** – Read-Out Integrated Circuit: CCD or CMOS register for the read-out process integrated on the detector chip
**SADA** – Standard Advanced Detector Assembly (SADA II): US detector program
**Scanner** – Opto-mechanical device (mostly an oscillating mirror) to shift the optical image laterally in the detector surface
**Silicon (Si)** – Semi-conductor: Material for lenses in MWIR; detector material in the visual range
**SMT** – Système Modulaire Thermographique: national French system of standard modules for thermal imaging
**SPRITE** – Signal PRocessing In The Element: British IR detector used in the TICM thermal imagers
**SPS** – Scan Position Sensor: Sensor for measuring the angular position of the scan mirror
**STAIRS C** – Sensor Technology for Affordable IR Systems: British program for Common Modules of the 2nd generation
**Stirling cooler** – Electrically driven cooler based on the Stirling cycle, a thermodynamic circular process invented in 1816 by the 25-year old Scottish referend Robert STIRLING
**Synergi** – French Common Module program of the der 2nd generation
**TDI** – Time Delay and Integration: multiple measurement of an image point by neighboring detector elements
**TEC** – Thermal Electric Cooler: based on cooling of a pair of current-carrying conductors of different material (Peltier effect)
**TED** – Tom Elliot's Device: another name of the British SPRITE detector
**Telescope** – From the Greek (looking far)
**Thermal image** – A visible reproduction of a scene based on the thermal radiation emitted by the objects
**Thermal imager** – Special camera for the infrared wavelength range to capture thermal images
**Thermal radiation** – All objects emit electromagnetic radiation according to their temperature. At room- temperature the maximum emission occurs at very long wavelengths in the infrared range which are invisible to the human eye.
**Thermocouple** – Electrical circuit for measuring temperature differences
**Thermography** – In general: Capturing of thermal images; – Today: Capturing of thermal images with an absolute temperature calibration, mostly displayed as false color image

**Thermopile** – Combination of several thermocouples
**TI** – Thermal Imager
**TI** – Texas Instruments Incorporate, USA
**TICM** – Thermal Imaging Common Modules: national British system of standard modules
**Relay system** – Telescopic system to increase the length of an optical system (e.g. mast of submarine periscope)
**V** – Volt: unit of electrical voltage
**W** – Watt: unit of mechanical/electrical/thermal power
**WBG** – Wärmebildgerät: German name for thermal imager
**Wheatstone bridge** – Electrical circuit to detect changes of the electrical resistance with very high sensitivity; made public by the British researcher WHEATSTONE
**WPG 15** – Wärme-Peil-Gerät: German thermal pointer with PbS detector and 150cm mirror
**ZnS** – Zinc sulfide: material for infrared windows and lenses; in multispectral grade high transmission in the LWIR, MWIR and visible region
**ZnSe** – Zinc selenide: material for infrared lenses; high transmission in MWIR and LWIR; dust is hazardous to health when in contact with water

## Bibliography


- Auerbach, Felix: Das Zeisswerk und die Carl-Zeiss-Stiftung in Jena, Verlag Gustav Fischer, Jena (1925)
- Bergmann-Schaefer: Lehrbuch der Experimentalphysik, Bd. III Optik, de Gruyter, Berlin (1978)
- Robert Bud, Philip Gummett: Cold War, Hot Science Harwood Academic Publishers, Australia (1999)
- Joseph Caniou : Les applications militaires de l'observation dans l'infrarouge, Congrès National de Thermographie THERMOGRAM' (2009)
- John P. Dakin, Robert G. W. Brown: Handbook of Optoelectronics, CRC Press (2006)
- Helmuth Giessler: Der Marine-Nachrichten- und Ortungsdienst, J. F. Lehmanns Verlag, München (1971)
- Henry L. Hackforth: Infrared Radiation, McGraw-Hill, New York (1960)
- Eugene Hecht: Optik, Oldenbourg Verlag, München (2001)
- Richard Holmes: The Age of Wonder, Harper Press, London (2009)
- Gerald C. Holst: Common Sense Approach to Thermal Imaging, SPIE Press, Bellingham (2000)
- Richard D. Hudson: Infrared System Engineering, John Wiley Publ., New York (1969)
- Jenaer Jahrbuch zur Technik- und Industriegeschichte, Verein Technikgeschichte, Jena (1999)
- Francis A. Jenkins, Harvey E. White: Fundamentals of Optics, McGraw-Hill, Kogakusha, Tokio (1976)
- Sean F. Johnston: A History of Light and Colour Measurement, Institute of Physics Publishing, Bristol (2001)
- Rudolf Lusar: Die deutschen Waffen und Geheimwaffen des 2. Weltkriegs, J.F. Lehmanns Verlag, München (1962)
- MDv. Nr. 291: Funkmeßgerätekunde, Oberkommando der Kriegsmarine (1944)
- John Lester Miller: Principles of Infrared Technology, Van Nostrand (1994)
- Mullard Ltd.: Application of Infrared Detectors, London (1971)
- Eric A. Newman, Peter H. Hartline: Infrarotsehen bei Schlangen, Spektrum der Wissenschaft, Heidelberg (1982)
- H. Plesse: Wärmepeilgeräte der Firma Carl Zeiß, Jena (1944)
- Joseph Mark Scalia: In geheimer Mission nach Japan: U 234, Ullstein Verlag (2005)
- Harry Schlemmer: History of Submarine Periscopes at Carl Zeiss, Mittler & Sohn, Hamburg (2011)
- Karl-Heinz & Michael Schmeelke: Fernkampfgeschütze am Kanal, Podzun-Pallas Verlag, Friedberg (1992)
- Michael Schmeelke: Alarm Küste, Dörfler Verlag, Eggolsheim (1993)
- Herbert Schober: Das Sehen, VEB Fachbuchverlag, Leipzig (1964)
- Norbert Schuster, Valentin G. Kolobrodov: Infrarotthermographie, John Wiley, Berlin (2004)
- Fritz Trenkle: Die deutschen Funkmessverfahren bis 1945, Motorbuch Verlag, Stuttgart (1979)
- Rolf Walter: Zeiss 1905 – 1945, Böhler Verlag, Köln (2000)
- Emil Wilde: Geschichte der Optik, Verlag Rücker & Püchler, Berlin (1838)

## List of Figures

# Notes

## Foreword

- **1** DASA was from 1989 to 2000 a German aeronautical and space group that belonged to Daimler-Benz; changing names were: Deutsche Aerospace Aktiengesellschaft, Daimler-Benz Aerospace Aktiengesellschaft, DaimlerChrysler Aerospace Aktiengesellschaft
- **2** EADS was founded on 10. July 2000 and had developed to the biggest European aeronautical and space group; it was created by merging of German DASA, the French Aerospatiale-Matra and the Spanish CASA.

## 1. Early History: Learning from Nature

- **3** By the word: Light from God and shadow
- **4** About the optical properties of the human eye, see e.g.:
  – Bergmann-Schaefer: Lehrbuch der Experimentalphysik, Bd. III Optik, Walter de Gruyter, Berlin (1978), S. 123 ff.
  – Lexikon der Optik, Bd. 1, Spektrum Akademischer Verlag, Heidelberg (2003)
  – Eugene Hecht: Optik, Oldenbourg Verlag, (2001)
  – David Macaulay: Das große Buch vom Körper, Ravensburger Buchverlag, Ravensburg (2009)
  – The neuronal image processing downstream of the eye is probably even more important than the mere optical properties. A comprehensive discussion may be found in
  – Herbert Schober: Das Sehen, VEB Fachbuchverlag, Leipzig (1964)
- **5** Before that, the evolution had made the leaves and the grass green.
- **6** SWIR stands for »Short Wave Infra-Red«
- **7** The small mantis shrimp (squilla mantis) probably has the most sophistic eye design evolution had created: it has two stalked eyes which are comprising 10,000 eye elements, and produce 3 images simultaneously, resulting in a panoramic view from 6 perspectives. The shrimps can distinguish 100,000 colors and uses up to 12 color receptors which covered the spectral region from ultraviolet to infrared. Furthermore, the shrimps can detect linear polarized light as well as circular.
- **8** Johan Marais: Die faszinierende Welt der Schlangen, Karl Müller Verlag, Erlangen (1995), S. 10, 13
- **9** Chris Mattison: Schlangen, Dorling Kindersley, München (2007), S. 21, 34, 185: Besides the pit vipers (crotalinae: 18 genera divided in 69 species) also the families of the boas (boidae) and the pythons (pythonidae) have pit organs; alas of lower performance compared to the vipers
  – S. 21: „Boas with pit organs carry these between scales at the rim of the mouth, at the pythons they lie in the mouth scales themselves. „
  – S. 186: The rattle snakes (crotalus), from which 29 species exist, belong to the best-known genera of pit vipers. Rattle snakes are viviparous, grow to a length of 50 cm to 2 m, and nourish from lizards, birds and small mammals. Their dispersal area comprises North, Latin and South America. Their haunt is very variable and comprises temperate and tropical forests as well as grasslands, stone and sand desserts and mountainous regions.
- **10** Eric. A. Newman, Peter H. Hartline: Infrarotsehen bei Schlangen, Spektrum der Wissenschaft (Mai 1982), S. 106
- **11** Ibid.
- **12** Chris Mattison: Schlangen, Dorling Kindersley, München (2007), S. 21
- **13** Volker Storch, Ulrich Welsch: Systematische Zoologie, Gustav Fischer Verlag, Stuttgart (1997), S. 619
- **14** Paul Raths, Gustav-Adolf Biewald: Tiere im Experiment, Urania Verlag, Leipzig (1970), S. 261
- **15** Eric. A. Newman, Peter H. Hartline: Infrarotsehen bei Schlangen, Spektrum der Wissenschaft (Mai 1982), S. 108
- **16** Andreas B. Sichert, Paul Friedel, J. Leo van Hemmen: Snake's Perspective on Heat: Reconstruction of Input Using an Imperfect Detection System, Physics Review Letters, PRL 97, 068105 (2006)
- **17** Eric. A. Newman, Peter H. Hartline: Infrarotsehen bei Schlangen, Spektrum der Wissenschaft (Mai 1982), S. 114
- **18** By mathematical definition, the AND and the OR operators of Boole's algebra produce the following results:
  AND: 1^1=1, 1^0=0, 0^1=0, 0^0=0
  OR: 1v1=1, 1v0=1, 0v1=1, 0v0=0
- **19** Können tote Schlangen beißen?, TV14, Nr. 5 (2012), S. 34

## 2. Fundamental Research until 1900: Worth a Nobel Prize

- **20** W. Herschel: Investigation of the Powers of the prismatic Colours to heat and illuminate Objects; with Remarks that prove the different Refrangibility of radiant Heat. To which is added, an Inquiry into the Method of viewing the Sun advantageously, with Telescopes of large Apertures and high magnifying Powers, Phil. Trans. Roy. Soc. London, 90, 284 (1800)
  – Richard D. Hudson: Infrared System Engineering, John Wiley, New York (1969), S. 3
  – W. Herschel: Experiments on the Refrangibility of invisible Rays of the Sun, Phil. Trans. Roy. Soc. London, 90, 255 (1800)

- **21** Max Theodor Felix von Laue [9.10.1879 in Pfaffendorf – 24.4.1960 in Berlin]; German physicist; contributions to optics, crystallography, quantum theory, supra conduction and relativity theory; 1914 Nobel prize for the diffraction of x-rays in crystals; professor in Frankfurt, Berlin und Gottingen; 1931–33 President of the Deutschen Physikalischen Gesellschaft; Director of the Kaiser-Wilhelm-Institute of Physics until 1946; from 1950 on Director of the Max-Planck-Institute of Physical Chemistry
- **22** Emil Wilde: Geschichte der Optik, Verlag Rücker & Püchler, Berlin (1838)
- **23** www.veterinary-thermal-imaging.com (10.05.2012)
- **24** The Hippocratic Oath is a guideline for ethic conduct in the field of medicine
- **25** Hippocrates of Kos [about 460 BC in Kos – about 377 BC in Larissa], Greek physician and scholar; founder of medicine as a science based on experience (Hippocratic school = observation & documentation) ; "Father of the western medicine"; diagnosis based on objective observation und description of the symptoms; partition of medicine from mystic, theology and philosophy; because of his opposition against typical thought patterns of his time, he spent 20 years in jail and wrote important books in this time ("The Complicated Body"); in total, 150 books seem to be connected to him; the awareness of moral responsibility in action and functioning is stated in the oath named after him, which most probably was formulated after his death; medical achievements: categorizing of diseases, first descriptions of several diseases, therapy to support the self-recovery-of the body, stretching bank to relieve broken bones, regulations with respect to illumination, instruments, bandages, cleanliness at operations, first operation of the chest, operative treatment of hemorrhoids, Hippocratic finger, Hippocratic face
- **26** www.gardenofpraise.com; www.historylearningsite.co.uk (29.07.2012);
- **27** Kimio Otsuka, Tatsuo Togawa: Hippocratic thermography, Physiol. Meas. 18, 227 (1997); Abstract: The thermal measurement method of Hippocrates, which has been referred to as the first for measuring body temperature, was reproduced and the image obtained was compared with an infra-red radiation thermogram. Hippocrates' method involves covering the patient's thorax with an earth-soaked cloth. As the warmer areas dry faster, the pattern of enlargement of the dry area shows the temperature distribution. Therefore, it appears that Hippocrates was the first to obtain thermograms.
  – www.thelancet.com (29.07.2012)
- **28** Lisa Jardine: Ingenious Pursuits: Building the Scientific Revolution, Little, Brown (1999)
- **29** Accademia del Cimento ("Academy for Experiments"); founded in 1657 by Giovanni Borelli and Vincenzo Viviani, both students of Galilei; financial support by Prince Leopoldo and Grand Duke Ferdinando II de'Medici; discontinued after 10 years; objectives: Experiments for all topics, speculations should be avoided, manufacturing of laboratory equipment, general rules for measurements; Motto: Provando e riprovando ("Sampling and repeated sampling" or "Experiment and Confirmation")
  – http://en.wikipedia.org/wiki/Accademia_del_Cimento
- **30** W. Middleton: The Experimenters: A study of the Accademia del Cimento, John Hopkins Press, Baltimore (1971), S. 225
  – Elian Kleger: Schwarzkörperstrahlung, Diplomarbeit PH St. Gallen (2008)
- **31** Sean F. Johnston: A History of Light and Colour Measurement, Institute of Physics Publishing, Bristol (2001), S. 24
- **32** F. Bacon: Neues Organ der Wissenschaften, Wissenschaftliche Buchgesellschaft, Darmstadt (1830), S. 111
  – Elian Kleger: Schwarzkörperstrahlung, Diplomarbeit PH St. Gallen (2008)
- **33** Sir Isaac Newton [4.1.1643 in Woolsthorpe – 31.3.1726 in Kensington], English physicist and mathematician; father of the classical physics and celestial mechanics; professor of mathematics in Cambridge; after a nervous breakdown in 1693 Ward of the in der Royal Mint, from 1699 on its director; from 1703 on President of the Royal Society;
  – Contributions to optics: developed and built a mirror telescope in 1672, investigated and explained the spectrum of white light, in 1704 his main work "Opticks" was published, formulated the particle theory of light in his paper "New Theory about Light and Colours";
  – Contributions to mathematics: developed the infinitesimal calculus independently from Leibniz, derived from this the Keplerian laws about the planetary orbits;
  – Contributions to classical mechanics: collisional laws, gravitation theory, Newton's axioms, developed methods to determine the mass of sun and planets;
  – 1687 main work "Philosophiae Naturalis Principia Mathematica", in addition alchemistic, natural philosophical & mystic work
  http://de.wikipedia.org/wiki/Isaac_Newton
- **34** Eugene Hecht, Alfred Zajac: Optics, Addison-Wesley, Reading (1980), S. 4
  – Eugene Hecht: Optik, Oldenbourg Verlag, München (2001), S. 4
- **35** http://en.wikipedia.org/wiki/Visible_spectrum
- **36** Richard Holmes: The Age of Wonder, Harper Press, London (2009), S. xvii
- **37** Wordsworth: The Prelude (1850), Book 3, lines 58-64
  – William Wordsworth [1770 – 1850], English poet of the Romantic, originated together with Samuel Coleridge the Romantic in England with the common publication "Lyrical Ballads"; main work is "The Prelude"; Britannia's "Poet Laureate" of 1843 -1850
- **38** Christiaan Huygens [1629 –1695], Dutch physicist, astronomer, cosmologist and manufacturer of astronomical instruments; wave theory of light; telescope studies of the Saturn rings and its moon Titan; idea of a densely populate space (with inhabitants on Jupiter who build space crafts)
  – Richard Holmes: The Age of Wonder, Harper Press, London (2009), S. 477

- **39** Gabrielle Émilie Le Tonnelier de Breteuil, Marquise du Châtelet (Émilie du Châtelet) [17.12.1706 in Paris – 10.9.1749 in Luneville, died after giving birth of her 4. child]; French mathematician, physicist and author;
- **40** Émilie du Châtelet derived that the kinetic energy E must be proportional to the square of the velocity V ($E \sim v^2$) while in her time it was generally supposed that a direct proportionality would hold ($E \sim v$).
- **41** http://en.wikipedia.org/wiki/ Émilie_du_Châtelet
- **42** http://en.wikipedia.org/wiki/ Émilie_du_Châtelet
  – David Bodanis: Passionate Minds: The Great Love Affairs of the Enlightenment, Crown, New York (2006)
- **43** Thomas Melvill [1726 – 1753]; Scottish natural philosopher; used kites in meteorology for the first time; father of the flame-emission spectroscopy; he postulated that light of different color also would propagate with different velocity.
- **44** Henry L. Hackforth: Infrared Radiation, McGraw Hill, New York (1960)
- **45** Carl Wilhelm Scheele [19.12.1742 in Stralsund – 21.5.1786 in Köping, Sweden]; German-Swedish pharmacist and chemist; he isolated many chemical compounds and elements; originator of the gas analysis; known by the discovery of oxygen and nitrogen; in 1777 a single book "Chemical Treatment of Air and Fire"
- **46** Marsilio Landriani [1751 in Mailand – 1815 in Wien]; Italian chemist, physicist and meteorologist; contributions to the chemistry of air and to electrochemistry; developed the Eudiometer for measuring the volume change of air that could also be used as barometer; published two scientific books
  – http://en.wikipedia.org/wiki/Marsilio_Landriani
- **47** Richard D. Hudson: Infrared System Engineering, John Wiley, New York (1969), S. 5
- **48** Marc-Auguste Pictet [23.7.1752 in Geneva – 19.4.1825 in Geneva]; Swiss scientist; contributions to meteorology; founded an observatory on the Great St. Bernhard; President of the Geneva Academy; from 1808 on member of the Bavarian Academy of Science
- **49** Antonin Vasko: Infra-red Radiation, Iliffe Books, London (1968), S. 16
- **50** Sir Frederick William Herschel [15.11.1738 in Hannover – 25.8.1822 in Slough, England]; German-British astronomer, telescope builder, musician and composer; A turbulent life:
  – 1752 Friedrich Wilhelm became military musician (Oboe) in Hannover like his father
  – 1757 Invasion of French troops in Hannover; Herschel escaped to England; worked as music teacher & organist
  – 1766 Music director in Bath; studies of music theory & mathematics; built & sale of first astronomical instruments; his sister Caroline supported his research work all her life
  – 1781 Discovery of the planet Uranus; member of Royal Society; annual pension by King George III.
  – 1788 Marriage with Mary Pitt, the widow of a neighbor;
  – 1792 Birth of their only son John Frederick William
  – 1816 Knightly accolade by the prince regent who later became King George IV
  – A comprehensive portrayal of the activities and the time can be found in the book of Richard Holmes: The Age of Wonder, Harper Press, London (2009), S. 60 ff, S. 163 ff, S. 381 ff
- **51** At that time, George II was king of England and Hannover.
- **52** Caroline Lucretia Herschel [16.3.1750 in Hannover – 9.1.1848 in Hannover]; German astronomer; followed her brother William to England in 1772; initially supported his astronomical work; then own discoveries of several comets, calculation of exact astronomical reductions and a zone catalog with several hundreds of star clusters and nebulae, in public she stands always modestly behind her brother; in 1822, after the death of her brother, she went back to Hannover; in 1828 gold medal of the Royal Astronomical Society; in 1835 honorary member; in 1838 member of the Irish Academy of Science in Dublin; in 1846 gold medal of the Prussian Academy of Science; after her the comet 35P/Herschel-Rigollet, a moon crater in Sinus Iridium and the planetoid (281) Lucretia were named
- **53** Richard D. Hudson: Infrared System Engineering, John Wiley, New York (1969), S. 5
- **54** Richard D. Hudson: Infrared System Engineering, John Wiley, New York (1969), S. 3
- **55** Thomas Johann Seebeck [9.4.1770 in Reval – 10.12.1831 in Berlin]; German-Baltic physicist; initial studies of medicine in Berlin and Gottingen and physician in Gottingen; then he turned to physics and contributed to optics and electro-magnetism; he discovered the thermoelectric effect that was named after him.
  – http://de.wikipedia.org/wiki/Thomas_Johann_Seebeck
- **56** Joseph Fraunhofer (since 1824 Ritter von Fraunhofer) [6.3.1787 in Straubing – 7.6.1826 in Munich]; German optician and physicist; contributions to scientific telescope design and spectroscopy; Fraunhofer achromat; new glasses; measurement of refractive indices; grinding machines for glass machining http://de.wikipedia.org/wiki/Joseph_von_Fraunhofer
- **57** William Hyde Wollaston [6.8.1766 in East Dereham – 22.12.1828 in London}; English physician, physicist and chemist; in 1802 he found 7 dark lines in the solar spectrum; refractometer based on total reflection; discovered the elements palladium and rhodium in 1803; in 1807 camera lucida (pinhole camera); discovered cysteine in 1810; goniometer in 1817; 1820 Wollaston prism comprising two sub prisms; since 1793 member of the Royal Society, 1804 - 1816 secretary; 1902 Copley Medal
- **58** Henry L. Hackforth: Infrared radiation, McGraw-Hill, New York (1960), S. 8
- **59** Norbert Schuster, Valentin G. Kolobrodov: Infrarotthermographie, John Wiley, berlin (2004), S. 16ff

- **60** Leopoldo Nobili [1784 in Trassilico – 5.8.1835 in Florenz]; Italian physicist; developed several electrical instruments; after finishing the military academy he became an artillery officer; veteran of Napoleon's Russian campaign; medal of the legion of honor; thereafter studies in physics, especially of electrical effects; since 1832 professor of physics at the Grand Ducal Museum in Florence; since Feb. 1835 member of the Berlin Academy of Science.
  – http://de.wikipedia.org/wiki/Leopoldo_Nobili
- **61** Macedonio Melloni [11.4.1798 in Parma – 11.8.1854 in Portici]; Italian physicist; known for his measurements of the thermal radiation of blackbodies; demonstrated that thermal radiation is refracted, diffracted and polarized like light; in 1839 Director of the Vesuvius Observatory; since 1845 foreign member of the Royal Swedish Academy of Science.
  – http://en.wikipedia.org/wiki/Macedonio_Melloni
- **62** Norbert Schuster, Valentin G. Kolobrodov: Infrarotthermographie, John Wiley, Berlin (2004), S. 16ff
- **63** Richard D. Hudson: Infrared System Engineering, John Wiley, New York (1969), S. 6
- **64** Thomas Young [13.6.1773 in Milverton – 10.5.1829 in London]; English multitalented researcher; deciphered parts of the Egyptian hieroglyphs earlier than Jean-Francois Champollion; further scientific contributions to vision, light, rigid body mechanics, energy, physiology, language and musical theory of harmony.
- **65** Andre-Marie Ampere [20.1.1775 – 10.6.1836]; French physicist and mathematician; the unit for the electrical current 'Ampere' was named after him
  – Leon Foucault [18.9.1819 – 11.2.1868]; French physicist; demonstrated the earth's rotation with a pendulum; measurement of the speed of light
  – Hippolyte Fizeau [23.9.1819 – 18.9.1896]; French physicist; measurement of the speed of light; co-discoverer of the Doppler effect
  – Paul-Quentin Desains [12.7.1817 – 3.5.1885]; French physicist
  – Pierre Curie [15.5.1859 – 19.4.1906]; French physicist; pioneer of crystallography, magnetism, piezo-electricity and radioactivity; Nobel laureate; husband to Marie Salomea Sklodowska-Curie.
  – Hermann Knobloch [11.4.1820 in Berlin – 30.6.1895 in Baden-Baden]; German physicist
  – Heinrich Rubens [30.3.1865 in Wiesbaden – 17.6.1922 in Berlin]; German physicist; contributions to electromagnetic radiation; demonstrated that Wien's Displacement Law is not valid in the longwave region; developed the Rubens flame tube
  – Ernest Fox Nichols [1.6.1869 – 29.4.1924]; US-American teacher and physicist; demonstrated the radiation pressure of light
- **66** Rock salt chemically consists of up to 98% table salt = sodium chloride NaCl
- **67** William Henry Fox Talbot [11.2.1800 in Melbury – 17.9.1877 in Lacock], British inventor and pioneer of photography; experiments with colored flames, monochromatic light and chemical changes of color; developed the Calotype process (incl. negative / positive) earlier than Daguerre, who showed his first pictures in 1839; Talbots process became the basis for the entire photography in the 19th and 20th century while Daguerre's process was virtually no longer used after 1865; http://en.wikipedia.org/wiki/Henry_Fox_Talbot
- **68** Henry L. Hackforth: Infrared Radiation, McGraw-Hill, New York (1960), S. 8
- **69** The knowledge about the atmospheric transmission was completed by many international contributions in the 100 years following:
  – 1839 Melloni (I): first absorption lines of water vapor in the solar spectrum
  – 1840 John Herschel (GB): three atmospheric transmission windows;
  – 1872 Tyndall (GB): quantitative measurements of the absorption coefficients of water vapor, carbon dioxide, methane etc.;
  – 1883 Langley (USA): all atmospheric absorption lines up to 5 µm;
  – 1917 Fowle (GB): all absorption lines up to 13 µm;
  – 1942 Adel: all absorption lines up to 24 µm; in the 1950s, several researchers have closed the gap to microwaves at 1000 µm
  – Richard D. Hudson: Infrared System Engineering, John Wiley, New York (1969), S. 7
- **70** Richard Holmes: The Age of Wonder, Harper Press, London (2009), S. 399
- **71** In total 132 plants were studied, and published as "Flora Herscheliana" in 1896
  – The camera lucida ("bright chamber") is a simple optical apparatus that helps the artist to view the object and the sketch on the paper simultaneously. Only in 1807 Herschel's compatriot William Wollaston had developed a special prism that could produce that function without any further optics. The camera lucida was known since Chinese high-cultures and was a popular aid for drawing before the era of photography
- **72** It was unknown to John Herschel that the term 'photography' was used before him by Hercules Florence in 1834
  – http://en.wikipedia.org/wiki/John_Herschel
- **73** Norbert Schuster, Valentin G. Kolobrodov: Infrarotthermographie, John Wiley, Berlin (2004), S. 16ff
- **74** http://en.wikipedia.org/wiki/John_Herschel
- **75** Richard D. Hudson: Infrared System Engineering, John Wiley, New York (1969), S. 299
- **76** Richard D. Hudson: Infrared System Engineering, John Wiley, New York (1969), S. 6
- **77** Richard D. Hudson: Infrared System Engineering, John Wiley, New York (1969), S. 299
- **78** Thermal Imaging for DUMIES: History of Thermal Imaging, Security Sales & Integration, July 2011
- **79** Antonin Vasko: Infra-red Radiation, Iliffe Books, London (1968), S. 21

- **80** Gustav Robert Kirchhoff [12.3.1824 in Königsberg – 17.10.1887 in Berlin]; German physicist; contributions to the fundamental understanding of electrical circuits, spectroscopy, emission of thermal radiation by black bodies (in all fields laws are named after him even today); Kirchhoff-Bunsen spectrometer; in 1875 first tenured professor of theoretical physics in Berlin
- **81** Written as a formula the Kirchhoff Law is: $P(\lambda,T) = a(\lambda) \cdot Ps(\lambda,T)$ with P = radiation power, a = absorption coefficient, $\lambda$ = wavelength, T = temperature) – The radiation power $P(\lambda,T)$ emitted by an area is proportional to the absorption coefficient $a(\lambda)$ and the radiation power $P_s(\lambda, T)$ of the ideally black body
- **82** John Tyndall [2.8.1820 in Leighlinbridge, Irland – 4.12.1893 in Haslemere]; Irish-British experimental physicist; studied and gained his doctorate in Marburg because he thought the German universities to be more practically orientated than the British; contributions to magnetism and magnetic-optical effects; investigations of the IR-absorption of gases; scattering in gases and fluids; contributions to atmospheric physics and glaciology; copious lecturing; numerous, and also didactic publishing (almost 18 books) http://en.wikipedia.org/wiki/John_Tyndall
- **83** Willoughby Smith [6.4.1828 in Great Yarmouth – 17.7.1891 in Eastbourne]; English electrical engineer; since 1848 with the Gutta Percha Company, London; headed in 1849 the manufacturing and the installation of the undersea cable between Dover and Calais; worked close together with Charles Wheatstone; discovered the photoelectric properties of selenium in 1873, publication *Effect of Light on Selenium during the passage of an Electric Current* in Nature, 20 February 1873 – http://en.wikipedia.org/wiki/Willoughby_Smith
- **84** David A. Kondas: Introduction to Lead Salt Infrared Detectors, Technical Report ARFSD-TR-92024, U.S. Army Armament Research, Development and Engineering Center (1993), S. 4
- **85** de.wikipedia.org/wiki/Max_Planck
- **86** Samuel Pierpont Langley [22.8.1834 in Roxbury – 27.2.1906 in Aiken]; American astronomer, physicist, inventor and widely-known aviation pioneer; since 1867 director of the Allegheny Observatory and Professor of Astronomy; since 1887 third secretary of the Smithonian Institution; in 1878 invention of the bolometer, application in the investigation of thermal radiation; results about the interaction of CO2 with thermal radiation motivated the Swedish physicist Svante Arrhenius to do first calculations of the greenhouse effect; in 1886 Henry Draper Medal for contributions to solar physics; since 1887 test flights with models; in 1903 the flight tests with his self-designed aircraft Aerodrome were discontinued after two accidents
- **87** Wheatstone bridge; invented by Samuel Hunter Christie [22.3.1784 – 24.1.1865]; British scientist and mathematician; worked on magnetism and measurement of magnetic fields; he proposed this method as 'diamond method' for measuring the resistance of thin wires; only after Charles Wheatstone seized on the method and presented it anew to the Royal Society, it became widely known; although Wheatstone introduced the method as Christie's invention, the term Wheatstone bridge became popular.
- **88** Sir Charles Wheatstone [6.2.1802 in Gloucester – 19.10.1875 in Paris]; English scientist and inventor; contributions to telegraphy; inventor of music instruments (English concertina), stereoscope and a encryption algorithm; today known by the bridge circuit which he presented in 1843, and which was named after him although it had been invented by Samuel Hunter Christie already in 1833. – http://en.wikipedia.org/wiki/Charles_Wheatstone
- **89** http://earthobservatory.nasa.gov/Features/Langley/langley_2.php
- **90** http://en.wikipedia.org/wiki/Bolometer
- **91** Joseph Stefan [24.3.1835 in St. Peter – 7.1.1893 in Wien]; Austrian physicist and mathematician of Slovenian heritage, poet in his Slovenian native tongue; since 1866 Director of the Physical Institute of the University Vienna; Vice president of the Vienna Academy of Science; 80 scientific publications; contributions to thermal radiation, thermal conductivity of gases, evaporation, diffusion and thermal conduction in fluids; Stefan-Boltzmann Law; thesis supervisor of Ludwig Boltzmann and Johann Josef Loschmidt
- **92** In the minutes of the sessions of the Vienna Academy of Science
- **93** Stefan-Boltzmann Law as a formula: $P = s \cdot T^4$, with the Stefan constant $s = 5.669 \cdot 10^{-12}$ W/(cm$^2$ K$^4$)
- **94** Ludwig Boltzmann [20.2.1844 in Wien – 5.9.1906 in Tybein]; Austrian physicist; student of Joseph Stefan; since 1869 Professor of Physics in Graz, then Vienna, Berlin and Vienna again; contacts to many well-known scientists of his time; contributions to the kinetic theory; Maxwell-Boltzmann distribution; Boltzmann equation for ideal gases; Stefan-Boltzmann Law; thesis supervisor of Paul Ehrenfest and Lise Meitner; suffered from bipolar disorder in his last years and committed suicide during a depression phase.
- **95** Sean F. Johnston: A History of Light and Colour Measurement, Institute of Physics Publishing, Bristol (2001), S. 25
- **96** Richard D. Hudson: Infrared System Engineering, John Wiley, New York (1969), S. 4
- **97** Wilhelm Wien [13.1.1864 in Gaffken – 30.8.1928 in München]; German physicist; 1886 promotion with Hermann von Helmholtz, and his assistant afterwards; since 1896 lecturer and professor at the Universities of Aachen, Gießen, Würzburg und Munich; essential contributions to the laws of thermal radiation; advocate of the electromagnetic approach to physical processes; equivalence of mass and energy; his differential equations for the electrodynamics of moving bodies in 1904 are counted among the preparations of the specific relativity theory; 1911 Nobel prize in physics for his work about thermal radiation. – http://de.wikipedia.org/wiki/Wilhelm_Wien

- **98** Wien's Radiation Law: $P(\lambda) = c1/\lambda^5 \cdot \exp(-c2/(\lambda \cdot T))$ with P = radiation power per area and wavelength unit in [$W/cm^2/\mu m$], λ = wavelength in [μm], T = absolute temperature in [K], c1 = 37418 $W{\cdot}\mu m^4/cm^2$, c2 = 14388 K μm
- **99** John William Strutt, 3rd Baron Rayleigh (The Lord Rayleigh) [12.11.1842 in Longford Grove – 30.6.1919 in Terling Place]; English physicist; discovered the element argon, for that Nobel prize in 1904; contributions to light scattering, Rayleigh scattering; explanation for the sky blue; contributions to acoustics and important textbook thereover; explanation of the localization of sound in the process of hearing.
  - http://en.wikipedia.org/wiki/John_Strutt,_3rd_Baron_Rayleigh
- **100** Sir James Hopwood Jeans [11.9.1877 in Ormskirk – 16.9.1946 in Dorking]; English physicist, astronomer and mathematician; Professor in Cambridge and New Jersey; important contributions to many fields of physics, quantum mechanics, rotating bodies, dynamics of star systems, co-founder of the British cosmology
  - http://de.wikipedia.org/wiki/James_Jeans
- **101** Rayleigh-Jeans Law: $P(\lambda) = c1 \cdot T/ (c2 \cdot \lambda^4)$ with P = radiation power per area and wavelength unit in [$W/cm^2/\mu m$], λ = wavelength in [μm], T = absolute temperature in [K], c1 = 37418 $W{\cdot}\mu m^4 / cm^2$, c2 = 14388 K μm
- **102** http://de.wikipedia.org/wiki/Rayleigh-Jeans-Gesetz
  - http://en.wikipedia.org/wiki/Ultraviolet_catastrophe
- **103** Max Karl Ernst Ludwig Planck [23.4.1858 in Kiel – 4.10.1947 in Göttingen]; German theoretical physicist; from a family with rich educational tradition; 1867-74 education in Munich; 1874-79 studies in Munich and Berlin; 1879 promotion, 1880 habilitation in Munich; 1880-85 private lecturer in Munich; 1885-89 Professor of Theoretical Physics in Kiel; 1889 Professor of Theoretical Physics in Munich; 1912 sekretary of the Kaiser-Wilhelm-Society (today: Max-Planck-Society); 1928-37 President; work about thermodynamics; insight about the connection between energy and entropy of radiation led to Planck's Radiation Law; the assumption of radiation packages with discrete frequencies, the frequency of which multiplied with Planck's constant h results in the energy (later called photons); co-founder of quantum theory; 1918 Nobel prize for his radiation law
- **104** There exist also other versions of the historical process; with respect to the intense work in radiation theory, this version is the most reasonable to the author.
  - http://de.wikipedia.org/wiki/Max_Planck
- **105** The exponential term of Planck is $1 / (\exp(C/(\lambda \cdot T)) - 1)$ with a constant C, wavelength λ and absolute temperature T in [K];
  - Limiting case for short wavelengths λ: $\exp(C/(\lambda \cdot T)) >> 1 \rightarrow 1 / (\exp(C/(\lambda \cdot T)) - 1) \approx \exp(-C/(\lambda \cdot T)) \rightarrow$ Wien's Law
  - Limiting case for long wavelengths λ: $\exp(C/(\lambda \cdot T)) \approx 1 + C/(\lambda \cdot T) \rightarrow 1 / (\exp(C/(\lambda \cdot T)) - 1) \approx 1 / (1 + C/(\lambda \cdot T) - 1) = (\lambda \cdot T)/C \rightarrow$ Rayleigh-Jeans Law
- **106** The 1st Planck radiation constant is $c1 = 2\pi hc^2$ and the 2nd Planck radiation constant is $c2 = hc/k$ with the three natural constants speed of light in vacuum $c = 2.99792458 \cdot 10^8$ m/s; Planck's constant $h = 6.626176 \cdot 10^{-34}$ J·s; Boltzmann constant $k = 1.380662 \cdot 10^{-23}$ J/K
- **107** Max Theodor Felix von Laue [9.10.1879 in Pfaffendorf – 24.4.1960 in Berlin]; German physicist; contributions to optics, crystallography, quantum theory, supra conduction and relativity theory; 1914 Nobel prize for the diffraction of x-rays in crystals; professor in Frankfurt, Berlin und Gottingen; 1931-33 President of the Deutschen Physikalischen Gesellschaft; Director of the Kaiser-Wilhelm-Institute of Physics until 1946; from 1950 on Director of the Max- Planck-Institute of Physical Chemistry
- **108** Albert Einstein [14.3.1879 in Ulm – 18.4.1955 in Princeton]; German-American theoretical physicist; studies at the TH Zürich; 1902-09 with the patent office in Bern; 1909-14 Prof. in Zürich and Prague; 1914-33 Prof. at the University Berlin, Director of the Kaiser-Wilhelm-Institute of Physics; emigrated in 1933; 1940 American citizen; professor at the Princeton University; developed the special theory of relativity in 1905 and in 1916 the general theory of relativity; in 1929 general field theory; in 1921 Nobel prize for the photoelectric effect
- **109** Paul Ehrenfest [18.1.1880 in Wien – 25.9.1933 in Amsterdam]; Austrian and Dutch physicist; contributions to statistical mechanics and quantum mechanics; Ehrenfest theorem; 1907 professor in Petersburg; 1912 in Leiden
- **110** Max Planck: Wissenschaftliche Selbstbiographie, Barth Leipzig (1948)

### 3. Thermal Imaging until 1945: Difficult Start

- **111** Richard D. Hudson: Infrared System Engineering, John Wiley, New York (1969), S. 464
- **112** http://en.wikipedia.org/wiki/Bolometer
- **113** http://earthobservatory.nasa.gov/Features/Langley/langley_2.php
- **114** Richard D. Hudson: Infrared System Engineering, John Wiley, New York (1969), S. 455
- **115** Examples: US 1099199 von 1914, US1158967 von 1915
- **116** Richard D. Hudson: Infrared System Engineering, John Wiley, New York (1969), S. 6
- **117** David A. Kondas: Introduction to Lead Salt Infrared Detectors, Technical Report ARFSD-TR-92024, U.S. Army Armament Research, Development and Engineering Center (1993), S. 4
- **118** Theodore Willard Case [12.12.1888 in Auburn – 13.5.1944); American inventor; contributions to photoelectric detectors; developed the „thallofide cell" and a communication system for the US Navy based thereon; famous and

commercially successful due to his contributions to sound film systems.
– http://en.wikipedia.org/wiki/Theodore_Case

- **119** David A. Kondas: Introduction to Lead Salt Infrared Detectors, Technical Report ARFSD-TR-92024, U.S. Army Armament Research, Development and Engineering Center (1993), S. 4
- **120** Henry L. Hackforth: Infrared Radiation, MaGraw-Hill, New York (1960), S. 152
- **121** Antoni Rogalski: History of infrared detectors, Opto-Electron. Rev., 20 (2012), S.282
- **122** http://en.wikipedia.org/wiki/Thallous_sulfide, http://en.wikipedia.org/wiki/Theodore_Case
- **123** Antoni Rogalski: History of infrared detectors, Opto-Electron. Rev., 20 (2012), S.282
- **124** Sean F. Johnston: A History of Light and Colour Measurement, Institute of Physics Publishing, Bristol (2001), S. 221
- **125** Richard D. Hudson: Infrared System Engineering, John Wiley, New York (1969), S. 8
- **126** Frederick Alexander Lindemann, 1st Viscount Cherwell [5.4.1886 in Baden-Baden – 3.7.1957 in Oxford]; English physicist and influential scientific consultant of the British government in the 1940s and 1950s; professor at the University of Oxford and director of the Clarendon Laboratory; consultant and close friend of Churchill; 1941 Baron Cherwell; since 1956 Viscount Cherwell
- **127** Sean F. Johnston: A History of Light and Colour Measurement, Institute of Physics Publishing, Bristol (2001), S. 222
- **128** Richard D. Hudson: Infrared System Engineering, John Wiley, New York (1969), S. 470, [3] S. O. Hoffman, "The Detection of Invisible Objects by Heat Radiation", Phys. Rev., 14, 163 (1919); [4] S. O. Hoffman, "Method of and Apparatus for Detecting and Observing Objects in the Dark", U.S. Patent No. 1,343,393
- **129** Emil Mechau [19.4.1882 in Seesen – 28.6.1945 in Koßdorf]; German designer and motion picture pioneer; from 1900 to 1908 at Carl Zeiss in the astronomical research workshop where he came in touch with the problem of flicker-free motion pictures via Prof. Siedentopf ; 1908 with Ernst Leitz in Wetzlar where he developed the flicker-free Mechau projector
- **130** Richard D. Hudson: Infrared System Engineering, John Wiley, New York (1969), S. 467, Zitat aus D. Terret: The United States Army in World War II: The Signal Corps, Office of the Chief of Military History, Department of the Army, Washington D. C. (1956)
- **131** Henry L. Hackforth: Infrared Radiation, MaGraw-Hill, New York (1960), S. 8
- **132** Marianus Czerny [17.2.1896 in Breslau – 10.9.1985 in München]; German experimental physicist; PhD in 1923 with the infrared physicist Heinrich Rubens; contributions to spectroscopy and measuring technology; 1937 – 1961 director of the Physical Institute in Frankfurt/Main;
- **133** Marianus Czerny: Die Fotographie im Infrarot betreffend, Zeitschrift für Physik, 53, 1 (1929)
- **134** Kalman Tihanyi [28.4.1897 in Üzbeg - 26.2.1947 in Budapest]; Hungarian physicist, electrical engineer and inventor; pioneer of electronic television technology; important contributions to the development of cathode ray tubes which were produced later by the Radio Corporation of America (RCCA) and the German companies Loewe and Fernseh AG; construction of the first automatic and unmanned aircraft in Great Britain.
http://en.wikipedia.org/wiki/Kalma_Tihanyi
- **135** Thermal Imaging for DUMIES: History of Thermal Imaging, Security Sales & Integration, July 2011
- **136** http://en.wikipedia.org/wiki/Kalma_Tihanyi
http://www.ctie.monash.edu.au/hargrave/tihanyi.html
- **137** Germanium (Ge) was discovered in 1886 by the German chemist C. A. Winkler who therewith could fill a known gap in the periodic system. Industrial production began in 1941 in the USA; the production of purest germanium, as it is needed for IR optics, is possible only since 1949.
– Anna Dammshäuser: Gallium & Germanium – Von der Lagerstätte bis zur Verwendung, Umweltgeochemisches Seminar WS 2003/2004, Universität Karlsruhe
- **138** C. Maxwell Cade: The industrial potential of the „heat camera", New Scientist, 499 (1964), S. 165
- **139** Herman Willenberg: Infrarote Photographie, Zeitschrift für Physik, 74, 663 (1932);
G. Moench & Herman Willenberg , Infrarote Photographie, Zeitschrift für Physik., 77, 170 (1932)
- **140** Edgar W. Kutzscher [? - ?]; German physicist; in the 1930s at the institute of physics of the University Berlin; 1939-45 head of development at Electroacustic (ELAC) in Kiel; in 1945 the Russians took his research equipment , Kutzscher himself was transferred to Great Britain; thereafter strong influence on Amerikan development projects as an employee of Lockheed Aircraft in California
- **141** Antoni Rogalski: History of infrared detectors, Opto-Electron. Rev., 20 (2012), S.283
- **142** David A. Kondas: Introduction to Lead Salt Infrared Detectors, Technical Report ARFSD-TR-92024, U.S. Army Armament Research, Development and Engineering Center (1993), S. 5
- **143** Sean F. Johnston: A History of Light and Colour Measurement, IOP Publishing (2001), S. 224
- **144** Henry L. Hackforth: Infrared Radiation, McGraw-Hill, New York (1960), S. 9

- **145** Richard D. Hudson: Infrared System Engineering, John Wiley, New York (1969), S. 470: [9] "Night Vision with Electronic Infrared Equipment", Electronics, 19, 192 (1946)
- **146** Bernhard Friedrich Adolf Gudden [14.3.1892 in Beuel – 3.8.1945 in Prag; German physicist; 1919 assistant of Robert Wichard Pohl; in 1921 habilitation; since 1924 professor of experimental physics at the Friedrich-Alexander-University in Erlangen; 1939 professor and head of the Physical Institute at the Karl-Ferdinands-University in Prag; fundamental work about photoelectric effects in semiconductors
- **147** Kai Christian Handel: Anfänge der Halbleiterforschung und –entwicklung, Dissertation, RWTH Aachen (1999)
- **148** Paul Robert Görlich [7.10.1905 in Dresden – 12.3.1986 in Jena]; German physicist and engineer; in 1932 promotion in Dresden; 1932 - 45 head of the lab of Zeiss-Ikon AG in Dresden; work about photo-detectors and secondary electron multipliers; in 1942 habilitation; 1946 -52 employed by the Soviet optical industry; in 1952 manager at VEB Carl Zeiss Jena; 1959-71 director of the Institute of Optics and Spectroscopy in Berlin; 1960-71 Research Director at VEB Carl Zeiss Jena
- **149** Zeiss Ikon AG: 1926 founded in Dresden by Carl Zeiss Jena as a merger of several German camera and optics manufacturer that were absorbed by Carl Zeiss in that time; among these were the Internationale Camera Actiengesellschaft (Dresden), Optische Anstalt C. P. Goerz AG (Berlin), Contessa-Nettel AG (Stuttgart) and the Ernemann-Werke AG (Dresden); later the Aktiengesellschaft Hahn für Optik und Mechanik (Ihringshausen/Kassel) and the Goerz Photochemische Werke AG (Berlin) were added; after 1945 the facilities in Dresden became states-owned enterprises and the headquarters of the Ikon AG were transferred to Stuttgart; in 1956 also Voigtländer and the Zett-Geräte-Werk (both in Brunswick) came in addition; since 1989 part of the Finnish-Swedish Assa-Abloy group.
- **150** Hans-Joachim Pohl: Paul Görlich – ein Leben für die Physik und für CARL ZEISS JENA, Sitzungsberichte der Leibniz-Sozietät, 83 (2006), S. 210-211
- **151** Henry L. Hackforth: Infrared Radiation, McGraw-Hill, New York (1960), S. 9
- **152** H. Plesse: Das Wärmepeilgerät der Firma Carl Zeiss, Carl Zeiss, Jena (1944), S. 1
- **153** The report was presented in parts at a session of the Work Commission UR in the frame of the Commission for Surveillance and Fire Control Equipment on 6.7.1944 in Berlin; perhaps this was the reason to file such a report.
  – H. Plesse: Das Wärmepeilgerät der Firma Carl Zeiss, Carl Zeiss, Jena (September 1944)
- **154** H. Plesse: Das Wärmepeilgerät der Firma Carl Zeiss, Carl Zeiss, Jena (1944), Titelblatt
- **155** SS Mauretania (GB) [1907 – 1934], length 240,8 m; SS Normandie (F) [1935 – 1943], length 313,6 m; SS Aquitania (GB) [1914 – 1950], length 274,6 m
- **156** Richard D. Hudson: Infrared System Engineering, John Wiley, New York (1969), S. 467, [11] cited from D. Terret: The United States Army in World War II: The Signal Corps, Office of the Chief of Military History, Department of the Army, Washington D. C. (1956)
- **157** Reginald Victor Jones [29.9.1911 in London – 17.12.1997 in Strathdon, Aberdeenshire]; British physicist and expert for scientific support of the military secret service, „father of S&T Intelligence“; specialist for new German weapons in WW II; active in the „Battle of Beams“ for the superiority in radar technology; decorated several time for his success
- **158** Sean F. Johnston: A History of Light and Colour Measurement, Institute of Physics Publishing, Bristol (2001), S. 222
- **159** Richard D. Hudson: Infrared System Engineering, John Wiley, New York (1969), S. 472, [24] R. V. Jones, "Infrared Detection in British Air Defense, 1935-38", Infrared Phys., 1, 153 (1961)
- **160** Sean F. Johnston: A History of Light and Colour Measurement, Institute of Physics Publishing, Bristol (2001), S. 222
- **161** M. Czerny & P. Mollet, Zeitschrift für Physik, 108, 85 (1937)
- **162** Nernst lamp: invented in 1897 by the German professor of chemistry Walther Nernst in Gottingen; generates white light being very similar to day light; more efficient than carbon fiber lamp; function is based on the electric conductivity due to ions in air of high temperature; typical operation temperature is 1600°C; the glow body is made from a mixture of zircon oxide and yttrium oxide; the lamp was used in spectroscopy until the 1980s.
- **163** Rolf Walter: Zeiss 1905 – 1945, Böhler Verlag, Köln (2000), S. 256
- **164** Rolf Walter: Zeiss 1905 – 1945, Böhler Verlag, Köln (2000), S. 253
- **165** H. Plesse: Das Wärmepeilgerät der Firma Carl Zeiss, Carl Zeiss, Jena (1944), S. 6
- **166** M. Czerny und H. Röder, Ergebnisse exakter Naturwissenschaften 17, 70 (1938)
- **167** H. Plesse: Das Wärmepeilgerät der Firma Carl Zeiss, Carl Zeiss, Jena (1944), S. 7
- **168** H. Plesse: Das Wärmepeilgerät der Firma Carl Zeiss, Carl Zeiss, Jena (1944), S. 9
- **169** Ibid.
- **170** Ibid.
- **171** H. Plesse: Das Wärmepeilgerät der Firma Carl Zeiss, Carl Zeiss, Jena (1944), S. 9-10
- **172** Lothar Kramer (Hrsg.): Jenaer Jahrbuch zur Technik- und Industriegeschichte, Band 8, Verein für Technikgeschichte in Jena (2006), S. 138
- **173** Henry L. Hackforth: Infrared Radiation, McGraw-Hill, New York (1960), S. 9
- **174** Otto Schott had developed his first borosilicate glass 3.3 already in 1887; in 1938, the marque Duran was filed at the Imperial Patent Office for this glass; Duran is representing also the international standard (DIN ISO 3585) for this type of glass.
  – http://de.wikipedia.org/wiki/Duran_(Glas)

- **175** In 1899 the German chemist and physicist Richard Küch at Heraeus in Hanau managed for the first time to melt rock crystal (= crystalline quartz $SiO_2$) in a oxyhydrogen flame at approx. 2000°C and to produce pure quartz glass; on 3.4.1912 the Heraeus Quarzglas GmbH in Griesheim was founded which produce high-value quartz glasses in different grades until today; since 1955 also synthetic quartz glass, that is distinguished by an extreme purity on the ppb-level, is produced directly in a oxyhydrogen flame
- **176** H. Plesse: Das Wärmepeilgerät der Firma Carl Zeiss, Carl Zeiss, Jena (1944), S. 2
- **177** L. Harris had used it in 1934 with his thermocouple arrangement (Phys. Rev. 45, 635 (1934));
  – In 1937, E. Lehrer at the I.G.-Farben had employed a fast bolometer with a modulation frequency of 68 Hz (Z. techn. Physik 18, 393 (1937))
- **178** H. Plesse: Das Wärmepeilgerät der Firma Carl Zeiss, Carl Zeiss, Jena (1944), S. 11
- **179** Evaporateed layers: L. Harris, Phys. Rev. 45, 635 (1934);
  – Metal foils: E. Lehrer, Z. techn. Physik 18, 393 (1937)
- **180** Today this effects at a sensor are known as microphony because the sensor unintendedly acts like a microphone, and converts external vibrations and sound waves into signals.
- **181** H. Plesse: Das Wärmepeilgerät der Firma Carl Zeiss, Carl Zeiss, Jena (1944) und Wärmepeilgeräte WPG, Broschüre von Carl Zeiss, Jena (1944?, not complete)
- **182** Funkmeßgerätekunde, Oberkommando der Kriegsmarine, MDv 291 vom Februar 1944; Angebotsschreiben von Carl Zeiss an das Verteidigungsministerium vom 1.12.1954, Leistungsblatt Nr. 17; Rudolf Lusar: Die deutschen Waffen und Geheimwaffen des 2. Weltkriegs, J.F. Lehmanns Verlag, München (1962)
- **183** H. Plesse: Das Wärmepeilgerät der Firma Carl Zeiss, Carl Zeiss, Jena (1944), S. 19
- **184** Wärmepeilgeräte WPG, Broschüre von Carl Zeiss, Jena (1944?, not complete), S. 7, 15-16
- **185** Wärmepeilgeräte WPG, Broschüre von Carl Zeiss, Jena (1944?, not complete), S. 8
- **186** David A. Kondas: Introduction to Lead Salt Infrared Detectors, Technical Report ARFSD-TR-92024, U.S. Army Armament Research, Development and Engineering Center (1993), S. 5
- **187** Robert Joseph Cashman [27.9.1906 in Wilmington – 27.9.1988]; American physicist; professor at the Northwestern University; work about photovoltaic and photo-emissive detectors before WW II; developed a stable thallium sulfide detector; manager of several military research programs during and after WW II; 7 detector patents; 1975 retired
- **188** Richard D. Hudson: Infrared System Engineering, John Wiley, New York (1969), S. 467: US Patent 2448517, filed 1944, granted 1948
- **189** Henry L. Hackforth: Infrared Radiation, McGraw-Hill, New York (1960), S. 152
- **190** Richard D. Hudson: Infrared System Engineering, John Wiley, New York (1969), S. 464
- **191** David A. Kondas: Introduction to Lead Salt Infrared Detectors, Technical Report ARFSD-TR-92024, U.S. Army Armament Research, Development and Engineering Center (1993), S. 5
- **192** H. Plesse: Das Wärmepeilgerät der Firma Carl Zeiss, Carl Zeiss, Jena (1944), S. 33
- **193** OEG Report No. 51: Antisubmarine Warfare in World War II, Operations Evaluation Group (1946), S. 155
- **194** Ibid., S. 156
- **195** ELAC in Kiel: Unofficially the history began in 1908 when a group of scientists around Professor Heinrich Hecht in Kiel did research work about sound propagation under water; on 1.9.1926 the ELECTROACUSTIC GmbH was founded; chiefly specialized on underwater sound technology; after 1945 the existence was secured by producing everyday objects; since 1948 the ELAC record player was famous world-wide; today , naval and phono applications are separated in two companies.
  – www.radiomuseum.org/dsp_hersteller_detail.cfm?company_id=812; http://de.wikipedia.org/wiki/Elac;
- **196** Helmuth Giessler: Der Marine-Nachrichten- und Ortungsdienst, J. F. Lehmanns Verlag, München (1971), S. 95
- **197** Funkmeßgerätekunde, Oberkommando der Kriegsmarine, MDv 291 (1944), Kap. Wärmepeilgeräte
- **198** Helmuth Giessler: Der Marine-Nachrichten- und Ortungsdienst, J. F. Lehmanns Verlag, München (1971), S. 95
- **199** Richard D. Hudson: Infrared System Engineering, John Wiley, New York (1969), S. 470: "Night vision with Electronic Infrared Equipment", Electronics, 19, 192 (June 1946)
- **200** Sean F. Johnston: A History of Light and Colour Measurement, Institute of Physics Publishing, Bristol (2001), S. 224
- **201** Henry L. Hackforth: Infrared Radiation, McGraw-Hill, New York (1960), S. 9
  – Richard D. Hudson: Infrared System Engineering, John Wiley, New York (1969), S. 466: [2] V. Krizek, V. Vand, "The Development of Infrared Technique in Germany", Electronic Eng. 18, 316 (1946)
- **202** OEG Report No. 51: Antisubmarine Warfare in World War II, Operations Evaluation Group (1946), S. 158
- **203** Ibid., S. 159
- **204** Robert Bud, Philip Gummett: Cold War, Hot Science, harwood academic publishers, (1999), Chap. 7: Ernest Putley "Thermal Radiation and its Applications", S. 189

- **205** Robert Bud, Philip Gummett: Cold War, Hot Science, harwood academic publishers, (1999), Chap. 7: Ernest Putley "Thermal Radiation and its Applications", S. 189
- **206** David A. Kondas: Introduction to Lead Salt Infrared Detectors, Technical Report ARFSD-TR-92024, U.S. Army Armament Research, Development and Engineering Center (1993), S. 5
- **207** Antoni Rogalski: History of infrared detectors, Opto-Electron. Rev., 20 (2012), S.283
- **208** Robert Bud, Philip Gummett: Cold War, Hot Science, harwood academic publishers, (1999), Chap. 7: Ernest Putley "Thermal Radiation and its Applications", S. 190
- **209** A comprehensive documentation of the last cruise of U 234 can be found in the book of Joseph Mark Scalia: In geheimer Mission: U 234, Ullstein Taschenbuch (2005).
  – This book contains a description and many relevant data of evaporography.
- **210** Further data about the X B type submarine: length 89.80 m, width 9.20 m, hight 10.20 m; displacement 1763 t surfaced and 2177 t submerged; maximum operating depth 120 m; 2 Diesel engines with 2100 PS each; surfaced 18450 sm at 10 kn, max. speed 17 kn ; submerged: 188 sm at 2 kn, max. speed 7 kn; containers for 66 mines outside the pressure hull; 2 rear torpedo tubes; crew: 5 officers / 47 men
  – U 234 was built at the Germania shipyard in Kiel (Built No. G 664) and was launched on 23.12.1943
- **211** Joseph Mark Scalia: In geheimer Mission: U 234, Ullstein Taschenbuch (2005), S. 74-75
- **212** Joseph Mark Scalia: In geheimer Mission: U 234, Ullstein Taschenbuch (2005), S. 241
- **213** Joseph Mark Scalia: In geheimer Mission: U 234, Ullstein Taschenbuch (2005), S. 257
- **214** Michael Schmeelke: Alarm Küste, Dörfler Verlag, Eggolsheim (1993), S. 10
- **215** Michael Schmeelke: Alarm Küste, Dörfler Verlag, Eggolsheim (1993), S. 18
- **216** Streoscopic rangefinder with 10 m base: The observer looks through a special type of binocular the front optics of which is laterally separated by 10 m. Due to this large base he gets a stereoscopic impression even of faraway objects. The distance to an interesting target can be read from a superimposed scale.
- **217** Karl-Heinz & Michael Schmeelke: Fernkampfgeschütze am Kanal, Podzun-Pallas Verlag, Friedberg (1992), S. 4-6
- **218** Michael Schmeelke: Alarm Küste, Dörfler Verlag, Eggolsheim (1993), S. 30
- **219** Karl-Heinz & Michael Schmeelke: Fernkampfgeschütze am Kanal, Podzun-Pallas Verlag, Friedberg (1992), S. 6
- **220** Michael Schmeelke: Alarm Küste, Dörfler Verlag, Eggolsheim (1993), S. 59-60
- **221** Margit Krake, Hendrik Rothe: Deutsche Infrarot- und Nachtsichttechnik: Von den Anfängen bis 1945, Lehrstuhl für Mess- und Informationstechnik, Helmut-Schmidt-Universität, Hamburg; Vortrag beim 5. DWT Workshop „Optik und Optronik in der Wehrtechnik", Meppen (2009)
- **222** Robert Bud, Philip Gummett: Cold War, Hot Science, harwood academic publishers, (1999), Chap. 7: Ernest Putley "Thermal Radiation and its Applications", S. 213, Ref. 6
- **223** Helmuth Giessler: Der Marine-Nachrichten- und Ortungsdienst, J. F. Lehmanns Verlag, München (1971), S. 95
- **224** Karl-Heinz & Michael Schmeelke: Fernkampfgeschütze am Kanal, Podzun-Pallas Verlag, Friedberg (1992), S. 51
- **225** Karl-Heinz & Michael Schmeelke: Fernkampfgeschütze am Kanal, Podzun-Pallas Verlag, Friedberg (1992), S. 8
- **226** Karl-Heinz & Michael Schmeelke: Fernkampfgeschütze am Kanal, Podzun-Pallas Verlag, Friedberg (1992), S. 16
- **227** Helmuth Giessler: Der Marine-Nachrichten- und Ortungsdienst, J. F. Lehmanns Verlag, München (1971), S. 95
- **228** This behavior is based on one of the three fundamental conservation laws of mechanics: the conservation of angular momentum (the other laws require the conservation of energy and momentum)
- **229** The inventor and company founder Hermann Anschütz [3.10.1872 in Zweibrücken – 6.5.1931 in München] has an impressive biography; while making plans to reach the North Pole with a submarine, he came across the issue to find north without a magnetic compass; in 1904 first patent for a gyroscopic compass; in 1905 company founding in Kiel; in 1907 the world's first north-seeking gyroscopic compass; 1916-20 world's first self-steering apparatus; in 1927 two-gyroscope compass which was groundbreaking for 6 decades; in 1930 Anschütz signed his company share over to the Zeiss foundation; in the 1930s, Anschütz gyroscopes were used on German battle ships to stabilized the line of sight of anti-aircraft guns
- **230** Jean-Marie Mathey, Alexandre Sheldon-Duplaix: Geschichte der Unterseeboote, Motorbuch Verlag, Stuttgart (2007), S. 53: On the Atlantic 41 convoy aircraft carriers and 300 long-range reconnaissance aircraft were deployed against approx. 240 German submarines; 43% of the submarine losses were due to air force.
  – Robert Hutchinson: Kampf unter Wasser, Motorbuchverlag, Stuttgart (2006), S. 111: German submarine losses: in total 782, due to aircraft and ship 50, due to aircraft 245, due to shipborne aircraft 45; the result is 43% again.
- **231** Helmuth Giessler: Der Marine-Nachrichten- und Ortungsdienst, J. F. Lehmanns Verlag, München (1971), S. 95
- **232** Helmuth Giessler: Der Marine-Nachrichten- und Ortungsdienst, J. F. Lehmanns Verlag, München (1971), S. 95

- **233** Richard D. Hudson: Infrared System Engineering, John Wiley, New York (1969), S. 471: E. J. Martin & G. G. Scott, "Sealed Heat Ray detector", US Patent 2491192, angemeldet am 11.11.1944, erteilt am 13.12.1949
- **234** Rudolf Lusar: Die deutschen Waffen und Geheimwaffen des 2. Weltkriegs; Lehmanns Verlag, München (1962), S. 168
- **235** H. Plesse: Das Wärmepeilgerät der Firma Carl Zeiss, Carl Zeiss, Jena (1944), S. 9 – Richard D. Hudson: Infrared System Engineering, John Wiley, New York (1969), S. 466: [3] F. E. Jones, "Infrared – Its Problems and Possibilities", Brit. Communications and Electronics, 4, 36 (1957)
  – [17] V. P. Moskovskii, P. T. Astashenkov, Modern War Technology, Moscow, Military Publishing House (1956)
- **236** Peter Hinchliffe: Luftkrieg bei Nacht 1939-1945, Motorbuch Verlag, Stuttgart (1998), S. 15
- **237** Fritz Trenkle: Die deutschen Funkmessverfahren bis 1945; Motorbuch Verlag, Stuttgart (1979), S. 138
- **238** Fritz Trenkle: Die deutschen Funkmessverfahren bis 1945; Motorbuch Verlag, Stuttgart (1979), S. 140
- **239** Heinz J. Nowarra: Die deutsche Luftrüstung 1933-1945, Band 4, Bernard & Graefe Verlag, Koblenz (1993), S. 142
- **240** Fritz Trenkle: Die deutschen Funkmessverfahren bis 1945; Motorbuch Verlag, Stuttgart (1979), S. 140
- **241** Sean F. Johnston: A History of Light and Colour Measurement, IOP Publishing (2001), S. 226
- **242** Richard D. Hudson: Infrared System Engineering, John Wiley, New York (1969), S. 469: [25] K. Lajos, "Infravoros Felderites es Alcazas", Budapest, Zrinyi Military Publishing House (1966)
- **243** Fritz Trenkle: Die deutschen Funkmessverfahren bis 1945; Motorbuch Verlag, Stuttgart (1979), S. 140
- **244** Th. Benecke, K.-H. Hedwig, J. Hermann: Flugkörper und Lenkraketen, Bernard & Graefe Verlag, Koblenz (1987), S. 34
- **245** Th. Benecke, K.-H. Hedwig, J. Hermann: Flugkörper und Lenkraketen, Bernard & Graefe Verlag, Koblenz (1987), S. 33
- **246** Ibid.
- **247** Th. Benecke, K.-H. Hedwig, J. Hermann: Flugkörper und Lenkraketen, Bernard & Graefe Verlag, Koblenz (1987), S. 34
- **248** Ibid.
- **249** Richard D. Hudson: Infrared System Engineering, John Wiley, New York (1969), S. 466: [3] F. E. Jones, "Infrared – Its Problems and Possibilities", Brit. Communications and Electronics, 4, 36 (1957)
  – Rudolf Lusar: Die deutschen Waffen und Geheimwaffen des 2. Weltkriegs; Lehmanns Verlag, München (1962), S. 207
- **250** Sean F. Johnston: A History of Light and Colour Measurement, IOP Publishing (2001), S. 224
- **251** Ibid., S. 225
- **252** Th. Benecke, K.-H. Hedwig, J. Hermann: Flugkörper und Lenkraketen, Bernard & Graefe Verlag, Koblenz (1987), S. 34
- **253** Henry L. Hackforth: Infrared Radiation, McGraw-Hill, New York (1960)
- **254** Richard D. Hudson: Infrared System Engineering, John Wiley, New York (1969)
- **255** Richard D. Hudson, S. 464
- **256** Henry L. Hackforth, S. 8
- **257** Henry L. Hackforth, S. 9
- **258** Richard D. Hudson, S. 6
- **259** Henry L. Hackforth, S. 10
- **260** Richard D. Hudson, S. 8
- **261** Richard D. Hudson, S. 464
- **262** Richard D. Hudson, S. 9
- **263** Richard D. Hudson, S. 9
- **264** Sean F. Johnston: A History of Light and Colour Measurement, IOP Publishing (2001), S. 224
- **265** Richard D. Hudson, S. 464

### 4. New Beginning in Germany in 1945: The Zero Hour

- **266** Armin Hermann, Und trotzdem Brüder, Piper Verlag, München (2002), S. 178: *„In wenigen Werken von Weltsichtbarkeit ist so das deutsche Schicksal markiert wie in diesem Zeiss-Werk"*
- **267** New York Times vom 13.4.45
- **268** Armin Hermann, Und trotzdem Brüder, Piper Verlag, München (2002)., S. 31
- **269** Matthias Judt, Burghard Ciesla: Technology Transfer out of Germany after 1945, harwood academic publishers, Amsterdam (1996), S. 37
- **270** Matthias Judt, Burghard Ciesla: Technology Transfer out of Germany after 1945, harwood academic publishers, Amsterdam (1996), S. 46
- **271** Matthias Judt, Burghard Ciesla: Technology Transfer out of Germany after 1945, harwood academic publishers, Amsterdam (1996), S. 37
- **272** Gesetz- und Verordnungsblatt für Groß-Hessen Nr. 21, Beilage Nr. 3 (3. Juli 1946)
- **273** The only goal of German rearmament: Bild am Sonntag, 1 March 2009

### 5. Thermal Imaging until 1975: How Pictures Learned to Move

- **274** Speaking of Pictures: Cameralike Device Sees a Purple Cow, LIFE (30.6.1956)
- **275** Gerald Holst: Common Sense Approach to Thermal Imaging, SPIE, Bellingham (2000), S. 16
- **276** Henry L. Hackforth: Infrared Radiation, McGraw-Hill, New York (1960), S. 11
- **277** Richard D. Hudson: Infrared System Engineering, John Wiley, New York (1969), S. 9
- **278** Richard D. Hudson: Infrared System Engineering, John Wiley, New York (1969), S. 465
- **279** S. J. Deitchman: Limited war and American Defense Policy, M.I.T. Press, Cambridge (1964)
- **280** Richard D. Hudson: Infrared System Engineering, John Wiley, New York (1969), S. 465
- **281** Richard D. Hudson: Infrared System Engineering, John Wiley, New York (1969), S. 465
- **282** IR Suchgerät FuG 280 »Kiel IV« eingebaut in Nachtjagdflugzeug JU 88 G
- **283** Robert Bud, Philip Gummett (Ed.): Cold War, Hot Science, Harwood Academic Publishers (1999), Chap. 7 Ernest Putley: Thermal Radiation and ist Application, S. 189-190
- **284** Robert Bud, Philip Gummett (Ed.): Cold War, Hot Science, Harwood Academic Publishers (1999), Chap. 7 Ernest Putley: Thermal Radiation and ist Application, S. 190
- **285** Robert Bud, Philip Gummett (Ed.): Cold War, Hot Science, Harwood Academic Publishers (1999), Chap. 7 Ernest Putley: Thermal Radiation and ist Application, S. 191
- **286** Gerald Holst: Common Sense Approach to Thermal Imaging, SPIE, Bellingham (2000), S. 15
- **287** Norbert Schuster, Valentin G. Kolobrodov: Infrarotthermographie, Wiley-VCH, Weinheim (2004), S. 17
- **288** David A. Kondas: Introduction to Lead Salt Infrared Detectors, Technical Report ARFSD-TR-92024, U.S. Army Armament Research, Development and Engineering Center (1993), S. 6
- **289** Marcel J. E. Golay [3.5.1902 in Neuchatel – 27.4.1989], Swiss mathematician, physicist and information theorist; application of mathematics on real military and industrial problems; contributions to Savitzky-Golay smoothing filter; development of the Golay code; inventor of the Golay-cell as infrared detector. http://de.wikipedia.org/wiki/Marcel_J._E._Golay
- **290** http://en.wikipedia.org/wiki/Golay-Zelle
- **291** US Patent 2,435,519 „Image-forming Heat Detector"; Inventor: William A. Tolson of Radio Corporation of America; filed on 14.2.1944; granted 3.2.1948
- **292** David A. Kondas: Introduction to Lead Salt Infrared Detectors, Technical Report ARFSD-TR-92024, U.S. Army Armament Research, Development and Engineering Center (1993), S. 6
- **293** William L. Wolfe, George J. Zissis: The Infrared Handbook, Environmental Research Institute of Michigan (1989), S. 13-72, 13-73
- **294** William L. Wolfe, George J. Zissis: The Infrared Handbook, Environmental Research Institute of Michigan (1989), S. 13-72, 13-73
- **295** http://en.wikipedia.org/wiki/Indium(III)_antimonide
- **296** Heinrich Johann Welker [9.91912 in Ingolstadt – 25.12.1981 in Erlangen]; German theoretical and applied physicist; 1948 invention of the 'transistron' a kind of transistor which was independently invented at Westinghouse parallel to Bell Laboratories; fundamental contributions to III-V semiconductors; 1940-45 at the Luftfunk Forschungsinstitut in Oberpfaffenhofen; 1947-51 at Westinghouse in Paris; 1951-61 head of the department of solid state physics of Siemens-Schuckert in Erlangen; 1961-69 director of the Research Laboratories of Siemens-Schuckert in Erlangen; 1968-77 director of all research laboratories until his retirement.
- **297** H. Gobrecht , W. Weiss: Über die Aufnahme von Bildern im Infraroten nach der Methode von Czerny (Evaporographie), Zeitschrift für Angewandte Physik, 5, 207-211 (1953)
- **298** US 2,855,522: Method and Apparatus for Long Wavelength Infra-red Viewing,
  – Inventors: David Z. Robinson, Arthur P. DiMattia, Gene W. McDaniel, Richard L. Collette von BAIRD Ass., Inc. Cambridge, MA, USA; angemeldet 30.4.1953; erteilt 7.10.1958
- **299** William L. Wolfe, George J. Zissis: The Infrared Handbook, Environmental Research Institute of Michigan (1989), S. 13-73
- **300** Radar Research Establishment in Malvern, Worcestershire, England; British radar research establishment; founded in 1953 by merging of the Tele-communications Research Establishment (TRE) and the Radar Research and Development Establishment (RRDE); work on radar, solid state physics and electronics; in 1957 renamed in Royal Radar Establishment after a visit of Queen Elisabeth II.; closed in 1976
- **301** C. Maxwell Cade: The industrial potential of the „heat camera", New Scientist 499 (1964), S. 165
- **302** http://en.wikipedia.org/wiki/Texas_instruments
- **303** Kenneth Lloyd-Williams: The "heat camera" in medicine, New Scientist, 499 (1964), S. 162
- **304** Speaking of Pictures: Cameralike Device Sees a Purple Cow, LIFE (30.6.1956)

- **305** Kenneth Lloyd-Williams: The „heat camera" in medicine, New Scientist, 499 (1964), S. 162
- **306** G. W. McDaniel, D. Z. Robinson: Thermal Imaging by Means of the Evaporograph, Applied Optics, Vol. 1, No. 3 (May 1962)
- **307** NETD = Noise Equivalent Temperature Difference
- **308** William L. Wolfe, George J. Zissis: The Infrared Handbook, Environmental Research Institute of Michigan (1989), S. 13-70
- **309** Henry L. Hackforth: Infrared Radiation, McGraw-Hill, New York (1960), S. 206-207
- **310** Henry L. Hackforth: Infrared Radiation, McGraw-Hill, New York (1960), S. 206-207
- **311** Robert W. Astheimer, Eric M. Wormser: "Image Transducer", US-Patent 2,895,049, filed 26.6.1957, granted 14.7.1959
  – Faster scanning procedure with rotating polygons: Robert Bowling Barnes „Infrared Thermogram Camera and Scanning Means therefor", US-Patent 3,287,559, filed 4.10.1963, granted 22.11.1966
  – Further improvements: Robert Bowling Barnes, Maggio Charles Banca, Nelson E. Engborg "Infrared Thermograph", US-Patent 3,169,189, filed 11.4.1963, granted 9.2.1965
- **312** Kenneth Lloyd-Williams: The „heat camera" in medicine, New Scientist, 499 (1964), S. 162-163
- **313** Radar Research Establishment in Malvern, Worcestershire, England; British radar research establishment; founded in 1953 by merging of the Telecommunications Research Establishment (TRE) and the Radar Research and Development Establishment (RRDE); work on radar, solid state physics and electronics; in 1957 renamed in Royal Radar Establishment after a visit of Queen Elisabeth II.; closed in 1976
- **314** W. D. Lawsen, S. Nielson, E. H. Putley, A. S. Young: Preparation and properties of HgTe and mixed crystals of HgTe-CdTe, J. Phys. Chem. Solids 9, 325 (1959)
- **315** The intrinsic conductivity is created by the tendency of a solid body itself to develop lattice defects, which allow the movement of charges. The number of defects increases with temperature and therewith the conductivity. Extrinsic conductivity, on the contrary, is the amount of conductivity that is due to impurity atoms built into the crystal lattice.
  http://de.wikipedia.org/wiki/Intrinsische_Leitfähigkeit
- **316** By a semiconductor with a direct band gap a photon is effectively absorbed if the photon energy is higher than the energy different between valence and conduction band (= band gap). On the contrary, together with the absorption of a photon by an indirect semiconductor, additionally to the energy also a momentum must be transferred. Due to the low momentum of a photon, the process with a single photon is less probable, and thus the indirect semiconductors show lower absorption.
  – http://de.wikipedia.org/wiki/Bandlücke
- **317** P. Norton: HgCdTe infrared detectors, Opto-Electron. Rev. 10, 3, 159 (2002)
- **318** Antoni Rogalski: History of infrared detectors, Opto-Electron. Rev., 20 (2012), S.289
- **319** P. Norton: HgCdTe infrared detectors, Opto-Electron. Rev. 10, 3, 159 (2002)
- **320** Epitaxy (Greek: epi = on, over; taxis = arrange, align) is a special method of crystal growth on crystal substrates. Thereby at least one crystallographic orientation of the growing crystals corresponds to an orientation of the substrate material.
  – http://de.wikipedia.org/wiki/Epitaxie
- **321** William L. Wolfe, George J. Zissis: The Infrared Handbook, Environmental Research Institute of Michigan (1989), S. 13-73, 13-74
- **322** William L. Wolfe, George J. Zissis: The Infrared Handbook, Environmental Research Institute of Michigan (1989), S. 13-73, 13-74
- **323** William L. Wolfe, George J. Zissis: The Infrared Handbook, Environmental Research Institute of Michigan (1989), S. 13-73, 13-74
- **324** William L. Wolfe, George J. Zissis: The Infrared Handbook, Environmental Research Institute of Michigan (1989), S. 13-73, 13-74
- **325** C. Maxwell Cade: The industrial potential of the "heat camera", New Scientist 499 (1964), S. 165
- **326** C. Maxwell Cade: The industrial potential of the "heat camera", New Scientist 499 (1964), S. 165
- **327** E. F. J. Ring: The historical development of thermal imaging in medicine, Rheumatology, 43 (2004), S. 800
- **328** C. Maxwell Cade: The industrial potential of the „heat camera", New Scientist 499 (1964), S. 166
- **329** Richard D. Hudson: Infrared System Engineering, John Wiley, New York (1969), S. 456
- **330** Raytheon Company: Fifty Years of Owning the Night: The History of Infrared Imaging (2013), www.raytheon.com/newsroom/feature/rtn13_flir
- **331** www.ti.com/corp/docs/company/history/timeline/defense/1960/docs/66-first_flir.htm vom 31.12.2012
- **332** Raytheon Company: Fifty Years of Owning the Night: The History of Infrared Imaging (2013), www.raytheon.com/newsroom/feature/rtn13_flir
- **333** E. F. J. Ring: The historical development of thermal imaging in medicine, Rheumatology, 43 (2004), S. 800
- **334** Gerald D. Dodd, John D. Wallace, Irwin M. Freundlich, Lee Marsh, Alhonso Zermino: Thermography and Cancer of the Breast, Thermology, 3 (1988), S. 74

- **335** Kenneth Lloyd-Williams: The „heat camera“ in medicine, New Scientist, 499 (1964), S. 162
- **336** Ibid.
- **337** Das Grosse Hobby-Lexikon, Ehapa Verlag, Stuttgart (1967), Band II, S. 199
- **338** C. Maxwell Cade: The industrial potential of the "heat camera", New Scientist 499 (1964), S. 167
- **339** C. Maxwell Cade: The industrial potential of the "heat camera", New Scientist 499 (1964), S. 167
- **340** FOA was merged in 1945 of three existing institutions: the Chemischen Anstalt für Verteidigung FKA, the Institut für Militärische Physik MFI and a part of the Swedish office for inventions SUN; 2001 absorbed by the Schwedische Verteidigungs-Forschungsagentur FOI.
- **341** U.S. Patent 3,253,498: Scanning Mechanism for Electro-magnetic Radiation; filed on 14.5.1962; granted on 31.5.1966
- **342** Joachim-Michael Engel, Udo Flesch, Günter Stüttgen (Hrsg.): Thermologische Meßmethodik, notamed, Baden-Baden (1983), S. 88
- **343** Gerald Holst: Common Sense Approach to Thermal Imaging, SPIE, Bellingham (2000), S. 16
- **344** Kazuhiko Atsumi (Edt.): Medical Thermography, University of Tokyo Press (1973), S. 4-6
- **345** Kazuhiko Atsumi (Edt.): Medical Thermography, University of Tokyo Press (1973), S. 18-29
- **346** www.ti.com/corp/docs/company/history/timeline/defense/1960/docs/66-first_flir.htm vom 31.12.2012
- **347** „Puff the magic dragon, A bird of days long gone, Came to fly the evening sky, in a land called … Vietnam“ (free after Peter, Paul & Mary)
- **348** The Douglas C-47 was a military version of the widely known and seasoned Douglas DC-3. The aircraft equipped as Gunships at first got the identifier FC-47. The Air Force, however, had difficulty to accept the label F (for fighter!) for such an outdated aircraft. The aircraft therefore got the label AC for Attack & Cargo. The call signal "Spooky" was chosen deliberately and reportedly originated from a dictum of a higher Air Force officer who mocked at "such spooky and old plane".
- **349** Vietnam War: Since 1965 the USA had massively deployed own forces in South Vietnam; in 1973 the US forces left the county; in 1975South Vietnam was occupied by North Vietnamese troops.
- **350** www.ti.com/corp/docs/company/history/timeline/defense/1960/docs/66-first_flir.htm on 31.12.2012
- **351** Ibid.
- **352** Gerald Holst: Common Sense Approach to Thermal Imaging, SPIE, Bellingham (2000), S. 16
- **353** www.ti.com/corp/docs/company/history/timeline/defense/1960/docs/66-first_flir.htm on 31.12.2012
- **354** Ibid.
- **355** H. R. Weinheimer: Zielerkennung bei Nacht, Studienbericht, Oberkochen (1977), S. 82
- **356** Carl Zeiss Sonderoptik: Datenblatt WBG 1/50 Wärmebildhandgerät, Oberkochen (undated)
- **357** Schlieren optics was developed for the purpose of schlieren photography in 1864 by the German chemist and physicist August Toepler; it is used to make changes of the optical density of fluids and gases visible, however. It can also be used in all transparent media (e.g. glass); applied to the testing of optical beam paths or the shape of optical surfaces, it is often known as Foucault-testing.
  - Principle: A small slit is used as light source; the emitted light is collimated, propagates as a parallel bundle along the measuring path and is image on a knife-edge stop; the stop blocks the regularly propagating light rays, and permits only the small fraction of erroneous rays to pass, which have been influenced e.g. by density variations or surface imperfections. Under usual illumination these effect, which are caused by very small errors, cannot be detected.
  - See also Francis. A. Jenkins, Harvey E. White: Fundamentals of Optics, McGraw Hill, Kogakusha, Tokio (1976), S. 604-607
- **358** U. Martens, F. Kneubühl: Dynamics of Thin-Film Thermal Detectors in Infrared Imaging Systems. 1: Basic Equations and Fourier Analysis, Applied Optics, Vol. 13, No. 6 (1974)
  - U. Martens, P. Jeannet, F. Kneubühl: Dynamics of Thin-Film Thermal Detectors in Infrared Imaging Systems. 2: Imaging of Fixed and Moving Objects, Applied Optics, Vol. 14, No. 5 (1975)
- **359** Datasheet of M.E.L. Equipment Company Limited, Crawley, Sussex, England (1974)
- **360** Private communication from Dr. Jens-Rainer Höfft who was in charge for the scientific development of the Ikotherm device

### 6. »Common Module« Devices until 1995: Success by Numbers

- **361** Historical documentation of Texas Instruments Incorporated on their website (1995-2007)
- **362** Contract between Carl Zeiss, Oberkochen and Texas Instruments Incorporated: *„Carl Zeiss ist ausschließlicher Lizenznehmer für den Vertrieb der FLIR Common Modules von TI auf dem Gebiet der BRD“ (1976)*
- **363** E. Putley in Cold War, Hot Science, S. 212 (1999)

- **364** Ibid.
- **365** DGA is the office of the French government for procurement and technology in the field of defense, and is responsible for program management, development and purchasing of weapon systems for the French military forces.
- **366** Joseph Caniou: Les applications militaires de l'observation dans l'infrarouge, Congres National de Thermographie THERMOGRAM' 2009
- **367** E. Putley in Cold War, Hot Science, S. 212 (1999)
- **368** Appendix 55 Memeorandum submitted by Pilkington Opronics, www.parliament.uk : Parliament home page > Parliamentary business > Publications and Records > Committee Publications > All Select Committee Publications > Commons Select Committees > Science and Technology > Science and Technology (22 October 1998)
- **369** Sir James Dewar [*20.9.1842 in Kincardine, Schottland; †27.3.1923 in London], Scottish chemist and physicist; 1875 professor at the University of Cambridge, 1877 member of the Royal Institution in London where he died in his office. He worked in the field of deep temperature physics and was the first to produce liquid oxygen and hydrogen. For this, he used a double-wall evacuated metal vessel as early as in 1874. In 1893, he invented a vessel for storing liquid gases that was built as a double-wall evacuated vessel with mirrored glass walls, and is named Dewar after its inventor until today. We also use the principle today from our Thermos bottles.
- **370** Reverend Dr. Robert Strirling [*25.10.1790 in Cloag, Methvin, Scotland; †6.6.1878 in Galston, Ayrshire, Scotland] Scottish reverend and inventor. He studied Latin, Greek, logic und mathematics as well as theology and law; in 1816 he was ordained a reverend of the Presbyterian Church and worked as such in his parish until the end of his life. On 27.9.1816 he filed a patent of a hot air machine which he built in 1818; therewith, the Stirling engine is the second oldest heat engine after the steam engine. From a further Scottish inventor and workshop owner Robert Morton he also learned how to grind and polish lenses, and he invented some optical instruments.
- **371** The term means something like "without focus"; this shall indicate that the system that the optical system does not create an image. Instead, a parallel ray bundle entering the objective leaves the ocular also as a parallel bundle; depending on the magnification, just direction and diameter are changed.
- **372** Harry Schlemmer: History of Submarine Periscopes at Carl Zeiss, S. 124, ISBN 978-3-8132-0937-2, Mittler & Sohn, Hamburg (2011)
- **373** As sea water is electrically conductive, between the germanium substrate of the window and the steel of the surrounding boat hull an electrochemical element is formed that acts like a battery with a low voltage. Due to the voltage, an electric current is flowing between window and boat which solves the germanium being of lower rank in the electrochemical series. Basically the same process is used vice versa to protect the submarine from corrosion: Everywhere on the hull thick copper pieces are mounted, as so-called sacrificial anodes, which are solved instead of the steel walls in the unavoidable chemical processes.
- **374** DLC means Diamond-Like-Carbon, a term that indicates that the coating shall be nearly as hard as diamond. However, it is not a true diamond layer (three-dimensional tetrahedron lattice) but a two-dimensional netted layer of carbon and hydrogen (C-H), and in literature it is sometimes termed as ACH = amorphous carbon-hydrogen. The coating is mechanically and chemically very resistive, and the refractive index of $n \approx 2$ is ideally suited for an anti-reflection coating on germanium.
- **375** Private communication: Participant of the team of Carl Zeiss-Sondertechnik for the equipment of the Norwegian ULA submarines
- **376** www.luftwaffe.de/portal über MRCA PA-200 Tornado (2017)
- **377** PROPOSAL FOR A FORWARD-LOOKING INFRARED (FLIR) SYSTEM FOR ECR TORNADO AIRCRAFT, Vol. 1 Technical Proposal, prepared by Carl Zeiss Produktbereich Sonderoptik, Oberkochen / D
- **378** Abschlußbericht "4 fach-Zeilensprung für CM – WBG" von G. Kürbitz, Carl Zeiss, Oberkochen (Juli 1985)
- **379** PROPOSAL FOR A FORWARD-LOOKING INFRARED (FLIR) SYSTEM FOR ECR TORNADO AIRCRAFT, Vol. 2 Introduction, prepared by Electro-Optics Division Equipment Group, Texas Instruments Incorporated, Dallas / USA (27 February 1985)
- **380** Privat communication: Carl Zeiss-Sondertechnik / Karl-Heinz Strauß
- **381** Manufacturer of the German Tornado was the PANAVIA AIRCRAFT GmbH, München
- **382** J. Nolting: Videokompatible Wärmebildgeräte, CCG-Kurs Se 3.06, Manuskript 6 (1991)
- **383** Gabor Orban: Feuertaufe bestanden, Militär & Geschichte Nr. 5/2017, S. 56
- **384** https://en.wikipedia.org/wiki/Panavia_Tornado
- **385** G. Kürbitz , D. Marx, H. Heinrich: Infrarot-Abbildungssystem mit einer Vorrichtung zum Ausgleichen des Temperatur-Einflusses auf den Fokussierzustand, EP 0412314 (1990)
- **386** KMW = Krauss-Maffei-Wegmann in Kassel is in charge for tank turrets, e.g. for Leopard and Puma; HDW = Howaldt Deutsche Werft in Kiel is today TKMS = Thyssen-Krupp Marine Systems, and build the most modern non-nuclear submarines of the 212 and 214 class with fuel cell drive.

## 7. Thermal Imagers of 2[nd] Generation until 2005: Simply Digital

- **387** Electro-optics, A special report on Thermal Imaging Systems, Jane's Information Group (1992)

- **388** Electro-optics, A special report on Thermal Imaging Systems, Jane's Information Group (1992)
- **389** https://de.wikipedia.org/wiki/Zweiter_Golfkrieg
- **390** Electro-optics, A special report on Thermal Imaging Systems, Jane's Information Group (1992)
- **391** Wikipedia
- **392** Broschüre „Wärmebildgeräte der 2. Generation OPHELIOS", Zeiss Optronik (II/1999)
- **393** G. Kürbitz: Zweite Generation der Wärmebildgeräte-Technologie, Zwischenbericht (1981)
- **394** John P. Dakin, Robert G. W. Brown: Handbook of Optoelectronics, Vol. 1, CRC Press (2006), S. 1306
- **395** Joseph CANIOU: LES APPLICATIONS MILITAIRES DE L'OBSERVATION DANS L'INFRAROUGE, Congrès National de Thermographie THERMOGRAM' (2009)
- **396** S. Kasap, P. Capper: Springer Handbook of Electronic and Photonic Materials, Springer, York (2006), S. 867
- **397** Project Agreement No. RTP-US-GE-A-97-0010 between the American and German departments of defense, S. 2
- **398** Manufacturer: VITRON Spezialwerkstoffe GmbH, Jena
- **399** CAN-Bus *(Controller Area Network)* is a serial bus system. The objective is to reduce the wiring harness and therewith cost and weight. CAN is an international standard ISO 11898 and defines the Layer 1 (physical layer) and Layer 2 (data backup layer).
- **400** BITE means Built In Test Equipment; by means of BITE a system tests itself to detect and display own malfunctions.
- **401** Militärisch, technische Forderung MTWF 7/92 der Bundeswehr (military, technical requirement of the German army)

## 8. Review and Outlook

- **402** Isaac Newton in a letter to Robert Hooke on 5 February 1676. The aphorism itself is much older and is attributed first to Bernhard of Chartres in 1120; it is often cited in science and cultural history, e.g. in conjunction with Albert Einstein. The words are also engraved on the edge of the British 2-pound coin.
- **403** Program AMASS (Autonomous Maritime Surveillance System), FP7/2008-2011
- **404** *Lightman, Alan P. (2005). The discoveries: great breakthroughs in twentieth-century science, including the original papers. Toronto: Alfred A. Knopf Canada. p. 8.* ISBN 0-676-97789-8.CRC Press (2006), S. 1306

We will gladly send you a complete list of available titles.
Please send an e-mail with your address to: vertrieb@mittler-books.de
Our web address is: www.mittler-books.de
Bibliographical information of the German National Library
The German National Library lists the original German
publication in the German National Bibliography;
detailed bibliographical data is available in the internet via
http://dnb.d-nb.de.

ISBN 978-3-8132-0980-8

Printed in Germany
Firmengruppe APPL, aprinta Druck, Wemding